Physical Basis of Cell-Cell Adhesion

Editor

Pierre Bongrand, M.D., Sc.D.

Professor of Immunology
Laboratory of Immunology
Hôpital de Sainte-Marguerite
Marseilles, France

CRC Press
Taylor & Francis Group
Boca Raton London New York

CRC Press is an imprint of the
Taylor & Francis Group, an **informa** business

T0174838

CRC Press
Taylor & Francis Group
6000 Broken Sound Parkway NW, Suite 300
Boca Raton, FL 33487-2742

First issued in paperback 2020

ISBN 13: 978-1-315-89647-2 (hbk)
ISBN 13: 978-0-367-65739-0 (pbk)

Library of Congress Cataloging-in-Publication Data

Physical basis of cell-cell adhesion.

   Includes bibliographies and index.
   1. Cell adhesion. I. Bongrand, Pierre.
[DNLM: 1. Cell Adhesion. 2. Cells--Physiology.
QH 623 P578]
QH623.P48   1988   574.87'5   88-9547
ISBN 0-8493-6554-6

A Library of Congress record exists under LC control number: 88009547

Visit the Taylor & Francis Web site at
http://www.taylorandfrancis.com

and the CRC Press Web site at
http://www.crcpress.com

INTRODUCTION

Cell adhesion is a ubiquitous process that influences many aspects of cell behavior. Indeed, the control of cell proliferation[1] and migration[2] through different tissues involves adhesive interactions. The invasion of a specific organ by a bacterium[3] or the penetration of a target cell by a virus[4] are initiated by adhesive recognition. The triggering of effector functions such as particle engulfment by a phagocyte[5] or target cell destruction by a cytotoxic T-lymphocyte[6] involves a binding step. Finally, metastasis formation by tumor cells was often postulated to include attachment and detachment steps, and statistical correlations were demonstrated between the adhesiveness and invasive potential of malignant cells in different experimental systems.[7]

Hence, it is not surprising that many authors studied adhesion on a variety of cellular models, with different experimental methods. In many cases, cell adhesion seemed to be driven by specific receptor-ligand interactions and much work was devoted to the characterization of "adhesion molecules".[8-10] However, several experimental data suggest that an exhaustive study of the structure and specificity of cell adhesion molecules would not allow a complete understanding of adhesion. Here are some specific examples.

Concanavalin A is a molecule with four binding sites specific for carbohydrate residues commonly found on different cell membranes. It is therefore not surprising that many cells are agglutinated by this substance through a cross-bridging mechanism. However, treating cells with glutaraldehyde, a well-known fixation procedure, may induce a drastic decrease of agglutinability without a parallel decrease of concanavalin A-binding ability.[11,12]

It is a common finding in many experimental systems[6] that cell adhesion may be decreased by the addition of divalent cation chelators, inhibitors of cell energy production or cytoskeleton assembly, or temperature decrease. In many cases, it is unlikely that these phenomena reflect a loss of cell surface receptors or alteration of these receptors with concomitant decrease of binding affinity.

Antibody-mediated erythrocyte agglutination is a widely used method of studying the presence of different antigens on the red cell surface. This is indeed a routine technique in blood transfusion centers and it is well known that in some cases antibodies cannot mediate agglutination unless some special procedures (such as modification of ionic strength or protein addition) are used. It is unlikely that these relatively mild procedures act by increasing antigen-antibody affinity.[13]

When cells are coated with low amounts of concanavalin A, they may be agglutinated provided they are subjected to gentle centrifugation. However, prolonged agitation of cell suspensions may not result in adhesion, despite the occurrence of numerous cell-cell encounters.[12]

Another point is that in some situations the concept of specific-bond mediated adhesion does not seem to hold. As an example, many cells may adhere to a variety of synthetic substrates that may hardly be considered as specific ligands for adhesion molecules.[14] Hence, in addition to well-characterized specific interactions, cell adhesion may involve a combination of nonspecific low affinity molecular associations.

Recent experimental and theoretical progress suggests that some results and physical methods may be used to deal with the aforementioned problems. Indeed, physics may help define and measure quantitative parameters of cell adhesion such as kinetics of bond formation, mechanical strength of adhesions, width of the cell-cell or cell-substrate gap in adhesive zones, adhesion-associated strain, and stress of the cell surface. Also, physical techniques may allow a quantitative description of the different cell properties relevant to adhesion, such as surface charge and hydrophobicity or mechanical properties. Finally, physical results obtained by studying model systems may yield some information on the forces experienced by membrane molecules during the cell-cell approach.

The present book is aimed at providing a readable account of physical methods and results required to measure cell adhesion and interpret experimental data. Since on the one hand readability seemed a major quality for a book, and on the other hand, the problems posed referred to a wide range of domains of physics, chemistry, and biology, completeness had to be sacrificed. Indeed, a whole book would not suffice to quote the relevant literature (and many more authors would be required to have read it). Hence, only a limited number of topics were selected for reliability of methods, availability of enough experimental results to illustrate basic concepts or potential use in the future. These were discussed in three sections.

Section I includes a basic physical background likely to help understanding of cell adhesion.

Intermolecular forces are reviewed in the first chapter; after a brief description of the structure of the cell surfaces, molecular interactions are described in systems of increasing complexity, from atoms in vacuum to macroscopic bodies suspended in aqueous ionic solutions. Also, selected examples of interactions between biological macromolecules are reviewed to convey a feeling for the concept of "binding specificity".

In Chapter 2, de Gennes gives a description of the latest principles underlying the interactions between polymer-coated surfaces. Although many problems remain unsolved, the language of polymer physics should provide a basically correct framework for the discussion of intercellular forces, since cells are essentially polymer-coated bodies surrounded by solute macromolecules.

In Chapter 3, some methods and results of surface physics are presented, since the wealth of experimental data gathered in this field may shed some light on the mechanisms of interaction between ill-defined surfaces such as cell membranes.

Chapter 4 is devoted to a description of recent results on the mechanical properties of cell membranes. Systematic use and development of the powerful micropipette aspiration technique allowed Dr. Evans to obtain a reliable picture of the cell response to mechanical stimuli. This kind of knowledge is an essential requirement for a correct understanding of the mechanisms by which cell surfaces are deformed to allow the appearance of extended cell-cell or cell-substrate contact areas.

The second section of the book includes a description of some quantitative methods of studying cell adhesion as well as selected experimental results.

Hydrodynamic flow methods are reviewed in Chapter 5. These methods provide a simple way of evaluating the minimal time required for the formation of stable intercellular bonds and the mechanical resistance of these bonds.

As a logical sequel to the description of procedures for generating and breaking intercellular bonds, David Segal presents in Chapter 6 the powerful and versatile methods of detecting and quantifying cell aggregation he developed with a flow cytometer. The increasing availability of this apparatus may make Dr. Segal's methodology a procedure of choice for those interested in the measurement of cell adhesion.

In Chapter 7, Evans describes an analysis of the adhesion-induced deformations he measured in different models. This work allowed quantitative evaluation of the work of adhesion between cell surfaces.

Colette Foa and colleagues present electron microscopical data on cell adhesion and describe a methodology allowing quantitative analysis of digitized micrographs in Chapter 8. Their results demonstrate the difficulty of a physical analysis of the adhesion between "usual" cells, since plasma membranes are studded with asperities of varying shape and unknown mechanical properties. A quantitative description of these features is needed to model adhesive processes involving these surfaces.

In Chapter 9, Curtis reviews a variety of experimental studies on cell adhesion with an emphasis on the danger of interpreting the obtained data with simplistic concepts. This shows how much caution is needed when detailed mechanisms are proposed to account for experimental data.

Finally, Chapter 10 is a description by George Bell of some models for the kinetics of intercellular bond formation and equilibrium contact area. It is shown that fairly simple and reasonable assumptions may lead to quite rich models, with predictions that are not excessively dependent on the details of underlying assumptions. These models may help understand available experimental data and suggest further studies.

REFERENCES

1. **Folkman, J. and Moscona, A.**, Role of cell shape in growth control, *Nature (London)*, 273, 345, 1978.
2. **Chin, Y. H., Carey, G. D., and Woodruff, J. J.**, Lymphocyte recognition of lymph node high endothelium. V. Isolation of adhesion molecules from lysates of rat lymphocytes, *J. Immunol.*, 131, 1368, 1983.
3. **Gould, K., Ramirez-Ronda, C. H., Holmes, R. K., and Sanford, J. P.**, Adherence of bacteria to heart valves in vitro, *J. Clin. Invest.*, 56, 1364, 1975.
4. **Ginsberg, H. S.**, Pathogenesis of viral infection, in *Microbiology*, Davis, B. D., Dulbecco, R., Eisen, H. N., and Ginsberg, H., Eds., Harper & Row, Philadelphia, 1980, 1031.
5. **Rabinovitch, M.**, The dissociation of the attachment and ingestion phases of phagocytosis by macrophages, *Exp. Cell Res.*, 46, 19, 1967.
6. **Golstein, P. and Smith, E. T.**, Mechanism of T-cell-mediated cytolysis: the lethal hit stage, *Contemp. Topics Immunobiol.*, 7, 269, 1977.
7. **Fogel, M., Altevogt, P., and Schirrmacher, V.**, Metastatic potential severely altered by changes in tumor cell adhesiveness and cell surface sialylation, *J. Exp. Med.*, 157, 371, 1983.
8. **Rougon, G., Deagostini-Bazin, H., Hirn, M., and Goridis, C.**, Tissue and developmental stage-specific forms of a neural cell surface antigen: evidence for different glycosylation of a common polypeptide, *EMBO J.*, 1, 1239, 1982.
9. **Rutishauser, U., Hoffman, S., and Edelman, G. M.**, Binding properties of a cell adhesion molecule from neural tissue, *Proc. Natl. Acad. Sci. U.S.A.*, 79, 685, 1982.
10. **Hynes, R. O. and Yamada, K. M.**, Fibronectins: multifunctional modular glycoproteins, *J. Cell Biol.*, 95, 369, 1982.
11. **Van Blitterswijk, W. J., Walborg, E. F., Feltkamp, C. A., Hilkmann, H. A. M., and Emmelot, P.**, Effect of glutaraldehyde fixation on lectin-mediated agglutination of mouse leukemia cells, *J. Cell Sci.*, 21, 579, 1976.
12. **Capo, C., Garrouste, F., Benoliel, A. M., Bongrand, P., Ryter, A., and Bell, G. I.**, Concanavalin A-mediated thymocyte agglutination: a model for a quantitative study of cell adhesion, *J. Cell Sci.*, 56, 21, 1982.
13. **Gell, P. G. H. and Coombs, R. R. A.**, Basic immunological methods, in *Clinical Aspects of Immunology*, Gell, P. G. H., Coombs, R. R. A., and Lachman, P. J., Blackwell Scientific, Oxford, 1975, 3.
14. **Gingell, D. and Vince, S.**, Substratum wettability and charge influence the spreading of Dictyostelium amoebae and the formation of ultrathin cytoplasmic lamellae, *J. Cell Sci.*, 54, 255, 1982.

THE EDITOR

Pierre Bongrand is Professor of Immunology at Marseilles Medicine Faculty and Hospital Biologist at Marseilles Assistance Publique, Marseilles, France.

He was educated in theoretical physics at the Ecole Normale Superieure, Paris and in medicine (Paris, Marseilles). He received the Sc.D. from Paris University in 1972, the M.Sc. in 1974 from Marseilles University, and state science doctorate in Immunology in 1979. He was appointed to the position of Assistant Professor in the Department of Immunology of the Marseilles Medicine Faculty in 1972. In 1982, he spent a sabbatical year at the Centre d'Immunologie, Marseilles-Luminy.

His main research interest is in the use of physical concepts and methods to quantify the behavior or mammalian cells, particularly phagocytes and lymphocytes, in experimental or pathological situations. His favorite models are cell adhesion and cell response to various biochemical or biophysical stimuli.

Dr. Bongrand is co-author of about 40 research publications in the field of immunology and cell biophysics.

CONTRIBUTORS

George I. Bell, Ph.D.
Division Leader
Theoretical Division
Los Alamos National Laboratory
Los Alamos, New Mexico

Anne-Marie Benoliel
Engineer
Laboratory of Immunology
Hôpital de Sainte-Marguerite
Marseilles, France

Pierre Bongrand, M.D., Sc.D.
Professor
Laboratory of Immunology
Hôpital de Sainte-Marguerite
Marseilles, France

Christian Capo, Sc.D.
Head of Research
Laboratory of Immunology
Hôpital de Sainte-Marguerite
Marseilles, France

Adam S. G. Curtis, Ph.D.
Professor
Department of Cell Biology
University of Glasgow
Glasgow, Scotland

P. G. de Gennes
Professor
Laboratoire de Matière Condensée
Collège de France
Paris, France

Evan A. Evans, Ph.D.
Professor
Departments of Pathology and Physics
University of British Columbia
Vancouver, BC, Canada

Colette Foa, D.Sc.
Head of Research
AIP "Jeune Equipe"
CNRS
Marseilles, France

Jean-Remy Galindo
INSERM
Marseilles, France

Jean-Louis Mege, M.D.
Assistant Hospitalo-Universitaire
Laboratory of Immunology
Hôpital de Sainte-Marguerite
Marseilles, France

David M. Segal, Ph.D.
Senior Investigator
Immunology Branch
National Institutes of Health/National
 Cancer Institute
Bethesda, Maryland

TABLE OF CONTENTS

Section I
Basic Physical Background

Chapter 1

INTERMOLECULAR FORCES

Pierre Bongrand

TABLE OF CONTENTS

I. INTRODUCTION

An obvious requirement for understanding cell-cell adhesion is to know what forces are experienced by membrane molecules when two cells encounter each other. Unfortunately, obtaining such knowledge seems a formidable task. Indeed, the detailed molecular structure of cell membranes has not yet been completely determined, although much progress was achieved during the last few years. Further, even if all required data were available, there would remain a need to predict the interaction between a very wide range of complex structures. This is by no means a simple problem since even the interaction between two small atoms or molecules in biological media is not easily amenable to rigorous theoretical treatment. The last point may be exemplified by the following data: the electronic energy of a small atom such as ^{12}C is on the order of 10^{-17} to 10^{-16} J.[1] The energy of a typical covalent bond such as C–C is much smaller since it is about 6×10^{-19} J.[2] However, intercellular adhesion usually involves a large number of much weaker noncovalent interactions; the energy of a typical hydrogen bond in vacuum is about 3×10^{-20} J.[3] Finally, in biological media, interactions between cell-surface molecules are often substantially screened by competition with water or ions. For example, this competition may decrease the energy of a hydrogen bond by a factor of 10.[4] Hence, the binding energies that are of interest for us are differences between much higher quantities. As a consequence, theoretical determination of intermolecular forces involves a cascade of calculations and approximations, which makes it very difficult to assess the validity of obtained results. Further, as is discussed later in the chapter, the experimental determination of binding energies is not a straightforward process and may also involve many assumptions. This situation is responsible for the many errors and unwarranted assumptions that appeared during the history of the study of intermolecular forces.[5]

The purpose of the present review is to provide a brief sketch of the available methods (both experimental and theoretical) of studying intermolecular forces that were applied to biological adhesion or that seemed of potential interest in this domain. Selected experimental results are also presented. First, a molecular description of the cell surface is given, in order to provide the reader with a well-defined model for applying different methods and results on intermolecular forces. Second, forces between atoms and molecules are described, both in vacuum and in aqueous ionic solutions. Third, forces between macroscopic bodies are discussed; although these are a sum of interatomic forces, specific theories were elaborated to bypass a detailed account of individual molecules. Fourth, selected examples of molecular interactions likely to occur near cell surfaces are described.

II. MOLECULAR STRUCTURE OF THE CELL SURFACE

Our aim is to provide a basically correct (although admittedly approximate) view of intermolecular forces that are likely to be generated when the distance between two cell surfaces is decreased. For this purpose, the cell surface has to be described at the nanometer level. Additional information on the organization of cell asperities with a size on the order of 0.1 to 1 μm may be found in Chapter 8. Also, much information of the mechanical properties of cell surfaces (as studied at the micrometer level) may be found in Chapter 4.

A. Core Structure of the Cell Plasma Membrane

The "fluid mosaic" model of the cell membrane as elaborated by Singer and Nicolson[6] is now widely accepted. More recent information may be found in excellent textbooks.[7,8] The basic structure is made of a lipid bilayer tightly associated to intrinsic membrane proteins (Figure 1).

1. Lipids

The lipid bilayer is made of amphipathic molecules comprising phospholipids (e.g., phosphatidylcholine or sphingomyelin), cholesterol, and glycolipids (such as gangliosides). This structure comprises a hydrophobic middle layer of about 40 Å thickness[9] that is made of hydrocarbon chains, and two external hydrophilic layers of about 15 Å thickness each[9]; these include the polar heads of lipid molecules and bear electric charges (e.g., the positive charges of choline groups and the negative charges of phosphate groups or the sialic-acid residues of gangliosides). It is well known that the bilayer is in a fluid state[10] and lipid molecules exhibit thermal motion with a diffusion constant on the order of 10^{-8} cm^2/sec.[11] However, molecular associations may result in coordinated movements of subclasses of lipid molecules (e.g., gathering of gangliosides to a pole of the cell after suitable cross-linking of a fraction of these molecules, a process called capping[12]). This possibility may be relevant to cell adhesion since local redistribution of membrane molecules may substantially alter intercellular forces. A final point is that several reports emphasized the possibility of phase separation with formation of distinct lipid domains.[13]

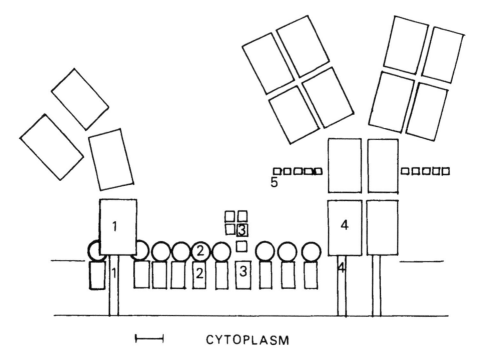

FIGURE 1.   Molecular shape of the cell surface at the nanometer level. Several molecules found on cell surfaces are shown. Emphasis is on molecular size rather than precise conformation. (1) Class I histocompatibility molecule, (2) phospholipid molecule, (3) ganglioside GM1, (4) surface immunoglobulin G (some lymphoid cells), (5) oligosaccharide chain. Bar is 2 nm.

2. Proteins

Cell-surface lipids are bound to a similar amount (by weight) of so-called "intrinsic membrane proteins" that cannot be removed without the aid of detergents and concomitant disorganization of the bilayer structure.[14] An erythrocyte membrane contains about 5×10^6 polypeptide chains[14] for a total area of about 140 μm^2.[15] A small number of protein species accounts for most of the membrane protein content; an erythrocyte membrane contains about 500,000 glycophorin molecules.[7] Lymphoid cells are endowed with several hundreds of thousands of class I histocompatibility molecules.[17] Also, many functionally important proteins are much less numerous on the cell surface (e.g., transport proteins, ionic channels, or hormone receptors).

Although the structure of membrane proteins is more difficult to study than that of soluble molecules, sufficient information is available to construct a reasonable picture of the cell-surface intrinsic proteins. Typical molecules may be anchored in the hydrophobic plasma membrane through a sequence of 10 to 20 hydrophobic amino acids,[17,18] and the largest portion is usually exposed to the extracellular aqueous medium (mostly intracellular proteins are of lesser importance for our purpose). Following are some specific examples.

A glycophorin molecule is made of 131 amino acid residues and about 100 saccharide residues arranged as 16 short carbohydrate chains with a length of order of 1 nm each.

The surface of a B lymphocyte may bear sIgM molecules that are made of four polypeptide chains with a molecular weight of about 180,000 (with 10% carbohydrate). The amino acid residues form 14 globular domains of 20 to 25 Å diameter. The molecule protrudes above the bilayer surface by about 100 Å. This structure has long been determined due to the easy availability of large amounts of soluble immunoglobulin molecules.

Class I histocompatibility molecules are made of "heavy" polypeptide chains of about 44,000 mol wt and β2 microglobulin chains of 12,000 mol wt. The discovery of sequence homologies between these proteins and immunoglobulins makes attractive the concept of an immunoglobulin-like domain structure for histocompatibility molecules. These molecules might thus be viewed as fairly globular entities of about 50 Å size. This example is important since essentially all nucleated cells bear histocompatibility molecules on their surface.

An additional result was obtained by electron microscopic observation of carbon or platinum replica obtained after freeze-fracture of plasma membranes, a procedure that may separate lipid monolayers and expose tangential membrane sections. These sections were found to contain particulate structures of about 75 Å diameter (the so-called intercalated particles) that are thought to represent membrane proteins. An erythrocyte membrane contains about 500,000 of these particles,[8] which is consistent with the view that they are made of a small number of polypeptide chains.

The above data are also consistent with several studies made on the surface distribution of different cell-surface molecules. Ferritin-labeled antibodies were used to label surface immunoglobulins[20] or histocompatibility antigens,[21] and electron micrographs were used to determine the radial distribution function of labeled molecules (in analogy with radial distribution functions used in liquid state theory). These experiments supported the concept that membrane molecules are scattered over the entire cell surface (polarized cells with specialized membrane areas, such as thyroid cells, may display different features[22]). However, the above results refer to undisturbed cells. Different experiments are required to understand the molecular redistribution phenomena that may occur during cell adhesion.

3. Movements of Cell-Surface Proteins

That cell membrane proteins can display extensive in-plane movements that has been checked experimentally for more than 10 years. Frye and Edidin[23] fused human and murine cells after labeling surface antigens with fluorescent antibodies of different colors; heterokaryons displayed rapid mixing of both labels. More recently, the diffusion constant of cell-surface molecules was quantified by measuring fluorescence recovery after photobleaching.[24,25] Briefly, a single cell bearing a fluorescent label bound to some class of membrane molecules is locally illuminated with a laser beam of a fraction of a micrometer width, and fluorescence is measured with a photomultiplier. The labeled molecules located in the illuminated area are then bleached by brief 1000-fold increase of the light intensity. The fluorescence recovery is then monitored under normal illumination. This recovery is ascribed to the replacement of bleached molecules with peripheral ones, due to diffusive movements. Simple quantitative models yielded diffusion coefficients on the order of 10^{-12} to 10^{-11} cm^2/sec. However, fluorescence recovery was often lower than 100%, indicating that a proportion of surface molecules did not exhibit long-range diffusion. Also, molecular diffusion was found to be anisotropic,[26] with a possible difference of a factor of 10 between the diffusion rates of a given molecule when measured along two perpendicular directions. Of course, an underlying assumption was that molecular movements were not altered by the bleaching process.

Another important point is that cross linking of a class of cell-surface molecules may result in extensive redistribution of these molecules with: (1) gathering into small patches scattered on the membrane (the patching phenomenon[27]), (2) gathering into a cap on a pole of the cell (this is the capping phenomenon, an active cellular process[27]), or (3) immobilization of cross-linked molecules.[28] These movements involve interactions between submembranous cytoskeletal constituents and cell-surface molecules.[29] Obviously, these phenomena may play a role in the reorganization of membrane contact areas during intercellular adhesion.

A final point is that redistribution of cell-surface molecules may be induced by a suitable external force. Subjecting embryonic muscle cells to an electric field higher than about 1 V/cm resulted in substantial redistribution of concanavalin A binding structures.[30]

The above data on the organization of intrinsic membrane components are summarized in Figure 1. However, this figure is not complete due to the presence on the surface of most cell species of more loosely bound polysaccharidic and proteic components collectively known as the cell "glycocalyx" or "fuzzy coat" that we shall now describe.

B. Structure of the Cell Glycocalyx

The word glycocalyx was coined by Bennett in 1963 to denote a polysaccharide-rich cell-surface structure that may be found on most cell species.[31] This may be visualized by staining cells with colloidal iron or ruthenium red and appears as a fuzzy coat the thickness of which may reach 1000 Å or more.[32,33]

Major constituents of the glycocalyx are proteoglycans. A typical proteoglycan such as bovine cartilage proteoglycan is made of a core protein (200,000 mol wt) bearing about 100 glycosaminoglycan chains of about 20,000 mol wt.[34] The length of a proteoglycan molecule may be as high as several micrometers.[35] Glycosaminoglycans are made up of disaccharide repeating units including a derivative of an aminosugar and at least a negatively charged carboxyl or sulfate group.[35]

In a quantitative study made on radioactive sulfate incorporation by fibroblastic cells, Roblin and colleagues[36] found that a typical 3T3 cell incorporated circa 10^{11} to 10^{12} sulfate groups and released this quantity within about 12 to 24 hr.

The cell coat also contains peripheral glycoproteins such as fibronectin. This cell adhesion molecule (also called LETS or cold insoluble globulin) is a glycoprotein (about 5% carbohydrate) made of two polypeptide chains of 225,000 mol wt each, appearing as two strands of 610 Å length and 20 Å diameter, with a limited flexibility restricted to specific regions.[37]

An important point is the relationship between the cell and the cell coat. First, measurement of precursor incorporation[33,36,38] demonstrated the endogenous origin of at least part of this coat. Second, detachment of the cell coat may be spontaneous,[36,39] or obtained by a mere washing[40] or mild exposure to proteolytic enzymes.[36] Hence, there must exist a wide heterogeneity in the tightness of association between the core membrane and the fuzzy coat. Third, peripheral cell macromolecules may be functionally associated with cytoskeletal elements.[41,42]

C. Approximate Quantitative Model of the Cell Surface

Now a realistic model of the cell surface must be built in order to evaluate the importance of the interactions that shall be reviewed in the next section. Due to the occurrence of wide differences between the surfaces of different cell types, some arbitrariness is required in the choice of our parameters, and only orders of magnitude can be obtained. The main points are as follows:

1. The cell surface is endowed with many asperities of various shape that were given different names such as microvilli, lamellipodia, blebs, veils, and ruffles. A typical surface asperity is a microvillus of about 0.1 μm thickness, corresponding to a protrusion of the plasma membrane surrounding a bundle of microfilaments (see, e.g., Reference 27 and Chapter 8).

2. The surface density of membrane-intrinsic proteins was taken as $10^4/\mu m^2$, and these were modeled as spheres of 50 Å diameter. Hence, the mean distance between these molecules on a plane section of the membrane is about 200 Å. This estimate is fairly consistent with the size of intercalated particles and the dimension of protein domains.

3. The glycocalyx was modeled as an assembly of polysaccharide chains with an extremity bound to the cell surface and a length of about 200 Å, which is somewhat intermediate between the length of side chains of intrinsic glycoproteins (i.e., about 5 residues or 25 Å) and a proteoglycan lateral chain of 100 residues (i.e., about 500 Å). Admittedly, this is only an order-of-magnitude estimate. The molecular density in the glycocalyx

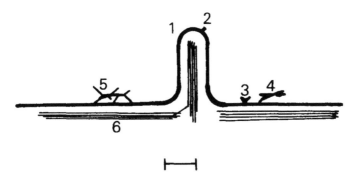

FIGURE 2. Cell surface asperities (10 nm level) with emphasis on the relative size of cell surface asperities. (1) Microvillus, (2) class I histocompatibility molecule, (3) surface immunoglobulin, (4) fibronectin, (5) proteoglycan, (6) microfilaments. Bar is 100 nm.

region may be estimated by taking as a typical membrane composition 10% carbohydrate and 45% lipid,[43] and assuming that lipids account for all the membrane area (the contribution of proteins to the cell-surface area depends on their state of aggregation, but this is not likely to alter our estimate by a factor lower than 1/2). Assuming an area of 70 Å^2 per phospholipid and 38 Å^2 per cholesterol molecule with a molar ratio of 1 between these species, it is readily found that there is a mean of 2.8 hexasaccharide residues per 108 Å^3. The mean density of saccharide residues within the glycocalyx is therefore 1.29 10^{-4} molecule/Å^3 or 0.2 M. Admittedly, the assumption of constant density within the glycocalyx is a crude approximation. However, we lack sufficient experimental data to propose an empirical law for the decrease of molecular density with respect to the distance to the bilayer.

4. The cell surface density of electric charges is likely to play a role in cell adhesion.[44,45] This charge was often studied with cell electrophoresis.[46] A problem with this technique is that the viscous drag due to the glycocalyx may drastically decrease the contribution to the measured charge of the molecules that are deeply buried in the cell coat.[47,48] This makes desirable a confirmation of experimental results with different techniques. Thus, Eylar and colleagues[49] studied the electrophoretic mobility of erythrocytes from different species and they assayed their sialic acid content with biochemical techniques. The mobility of human red cells was essentially abolished by enzymatic removal of negatively charged sialic acid molecules, and the surface density of these groups was estimated at 1/1369 Å^2. Biochemical assays revealed one neuraminidase-sensitive sialic acid group per 679 Å^2. The latter estimate leads to a charge density of 1.2 × 10^6 C/m^3. According to the above estimate for cell-surface carbohydrate density, about 6% of monosaccharide residues would bear a negative charge.

The above estimates are summarized in Figure 2. They will be used to calculate the order of magnitude of different interactions likely to occur on cell surfaces.

Now we shall review some basic results on interatomic or intermolecular forces in vacuum or in aqueous ionic solutions.

III. FORCES BETWEEN ATOMS AND SMALL MOLECULES

A comprehensive coverage of the field of intermolecular forces is clearly out of the scope of this book (and, indeed, many books would be required to do justice to this topic). Reference 50 may be of interest for a short historical perspective. A comprehensive review of older work may be found in the celebrated treatise by Hirschefelder et al.[51] that appeared in 1954.

More recent work is described in the books by Margenau and Kestner (1969)[52] and Maitland and colleagues (1981).[5] First we shall describe the interaction between atoms or small molecules in vacuum. This domain is amenable to rigorous treatment, and fairly direct experimental check of calculated potentials can be achieved. We shall then review intermolecular forces in aqueous ionic solutions. The evaluation of these forces relies on several assumptions, the most important of which may be the additivity of interactions. Also, in most cases a direct check of theoretical results is not feasible, which makes it very difficult to assess the validity of the cascade of approximations required to obtain these results.

A. Intermolecular Forces in Vacuum

First we shall describe some experimental methods of studying intermolecular forces. Then we shall consider sequentially long- and short-range interactions.

1. Experimental Study of Intermolecular Forces

We shall briefly quote several methods of interest, referring the interested reader to specialized reviews.[5]

a. Gas Imperfections

The most widely used means of evaluating intermolecular forces or testing a calculated pair energy function $W(r)$ may be to compare the experimental and calculated values of the second virial coefficient. It is well known that the pressure P of a real gas may be expanded as follows:

$$P\hat{V}/RT = 1 + B(T)/\hat{V} + C(T)/\hat{V}^2 + \ldots \qquad (1)$$

where $\hat{V}$ is the molar volume, T is the absolute temperature, R the perfect gas constant, and B, C... are the second, third... virial coefficients.

The link between virial coefficients and intermolecular forces is provided by statistical mechanics. The partition function Z of a closed system of N particles maintained in equilibrium with a heat reservoir at temperature T is

$$Z = \sum_i \exp(-E_i/kT) \qquad (2)$$

where the summation is extended over all quantum states (i) of energy E_i, and k is Boltzmann's constant. Z is linked to the free-energy F and pressure P by the following formulas:

$$F = -kT\ln Z$$
$$P = -(\partial F/\partial V)_T \qquad (3)$$

For a classical gas of indistinguishable particles of mass m, the classical expression for Z is

$$Z = (1/N!) \int \exp[-U(r_1,\ldots r_N)/kT]d\vec{r}_1..d\vec{r}_N.$$
$$\exp[-(p_1^2 + \ldots p_N^2)/2m]d\vec{p}_1\ldots d\vec{p}_N/h^{zN} \qquad (4)$$

where z is the number of degrees of freedom of a molecule, $\vec{r}_i$ and $\vec{p}_i$ denote the spatial coordinates and momentum of the ith particle, and h is Planck's constant.

Assuming pairwise additivity of interactions, and provided that the gas is diluted enough that three-body interactions are negligible, Equations 1 to 4 yield:

$$B(T) = (-N/2) \int [\exp(-W(\vec{r}_1, \vec{r}_2)/kT] - 1)d\vec{r}_1 d\vec{r}_2/V \tag{5}$$

where $W(\vec{r}_1, \vec{r}_2)$ is the intermolecular potential between particles of coordinates $\vec{r}_1$ and $\vec{r}_2$. If W depends only on the distance between particles 1 and 2, Equation 5 yields:

$$B(T) = -2\pi N \int_0^\infty [\exp(-W(r)/kT) - 1]r^2 dr \tag{6}$$

A theoretical expression for the intermolecular potential W can thus be checked by comparing the experimental and calculated functions B(T). A problem is that different potentials may yield fairly comparable values for B(T) on a finite temperature range.

Gas viscosity was also related to intermolecular forces through equations derived from statistical mechanics.[5]

b. Molecular Beam Scattering

This method consists of studying the scattering of a monoenergetic beam of molecules by other molecules. Scattering cross sections can be related to intermolecular forces, although calculations are somewhat more complex than those leading to Equation 6. Also, experiments are more difficult to perform than the measurement of bulk gas properties.

c. Spectra of van der Waals Dimers

Fairly involved experimental methods recently allowed quantitative determination of the spectra generated by the vibrations and rotations of dimers of molecules linked by noncovalent forces. This makes possible a fairly direct check of theoretical results on these interactions.

d. Experimental Determinations of Liquid Properties

The liquid state theory provides a link between the thermodynamic properties of a liquid and intermolecular forces. The methods that were evolved were used both to predict liquid properties[53] and to compare the value of different potential functions (see a recent study on water[54]). An important tool in this field is computer simulation. This methodology may prove of interest in biology due to the increasing availability of high-speed computers, and the possibility of dealing with complex systems. Two main methods are currently used.

The Monte Carlo method, first described by Metropolis and colleagues,[55] consists of using random samplings to generate a subset S of the states of a system such that the probability of including a given state i be proportional to $\exp(-E_i/kT)$ (see Equation 2) in accordance with results from statistical mechanics. The set S may then be used to average quantities such as energy or pressure (see Reference 56 for methodological details), which writes (for energy U):

$$U = \sum_{i \in S} U_i / \sum_{i \in S} 1$$

The molecular dynamics method of Alder and Wainwright[57] consists of simulating the evolution of a thermodynamic system by starting from an arbitrary configuration and calculating the changes of the particles coordinates and velocities for a sequence of short steps. As an example, Stillinger and Rahman[58] modeled liquid water by considering a few tens of thousands of steps of a duration on the order of 10^{-16} sec.

A basic limitation of these methods is that only small systems (a few hundred molecules) can be handled, which makes edge effects important and requires special procedures to ensure the constancy of the particle number and to take care of long-range interactions.

Another problem with the molecular dynamics technique is that it is difficult to deal with systems the evolution of which involves widely different time constants. Hence, these methods have to be checked with reliable theoretical results.[53,59]

Finally, there exists a variety of methods of extracting information on intermolecular potentials from experimental data. We shall now review some theoretical tools that may be combined with these methods to obtain pair potential energy functions. As will be emphasized, fairly reliable intermolecular potentials are now available only for a limited number of atoms or small molecules.[5] However, these data may be a starting point for further studies on macromolecules of biological interest.

2. Simple Treatment of Long-Range Intermolecular Forces

The starting point is a classical expression of the electrostatic interaction between molecules considered as polarizable multipoles. Three main interactions were defined.

a. Interaction between Permanent Dipoles

This interaction was first calculated by Keesom. Let us consider a cartesian frame Oxyz with $\vec{O}z$ as axis for polar coordinates and a molecule with a dipole moment $\vec{p}_1$ parallel to $\vec{O}z$ fixed at point 0. The electric field $\vec{E}$ generated by the dipole at any point M of coordinates (r, Θ, ϕ) is

$$\vec{E} = -\vec{p}_1/4\pi\epsilon_o r^3 + 3(\vec{p}_1 \cdot \vec{r})\vec{r}/4\pi\epsilon_o r^5 \tag{7}$$

where $\vec{r}$ is $\overrightarrow{OM}$ and ϵ_o is the dielectric constant of vacuum. If a molecule with dipole moment $\vec{p}_2$ is located at M, the interaction energy between both dipoles is

$$W = -\vec{p}_2 \cdot \vec{E} = [\vec{p}_1 \cdot \vec{p}_2 - 3(\vec{p}_1 \cdot \vec{r})(\vec{p}_2 \cdot \vec{r})/r^2]/4\pi\epsilon_o r^3 \tag{8}$$

The Keesom potential W_K was derived by averaging the righthand side of Equation 8 over all spatial orientations of $\vec{p}_1$ and $\vec{p}_2$ and weighting with Boltzmann's factor, which yielded:

$$W_K = -(p_1^2 p_2^2/24\pi^2\epsilon_o^2 kT)/r^6 \tag{9}$$

(W/kT was assumed to be much smaller than one, and Equation 9 represents the first order approximation for W_K).

The main conclusion is that the interaction energy is temperature dependent and varies as r^{-6}. As an example, the numerical coefficient of r^{-6} for two water molecules (p is 6.1 $\times$ 10^{-30} C.m[60]) at 300 K is 1.81 $\times$ 10^{-77} J.m^6. The Keesom interaction between two neutral molecules without any permanent dipole is zero in our first order approximation.

b. Interaction between a Permanent Dipole and a Polarizable Molecule

This potential was first calculated by Debye. A molecule exposed to an electric field $\vec{E}$ acquires a dipole moment $\vec{p}$ proportional to $\vec{E}$:

$$\vec{p} = \alpha\vec{E} \tag{10}$$

which defines the molecular polarizability α. The work w done on the molecule when an electric field E is applied is therefore:

$$w = \int \vec{E} \cdot d\vec{p} = \alpha \int \vec{E} \cdot d\vec{E} = \alpha E^2/2 = p^2/2\alpha \qquad (11)$$

This is the molecular polarization energy. The interaction potential between the molecule and the electric field is the algebraic sum of the electrostatic potential energy and polarization energy:

$$W = -\alpha E^2/2 \qquad (12)$$

Using Equations 7 and 12, the induction energy corresponding to the polarization of molecule 2 with polarizability α_2 by the dipole moment p_1 is

$$W = -(\alpha_2/2)(p_1/4\pi\epsilon_o r^3)^2 (1 + 3\cos^2\theta) \qquad (13)$$

Averaging over all spatial orientations of p_1, we obtain the high-temperature limit:

$$W_{ind} = -\alpha_2(p_1/4\pi\epsilon_o)^2/r^6$$

Adding the reciprocal term corresponding to the polarization of molecule 1 by the dipole moment of molecule 2, we obtain the Debye interaction potential W_D:

$$W_D = -(\alpha_1 p_2^2 + \alpha_2 p_1^2)/(4\pi\epsilon_o)^2/r^6 \qquad (14)$$

The numerical coefficient of r^{-6} for two water molecules (α is 1.6×10^{-40} SI units[60]) is 9.6×10^{-79} J.m^6.

c. London Interaction between Apolar Molecules

Whereas the Keesom and Debye interaction potentials between two molecules are zero when these have no permanent dipole moment, a third kind of interaction was shown to occur between all molecule species due to a typical quantum effect. The basic idea is that even if a molecule has no dipole moment ($<\vec{p}> = \vec{0}$) the mean square ($<p^2>$) may be nonzero.

A simple approximation that was often used was to model atoms as harmonic oscillators made of two charges, q and $-q$. The positive charge may be considered as fixed at point 0, and the negative charge remains on axis Oz and oscillates about the positive charge.

Now, a well-known result from quantum mechanics[61] is that the ground-state energy w_o of a harmonic oscillator is nonzero:

$$w_o = h\nu_0/2 \qquad (15)$$

where h is Planck's constant and ν_o is the oscillator frequency. This result is clearly in contrast with classical mechanics. Tentatively combining Equations 11 and 15, the mean square dipole moment of a ground-state oscillator is

$$<p^2> = h\nu_0\alpha$$

The interaction between the fluctuating dipole moment and a polarizable molecule may now be calculated with Equation 14, which yields:

$$W = [-2\alpha^2 h\nu_0/(4\pi\epsilon_o)^2]/r^6$$

The important points are (1) the quantum origin of the nonzero mean square dipole moment, which explains the appearance of Planck's constant in the theoretical formulas for the dispersion energy, and (2) the link between London interactions and polarization fluctuations, which is a starting point for macroscopic theories of electrodynamic interactions between material bodies.[62] However, the approximate reasoning we described cannot be expected to yield reliable formulas. A quantum mechanical treatment of the interaction between atomic systems may be achieved by means of the perturbation procedure, using Equation 8 for the perturbation potential. The following formula was obtained[52,63] for the "London interaction":

$$W_L = -h[3\alpha_1\alpha_2\nu_1\nu_2/32\pi^2\epsilon_o^2(\nu_1 + \nu_2)]/r^6 \tag{16}$$

where ν_1 and ν_2 are frequencies corresponding to the "dominant" oscillator component of atoms 1 and 2. It was often found that reasonable numerical values were obtained by replacing $h\nu_i$ with the ionization energy E_i of atom i, yielding:

$$W_L = -[3\alpha_1\alpha_2 E_{I1}E_{I2}/32\pi^2\epsilon_o^2(E_{I1} + E_{I2})]/r^6 \tag{17}$$

The advantage of this formula is that it involves only experimentally accessible parameters. When molecules 1 and 2 are identical, Equation 17 yields:

$$W_L = -(3\alpha^2 E_I/64\pi^2\epsilon_o^2)/r^6 \tag{18}$$

Using 12.6 eV for I_1, the numerical coefficient of r^6 for water is found to be 3.1×10^{-78} $J.m^6$. More refined formulas were obtained by replacing Equation 8 with more accurate formulas including quadrupolar and other higher-order electrical moments. However, a complete discussion of these methods is outside the scope of this review. We shall emphasize only two important points.

d. Many Body Interactions

An important problem is the additivity of intermolecular potentials. Clearly, assuming pairwise additivity allows easy expression of the potential energy of a large assembly of molecules. However, recent theoretical progress in the understanding of the liquid state led to the conclusion that many body interactions played a significant role in molecular organization.[59] The leading correction term is the Axilrod-Teller triple dipole term. The interaction potential between three identical atoms i, j, and k reads:

$$W_L = -C(1/r_{ij}^6 + 1/r_{jk}^6 + 1/r_{ki}^6)\nu_{ijk} (3\cos\theta_i\cos\theta_j\cos\theta_k + 1)/(r_{ij}r_{jk}r_{ki})^3 \tag{19}$$

where r_{ij} is the distance between atoms i and k, Θ_i is the angle $\widehat{O_kO_iO_j}$ (where O_i is the center of atom i), C is the coefficient of r^{-6} in Equation 18, and ν_{ijk} may be approximated as $-3 C/(16\alpha_o)$. Note that $(\alpha/4\pi\epsilon_o)$ is of the same order of magnitude as the atomic size.

e. Combining Rules

As is detailed in Chapter 3, it would be of high interest to derive the interaction between two different molecules of type 1 and 2 (i.e., $-C_{12}/r^6$) from the numerical coefficients C_{11} and C_{22} corresponding to the London interaction between two identical molecules. As readily shown with Equation 9, if the dipole-dipole interaction is dominant, a simple formula holds:

$$C_{12} = (C_{11}C_{22})^{1/2} \tag{20}$$

A similar equation may be used as a first approximation when London forces are dominant and molecules 1 and 2 have similar ionization energy (see Equation 17). The validity of these approximations in complex systems is obviously questionable (see Chapter 3).

A final point is about the relative order of magnitude of the above interactions. London interaction is clearly dominant when apolar molecules are studied. Keesom energy is important in highly polar molecules (such as H_2O) and Debye energy is often less than 10% of the sum of the other two interaction potentials.[5,51]

3. Ab Initio Calculation of Intermolecular Forces at Short and Intermediate Distances

The quantitative treatment of intermolecular forces at short and intermediate distances involves very heavy calculations that were made possible only with the availability of high-speed computers. Indeed, only the interaction between a proton and a hydrogen atom could be solved analytically.[64] Presently, the interaction between two small molecules such as H_2O[65] or CO_2[66] may be calculated with fairly high accuracy and satisfactory consistency with available experimental data. However, nonadditivity of intermolecular forces makes it very dangerous to apply numerical results obtained on small molecular systems to more complex molecules, although cancellation of different errors occurred several times during the history of sciences. Hence, we shall give only a very brief sketch of currently available theoretical tools of quantum chemistry, while emphasizing that it may be too early to try to apply them to the complex problem of macromolecule interactions in aqueous ionic solutions.

The interested reader is referred to several excellent textbooks for further details: while earlier developments were described in the well-known treatise by Eyring et al.,[67] followed by more elementary introductions,[68,69] the recent book by Szabo and Ostlund[70] provides a self-contained description of the present methods of quantum chemistry. The present section may be skipped without impairing the understanding of later paragraphs.

a. Schrodinger Equation

The starting point of the ab initio treatment of the interaction between two molecules 1 and 2 is to write the complete Hamiltonian of the system. A first approximation is then to fix the nuclear coordinates and calculate the electronic energy levels for a given nuclear configuration. The justification of this procedure, known as the Born-Oppenheimer approximation, is that electrons move much faster than nuclei and adopt a configuration corresponding to the instantaneous location of these heavier particles. The useful part of the Hamiltonian operator H is therefore the sum of the kinetic contributions of individual electrons and electrostatic electron-nucleus and electron-electron interactions:

$$H = \sum_{i=1}^{N} h(i) + \sum_{i,j}^{N} q^2/4\pi\epsilon_o r_{ij}$$

$$h(i) = -\hbar^2\Delta_i/2m + \sum_{\alpha} - qq_\alpha/4\pi\epsilon_o r_{i\alpha} \qquad (21)$$

where N is the number of electrons, $-q$ is the electronic charge, q_α is the charge of nucleus α, and r_{ij} and $r_{i\alpha}$ are the distances between electron i and electron j and nucleus α, respectively. Δ_i is the Laplacian operator applied to the spatial coordinates of electron i. $\hbar$ is Planck's constant divided by 2π.

b. The Antisymmetry Principle

A second point that was neglected in the approximate treatment of long-range forces is that electrons follow Fermi-Dirac statistics. The wave function of a many-electron system must be an antisymmetric function $\Phi(1, \ldots i, \ldots N)$ where N is the number of electrons and i stands for the coordinates of the ith electron. Hence:

$$\phi(1,..,i,..j,..N) = -\phi(1,..j,..i,..N) \tag{22}$$

for every permutation of the coordinates of electron i and j.

c. Use of a Basis of Functions

Now the problem is to solve the Schrodinger equation and find the eigenfunctions ϕ_u and the corresponding eigenvalues E_u of H:

$$H\phi_u = E_u \phi_u \tag{23}$$

The usual way to proceed is to express function ϕ_u as a linear combination of well-chosen functions U:

$$\phi_u = \sum_\alpha c_{u\alpha} \psi_\alpha \tag{24}$$

The theoretical justification for this procedure is that in a set of functions with suitable properties (concerning derivability and vanishing at infinity) there exists particular sets of functions (constituting bases) such that any function Φ may be uniquely expressed as a linear combination of the basis functions. The most convenient bases are orthonormal, which reads:

$$<\psi_\alpha|\psi_\beta> = \int \psi_\alpha^* \psi_\beta \, d\tau = \delta_{\alpha\beta} \tag{25}$$

where the first symbol denotes a hermitian product between functions considered as elements of a vector space. This product may be calculated by integration over all variables $(1, \ldots n)$ represented as $d\tau$, $\delta_{\alpha\beta}$ is Kronecker's symbol that is O if $\alpha \neq \beta$, and 1 if $\alpha = \beta$. ψ_α^* is the complex conjugate of ψ_α. Now, if the basis (ψ_α) is orthonormal, Equations 24 and 25 yield:

$$c_{u\alpha} = <\psi_\alpha|\phi_u>$$

Although a basis is an infinite set of functions, the basic idea is that an approximate value of O_u may be obtained by using a well-chosen finite set of functions constituting an incomplete basis. Clearly, the more extensive this incomplete basis, the more accurate the approximation of ϕ_u. Hence, solving the Schrodinger equation amounts to diagonalizing a finite matrix, which is a simpler problem than solving a many-variable, partial differential equation. There remains the finding of suitable basis functions.

d. Slater Determinants

A classical way of constructing a properly antisymmetrized N-electron wave function is to write a so-called Slater determinant:

$$\psi = (N!)^{-1/2} |X_i(j)| \tag{26}$$

where $|X_i(j)|$ is an $N \times N$ determinant with $X_i(j)$ on line i and row j. X_i is a one-electron function (it is a product of a function of three spatial coordinates, called an orbital, and a spin function that is an element of a two-dimensional vector space. X is called a spin orbital. The factor $N!^{-1/2}$ is to ensure that $<\psi|\psi>$ is equal to one, provided functions X_i are orthonormal. A physical interpretation of ψ is that this function represents a set of N electrons occupying N spin orbitals. Since ψ is obviously zero if $X_i = X_j$ for some couple (i,j), it appears that two electrons cannot occupy the same spin orbital. Antisymmetry may thus be considered as a mathematical expression of Pauli's exclusion principle.

Now, there remains to find functions X_i in order to construct the basis (ψ_α). The simplest procedure is to take as X_i linear combinations of eigenstates of isolated molecules of type 1 or 2. This was done by Heitler and London in their classical study of the interaction between two hydrogen atoms, which yielded:

$$X_1 = [2(1 + S_{12})]^{-1/2} (\phi_1 + \phi_2)|+>$$
$$X_2 = [2(1 + S_{12})]^{-1/2} (\phi_1 + \phi_2)|-> \qquad (27)$$

where Φ_i and Φ_2 are spatial 1s orbitals centered on the nuclei of atoms 1 and 2, S_{12} is $<\Phi_1|\Phi_2>$ and $|+>$ and $|->$ are spin functions. The analytical expression for Φ_i is

$$\phi_i(r) = \exp(-|\vec{r} - \vec{R}_i|)/\pi^{1/2}$$

where $|\vec{r} - \vec{R}_i|$ is the distance between any point $\vec{r}$ and the nucleus of atom i. It may be noted that the spatial part of X_i is a molecular orbital (MO) that is a linear combination of atomic orbitals (LCAO), which explains the abbreviation of MO-LCAO for the method of building basis orbitals we have just described.

However, the simple combination of ground states we have just written is only an approximation. We shall now define more accurate procedures for calculating molecular energies.

e. The Hartree-Fock Approximation

The Hartree-Fock approximation is a starting point for many modern calculations. This may be sketched as follows: the basic approximation is to write the N-electron wave function of the studied system as a simple Slater determinant (Equation 26). The choice of the one-electron function X is based on the "variational principle"; this consists of noticing that according to Equation 23 the eigenvalue associated to any eigenfunction Φ_u of the Hamiltonian operator H is

$$E_u = <\phi_u|H|\phi_u>/<\phi_u|\phi_u> \qquad (28)$$

Now, all eigenvalues of H may be shown to be positive, and the eigenfunctions constitute a complete basis. As a direct consequence of these properties, the energy E associated to any approximate form ψ of the ground state through Equation 28 is higher than the actual ground energy E_o. Hence, if the system wave function is approximated by using a trial expression ψ involving unknown parameters, these may be determined by choosing the values that make minimum the corresponding energy E. Also, using functional variations, it may be shown[70] that the spin orbitals X that minimize the energy associated to the Slater determinant ψ satisfy one-electron Schrodinger equations, the potential energy of which may be interpreted as the average potential exerted by other electrons and nuclei. These one-electron Schrodinger equations are solved approximately by expressing X as a combination of a finite set of functions. These may be atomic orbitals, but Gaussian orbitals (with an exponential term $\exp[-kr^2]$ instead of $\exp[-k'r]$) are often preferred for computational purposes. Finally, the partial differential Schrodinger equation may be replaced with a more easily solvable matricial equation, yielding a number of solutions. A Slater determinant ψ is then built out of the N spin orbitals with lowest energy. Since the basic solution procedure is an iteration consisting of guessing trial functions to derive average potentials and obtain better functions until fair consistency is obtained, this was called the self-consistent field (SCF) method. Obviously, the more extensive the basis, the more exact the solution. However, approximating ϕ as a single Slater determinant results in a neglect of electron correlations. A more refined procedure that may allow overcoming this limitation is the configuration-interaction method (CI).

f. Configuration Interaction

Whereas the Hartree-Fock N-electron wave function ϕ is written as a single determinant including the N one-electron spin orbitals with the lowest energy, the configuration-interaction method consists of expanding ϕ as a combination of "excited" Slater determinants where a number of ground state spin orbitals are replaced with excited states. Since excited states form a complete basis, the accuracy of the method is limited only by the number of basis functions and determinants.

The difference between the "accurate" value of the ground state energy and the Hartree-Fock limit is called the correlation energy.

4. Results and Problems

a. Water-Water Interaction: A Specific Example

In order to illustrate the methods we have just described, we shall give a specific example of particular interest. Clementi and Habitz[65] reported ab initio calculations on water dimers. They studied 169 geometrical arrangements, and used for their basis set 20 orbitals for oxygen atoms and 7 orbitals for hydrogen atoms. The calculation included 108,000 configurations for the dimer. Their estimate of the minimum interaction energy (5.5 kcal/mol or 23 kJ/mol) and of the corresponding geometry was consistent with experimental data obtained by molecular-beam electric resonance spectroscopy (the same value of 2.97 Å was obtained for the distance between oxygen atoms, and experimental and theoretical molecular orientations did not differ by more than a few degrees). Also, the calculated dipole moment of the isolated water molecule did not differ from the experimental value by more than 1%. However, the calculated second vivial coefficient was about half the experimental value.

The authors pointed out that about one third of the total two-body energy at the dimer equilibrium geometry was due to correlation energy corrections. They also emphasized that three-body corrections were not negligible, thus invalidating the assumption of pairwise additivity of interactions.

b. Empirical Potentials

Clearly, writing a set of discrete numerical values of the intermolecular potentials corresponding to particular geometries is not a convenient way of expressing theoretical results. Hence, much work was devoted to the derivation of empirical or semiempirical potential functions to summarize available results. A wide variety of functions with different complexity and accuracy were suggested,[5] but we shall restrict ourselves to the simplest cases.

The simplest and most common way of expressing the long-range attraction between simple neutral atoms is the well known -6 potential ($-c/r^6$ where c is a constant). However, a short-range repulsion term is required to account for the impossibility of atomic interpenetration and occurrence of a minimum energy at some equilibrium distance. The most widely used repulsion function may be the empirical r^{-12} potential, yielding the celebrated 6-12 Lennard-Jones potential:

$$W = 4\epsilon[(\sigma/r)^{12} - (\sigma/r)^6] \tag{29}$$

where ϵ and σ are constants. The 12 exponent was chosen for computational purposes. An alternative repulsion function that may be more justified theoretically is the exponential repulsion (see Reference 71 for a model study and Reference 5 for a review). However, it must be pointed out that the resulting potential:

$$W = A \exp(-\eta r) - c/r^6 \tag{30}$$

cannot be used at all distances since it exhibits an unphysical decrease to $-\infty$ at $r = 0$, which requires suitable correction. Finally, the simplest repulsion potential is the so-called hard-core potential (W is infinite when r is less than some parameter d and it is zero elsewhere).

Now, Equations 29 or 30 cannot be applied to asymmetric molecules. A convenient way of dealing with such systems is to write the interaction between small molecules as a sum of site-site interactions. For example, Bohm and colleagues[66] represented the short-range repulsion and the coulomb electrostatic forces between simple molecules A and B (such as CH_4, CH_3CN, CO_2) with the following formula:

$$E = \sum_{i/j} \exp[-(R_{ij} - \sigma_i - \sigma_j)/(\sigma_i + \sigma_j)] + q_i q_j / R_{ij}$$

where R_{ij} is the distance between sites i and j, the exponential term represents short range repulsion, and $q_i q_j / R_{ij}$ is the long-range coulombic interaction. Interaction sites may be atomic centers or arbitrary points (see also the fitting of water-water interaction potentials to empirical expressions[65,72]).

An obvious temptation is to express the interaction between macromolecules as a sum of interactions between small constituent groups and use for the latter the potentials obtained from studies made on small molecules. Although this procedure seems reasonable, it is very difficult to check experimentally or theoretically[73] the validity of the assumption of additivity between intramolecular interactions.

c. Numerical Examples of Biological Interest

Although the above results apply only to very simple molecules, it may be instructive to describe presently available numerical data relevant to chemical groups likely to be found on biological molecules, such as $-CH_3$, $-OH$, $-CO$, or $-NH_2$. These data are often obtained by combination of theoretical and experimental information.

A first point is that whatever the exact short-range repulsive potential, atoms behave as low-penetrability spheres with a fairly well-defined radius (the van der Waals radius). Numerical values for this radius may be found in standard chemistry textbooks.[74]

A second point is that polar interactions are much more important than dispersion forces in the biological milieu. The minimum intermolecular energy between two CH_4 groups is about 2.1 kJ/mol,[71] which is about tenfold lower than the interaction between two water molecules.

A third point is that the interaction potentials between polar molecules is clearly anisotropic. Okazaki and colleagues[75] constructed an intermolecular potential for H_2O–CH_3OH dimers out of quantum chemical calculations made on 291 geometrical configurations. They found that when water was in the direction of the hydroxyl group of the methanol molecule, the minimum energy configuration corresponded to an interaction energy of about 24 kJ/mol (representing a hydrogen bond where water was a proton acceptor) whereas the minimum interaction energy was about 3 kJ/mol when water was in the direction of the methyl group.

A fourth point is about the "specificity" of polar interactions. It is of high interest to determine the dependence of the interaction energy on slight changes of intermolecular distances, relative orientation of interacting groups, or replacement of a chemical group with another one. Many numerical data on hydrogen bonds are reviewed in Reference 3 and we give only a few orders of magnitude. A 0.5 Å variation of the distance between hydrogen-bonded water molecules may result in a loss of 50% of the binding energy. However, some rotations of interacting molecules by several tens of degrees involved much smaller energy increases.[65]

Another point is the dependence of bond energies on the nature of the proton donor and acceptor. The experimental dimerization energies of H_2O and NH_3 are about 22 and 17 kJ/

mol, respectively.[3] Also, Okazaki and colleagues[75] found that bonding energy was only 17 kJ/mol (instead of 25 kJ/mol) when methanol acted as a proton acceptor instead of water in the water-methanol interaction.

Clearly, hydrogen bonding between chemical groups of biological interest in vacuum may involve fairly high energies. However, the influence of the biological milieu on intermolecular forces may be very important, as discussed below.

B. Intermolecular Forces in Aqueous Ionic Solutions

The results described in the preceding section cannot be used to predict the mutual interaction between cell-surface macromolecules because of the dramatic influence of the surrounding medium on these interactions. Due to the complexity of involved phenomena, it may be useful to separately describe different basic processes. Further theoretical and experimental progress may lead to a more synthetic view of molecular interactions. We shall review sequentially the so-called hydrophobic effect and the influence of biological media on electrostatic forces and hydrogen bonds.

1. The Hydrophobic Effect
a. Qualitative Description

The starting point is that water is a very peculiar fluid as compared to substances of similar structure and molecular size.[76] Indeed, the boiling point of H_2O is much higher than that of H_2S or NH_3, and the surface tension of liquid water is much higher than that of, e.g., CH_3OH or alcanes (see Chapter 3). Also, the solubility of an apolar gas in water is usually lower than in typical organic liquids. Further, the enthalpy and entropy of solution of a polar gas in water are both negative, resulting in a decrease of solubility when temperature is increased.[76]

Another interesting point is that when different authors measured the free energy involved in the transfer of different amino acids from water to different organic liquids, they found a striking correlation (with a correlation coefficient of order 0.99) between this free energy and the molecular area of these amino acids (see Reference 77 for references). An attractive interpretation of these findings was that the dominant component of the transfer free energy was the free energy required to create an empty cavity in water in order to accommodate the solute molecule, the free energy decrease by solute-solvent interaction being negligible in a first approximation, in the absence of any hydrogen bond formation. It was thus of interest to examine these hypotheses quantitatively. This required a clear definition of molecular areas (see Reference 78 for a discussion of this problem). The "accessible surface" of a molecule X was defined as the locus of the center of a spherical probe representing a water molecule (about 2.8 Å diameter) touching the molecule X represented as a set of spheres obtained by taking the van der Waals radii of individual atoms. The "contact surface" is the locus of contacts between the wandering spherical probe and the van der Waals spheres constituting molecule X; this is a disconnected surface. This contact surface is about 3.5 times lower than the accessible surface for different protein structures[77] and the free energy of transfer of small molecules from an organic solvent to water is about 105 $J/mol/Å^2$ of accessible surface (or 17 mJ/m^2) and it is about 420 $J/mol/Å^2$ of contact area (or 70 mJ/m^2). These orders of magnitude are comparable with the surface energy of water (about 73 mJ/m^2 at room temperature).

It may thus be expected that the coalescence of two cavities will lower the system free energy, as a consequence of the decrease of the total cavity surface. This phenomenon might be responsible for the generation of an attractive force between two solute molecules without any requirement for "direct" intermolecular forces. Clearly, this interaction will be maximal with apolar substances that cannot form any hydrogen bond with water.

b. Quantitative Theories and Computer Simulations

Although much indirect information on hydrophobic interactions was gathered, mainly by thermodynamic analysis of experimental solvation enthalpies and entropies in different molecular systems, it was only recently that the modern tools of liquid-state science provided a detailed picture of this phenomenon with reasonable reliability.

The first point is that the interaction potential between two apolar particles embedded in water is related to the pair-distribution function g that may be defined as follows: let ρ be the mean number of particles per unit volume. The probability of finding a particle (2) in an elementary volume dv at a distance r from a given particle (1) is $\rho.dv.g(r)$ (we assume spherical symmetry). According to this definition, g(r) would be equal to 1 in the absence of correlation between particle positions, and the pair correlation function $h(r) = g(r) - 1$ would be zero. Further, g plays a central role in liquid-state theory since it is closely related to thermodynamic parameters, and it can be studied in a fairly straightforward way with light- or neutron-scattering experiments.[53] Thus, the "potential of mean force" w between two particles is given by:

$$w(r) = -kT \ln g(r)$$

where k is Boltzmann's constant. Since the probability of occurrence of any state of free energy F of a system at temperature T is proportional to $\exp(-F/kT)$, it is easily shown that w(r) is the reversible work required to bring the two particles from infinite separation to a distance r.[79] A widely used way of obtaining an approximate form of g(r) is to use the Ornstein-Zernicke equation that writes, in the case of a monomolecular fluid of symmetrical molecules, as follows:

$$h(\vec{r}) = c(\vec{r}) + \int h(\vec{r}') c(\vec{r} - \vec{r}') d\vec{r}'$$

where c is called the direct correlation function. The intuitive interpretation of the above equation is that the correlation between the molecular density at two points separated by a distance r is the sum of:

1. A direct component c(r) corresponding to the potential generated by a particle located at point 0 on point r
2. An indirect component corresponding to the change of the potential generated at point r by the charge density at points r', due to the change of the particle density caused by the presence of a fixed particle at 0

The Ornstein-Zernicke equation serves to define c(r). It would be of no use if it had not been demonstrated that simple approximations might "close" this relation. A simple example of such a closure is the following "mean spherical approximation":

$$g(r) = 0 \ (r < d)$$

$$c(r) = -V(r)/kT \ (r > d)$$

where d is the diameter of particles considered as hard spheres and V(r) is the attractive potential at distance $r > d$.

Pratt and Chandler[80] elaborated a semiempirical theory of the hydrophobic effect by relating the correlation between Lennard-Jones apolar molecules embedded in water to the experimental correlation function between water molecules. Further, Pangali and colleagues[79] performed a Monte-Carlo simulation of the hydrophobic interaction. Also, Nakanishi and colleagues[81] simulated the arrangement of water around a hydrophobic tertiary butanol molecule. Finally, Lee and colleagues[82] performed a molecular dynamics simulation of water near an extended hydrophobic surface. The main results may be summarized as follows:

1. The formation of an apolar cavity results in a local increase of water-water interaction with increased order and decreased potential energy, in accordance with classical concepts.[81]
2. When the distance between two apolar spheres embedded in water is decreased, there are two relatively stable configurations, in which (1) spheres are in close contact without any water molecules in between, and (2) each sphere is in a water cage, with a water molecule between spheres[79] (see also Reference 83).
3. An important point is that the hydration structure of large hydrophobic surfaces can be very different from that of small hydrophobic molecules, with an important dependence on geometrical parameters.[82]

Indeed, the overall conclusion is that there is a balance between the tendencies of the liquid to maximize the number of hydrogen bonds on the one hand, and to maximize the packing density on the other.[82] Precise molecular shapes and dimensions are therefore expected to play an important role in hydrophobic interactions.

2. Electrostatic Interactions in the Biological Milieu

We shall describe sequentially the influence of the solvent and that of solute ions on coulombic forces.

a. Effect of Water on Electrostatic Interactions

We shall first describe the macroscopic concept of dielectric constant, purposely ignoring the problems posed by this approach. The significance of this treatment will then be discussed and more recent concepts and results will be reviewed.

i. Macroscopic Treatment of Solvent Polarization

Let us consider a volume distribution of electrostatic charge density $\rho(r)$ in vacuum. The potential V_o at some point M is

$$V_o(M) = \int [\rho(\vec{r}')/4\pi\epsilon_o r] d^3\vec{r}' \tag{31}$$

where the integration constant is extended over all points M' of space and $\vec{r}'$ is $\overrightarrow{MM'}$. Now, suppose we fill a portion of space with a continuous material medium. The basic hypothesis is that the electric field $\vec{E}_o = -\overrightarrow{grad}V_o$ will polarize the medium, thus inducing a dipole moment density $\vec{P}(r)$. According to Equation 31 and results from classical electrostatics, the total potential at M will be

$$V(M) = \int [\rho(\vec{r}')/4\pi\epsilon_o r] d^3\vec{r}' + \ldots$$

$$\ldots \int [\vec{P}(\vec{r}') \cdot \overrightarrow{grad}(1/r)/4\pi\epsilon_o] d^3\vec{r}' \tag{32}$$

Now, using classical formulas from vector analysis, we obtain:

$$V(M) = (1/4\pi\epsilon_o)\left[\int (\rho(\vec{r}') - div\,\vec{P})/r\,d^3\vec{r}' + \int_s \vec{P} \cdot \vec{n} \cdot dS/r\right]$$

where the second integral is calculated over the dielectric surface S. This equation shows that the electric field generated by the dielectric is equivalent to that produced by a volume charge distribution of density $-\text{div}\,\vec{P}$ combined with a surface density of charges located on the dielectric surface equal to $\vec{P}\cdot\vec{n}$, where $\vec{n}$ is a unit vector normal to the dielectric surface and pointed toward vacuum.

Now, let us assume that the dielectric fills all space and that the polarization P is proportional to the total electric field $\vec{E} = -\overrightarrow{\text{grad}\,V}$:

$$\vec{P} = (\epsilon - \epsilon_o)\cdot E \tag{33}$$

where ϵ is defined as the dielectric constant of the medium. Since it is well known from classical electrostatics that:

$$\text{div}\,\vec{E}_0 = \rho_o/\epsilon_o$$

This is Poisson's law. If we assume that the total electric charge is zero and charges are restricted to some finite region of space, the above equations are satisfied by the following formulas:

$$\vec{E} = \vec{E}_0\cdot\epsilon_o/\epsilon$$

$$\text{div}\,\vec{P} = \rho_o(1 - \epsilon_o/\epsilon)$$

Assuming that the solution to the basic equations is unique, we conclude that the effect of the dielectric is to replace the charge distribution ρ by an effective distribution $\rho.\epsilon_o/\epsilon$. The dielectric thus screens all electrostatic interactions and divides them by a factor of ϵ/ϵ_o. This factor is called the relative dielectric constant of the medium, ϵ_r. In water, ϵ_r is about 80. The interaction potential between two electronic charges at distance r is therefore found to be

$$w = \frac{17.3}{r}\text{ kJ/mol (water)}$$

instead of:

$$w = \frac{1387}{r}\text{ kJmol (vacuum)}$$

where r is expressed in angstrom units.

ii. Link between Macroscopic and Microscopic Theories

Let the dielectric be considered as an assembly of r molecules of polarizability α per unit volume. The dipole moment density in a region of constant electric field $\vec{E}$ is

$$\vec{P} = r\alpha\vec{E}' \tag{34}$$

where $\vec{E}'$ is the electric field "seen" by a molecule. The usual way of calculating this field is to consider that $\vec{E}'$ is the average electric field $\vec{E}$ minus the molecule contribution to $\vec{E}$. This contribution is calculated as the electric field within a sphere with constant polarization density $\vec{P}$. It is easily shown that this field is equal to the field generated by two spherical charge densities ρ_q and $-\rho_q$ separated by a distance P/ρ_q. The electric field is constant and equal to $\vec{P}/3\epsilon_o$, which yields:

$$\vec{E}' = \vec{E} + \vec{P}/3\epsilon_o \tag{35}$$

combining Equations 33, 34, and 35, we obtain:

$$(\epsilon - \epsilon_o)/(\epsilon + 2\epsilon_o) = r\alpha/3\epsilon_o \tag{36}$$

which is the well-known Clausius-Mosotti relationship. Now, the polarizability α of a fluid made of polarizable molecules with a permanent dipole moment $\vec{P}$ is the sum of an induction term (α_d) corresponding to the electronic cloud polarization by the electric field and a temperature-dependent orientation term α_p corresponding to the orientation of the permanent dipole moment, according to Boltzmann's law:

$$\alpha_p = (1/E) \int_0^{2\pi} d\phi \int_0^{\pi} d\theta \cdot p \cdot \cos\theta \cdot \exp(-pE\cos\theta/kT) \Big/ \int_0^{2\pi} d\phi \int_0^{\pi} d\theta \exp(-pE\cos\theta/kT)$$

Assuming that pE/kT is much smaller than 1, we obtain:

$$\alpha_p = p^2/3kT \tag{37}$$

the dependence of α_p on the second power of p has an important consequence: if intermolecular forces induce some correlation between the orientation of neighbor molecules, the "effective" α_p will be increased. Hence, the high value of the dielectric constant of water ($\epsilon_r = 78.5$) as compared to NOCl (ϵ_r is 18 at 285 K) that has a comparable dipole moment[84] may be ascribed to the high correlation between the orientations of neighboring molecules, which is a consequence of hydrogen bonding. This is the basis of Kirkwood's theory of the dielectric constant of water.[85] However, the macroscopic approach we have just described is fraught with severe difficulties. Indeed, the concept of a macroscopic electric field as an average field calculated on a "sufficient volume" is not very useful at the molecular level. Further, the introduction of spherical cavities with a macroscopic derivation of the inside effective field is not warranted (indeed, the shape of the cavity may influence the result of calculations). Recently, several authors made use of the concepts of pair correlation functions and Ornstein-Zernicke-like equations with approximative closures to estimate the dielectric constant of polar solvents[86,87] or molecules with multiple interaction sites.[88] Also, computer simulations were used to study ionic interactions in polarizable fluids.[89,90] Although more information on interionic forces in water are needed, the following conclusions may be drawn:

1.   The concept of dielectric constant is of little use to account for short-range interionic forces[90] and should be replaced with a distance-dependent screening coefficient.

2. The screening of electrostatic interactions decreases very rapidly when the distance between interacting charges is decreased and becomes comparable to the size of the solvent molecules,[86] and the detailed dependence of interionic forces on distance is highly influenced by molecular shapes and the relative size of ions and solvent molecules.[88,89]

3. The discrepancy between the predictions of continuum and more refined theories for interionic potential at contact may amount to 100%.[88]

However, the aforementioned theories essentially referred to the low dilution limit of ionic solutions. Another important parameter of electrostatic interactions is the screening effect of solute ions.

b. Electrostatic Interactions in Ionic Solutions

In the following paragraphs, we shall only describe the Debye-Huckel theory of dilute ionic solutions. The treatment of ion or dipole interaction in concentrated solutions should involve the same methods as were described in the preceding sections, with somewhat increased complexity.

We refer the reader interested in more detailed experimental data to general textbooks on ionic solutions.[91] We shall only describe a macroscopic theory that is a basis for the DLVO theory (Section IV).

Let us consider a dilute aqueous solution of a 1:1 electrolyte such as NaCl. The mean concentrations of cations (c_o^+) and anions (c_o^-) are equal. Now, let us assume that there is a varying potential field $V(r)$ in some region of the solution. Neglecting ion-ion correlations, we may use Boltzmann's law to estimate the mean charge density ρ at any point r:

$$\rho = c_{o+}q \exp[-qV(r)/kT] - c_0^- q \exp[qV(r)/kT]$$

$$\rho = -2 c_o q \sinh[qV(r)/kT] \tag{38}$$

where q is the electron charge, k is Boltzmann's constant, T is the absolute temperature, and V is taken as zero at infinity. Combining with Poisson's law, we obtain the well-known Poisson-Boltzmann equation:

$$\Delta V = 2 c_o q \sinh[qV(r)/kT]/\epsilon \tag{39}$$

Assuming that $V(r)/kT$ is much smaller than 1, we obtain the linearized Poisson-Boltzmann equation:

$$\Delta V = (2c_o q^2/kT\epsilon)V \tag{40}$$

Now, suppose there is a fixed charge Q at some point O. Due to spherical symmetry, the mean potential V at any point M will be dependent only on the distance r = OM. The Laplacian operator Δ then becomes very simple and Equation 40 reads:

$$d^2/dr^2(rV) = (2c_o q^2/kT\epsilon)(rV) \tag{41}$$

Since $V(r)$ must vanish at infinity and tend to $Q/4\pi\epsilon r$ when r is small, we obtain:

$$V(r) = (Q/4\pi\epsilon r) \exp[-(2c_o q^2/kT\epsilon)^{1/2}r]$$

$$V(r) = (Q/4\pi\epsilon r) \exp(-r\kappa), \quad 1/\kappa = (kT/2c_o q^2)^{1/2} \tag{42}$$

$1/\kappa$ is the so-called Debye-Huckel length. This is 8.1 Å in a 0.15 M NaCl solution. The simple interpretation of Equation 42 is that in an ionic solution the potential field generated by a given ion exhibits an exponential decay due to the screening exerted by other ions. However, the simple treatment we described is subject to the severe criticism concerning the use of a macroscopic dielectric constant to describe ionic interactions. Also, correlations between the fluctuations of ionic displacements was neglected. Experimental studies made on the thermodynamic properties of aqueous ionic solutions suggest that the Debye-Huckel theory fails when c_o is higher than 10^{-2} to 10^{-1} M.[91] The modern theoretical treatment of concentrated ionic solutions relies on pair correlation functions[92,93] and computer simulations, as was explained in the review on the dielectric constant. However, present studies mostly address the problem of ionic interactions in water at infinite dilution. Further progress is needed before a reliable theory relevant to biological systems is provided.

In conclusion, it appears that electrostatic interactions may be several orders of magnitude lower in aqueous electrolyte solutions than in vacuum due to a screening by water and ions. Whereas long-distance interactions are described with a very simple formula (Equation 42), the "macroscopic" or "continuum" theories we described failed to account for the electrostatic interaction between two charges separated by a few water-molecule diameters. In this case, ionic shapes and dimensions may play an important role in the local arrangement of solvent molecules. This phenomenon may play a prominent role in the selectivity of ionic interactions (e.g., this may explain why K^+ and Na^+ or Ca^{2+} and Mg^{2+} may be easily discriminated by biological macromolecules).

3. Hydrogen Bonds in Aqueous Ionic Solutions

As was indicated above, the hydrogen bond is a fairly strong interaction (the binding energy is of order of 20 kJ/mol). However, since water can act as both a proton donor and acceptor, the formation of a hydrogen bond between two groups R–H and :X–R' dissolved in aqueous medium may read:

$$R\text{–}H...OH_2 + H\text{–}O\text{–}H...X\text{–}R' \leftrightharpoons H\text{–}O\text{–}H...OH_2 + R\text{–}H...:X\text{–}R' \qquad (43)$$

where . . . is for a hydrogen bond. As a consequence, the association between these groups results in no change of the overall number of hydrogen bonds. The apparent energy of the hydrogen bond is therefore the net result of formation and destruction of four H bonds. Two factors may influence the final equilibrium:

1. The nature of proton donors and acceptors may influence the hydrogen bond energies (see Section III.A.4.c).
2. When a potential proton donor and acceptor are brought in close distance following the approach between two molecules, the formation of a hydrogen bond between these groups may be favored by an entropic parameter that is discussed below. This may be as high as 20 kJ/mol.[94] However, if the relative orientation of these groups is not compatible with the formation of an unbent hydrogen bond, association of these groups with water may be favored.[94]

Also, the competition of water molecules for bond formation is obviously decreased in regions that are poorly accessible to water (e.g., the protein interior or some regions of contact between two macromolecules). Hence, no general rule can be obtained to estimate the effective hydrogen bonding energies in biological systems, and experimental data are required to estimate these energies. In this respect, a particularly attractive method consisted of using the techniques of protein engineering to study the effect of single amino acid changes on enzyme-substrate interaction.[95] It was thus found that the removal of a protein donor or acceptor group in the interaction area between tyrosyl tRNA synthetase and tyrosyladenylate resulted in a decrease of the binding energy ranging between about 2 to 6 kJ/mol (for an uncharged group) and 18 kJ/mol (for a charged group). Older estimates obtained by considering small molecules may be found in several articles.[96]

C. Selected Examples of Intermolecular Forces between Molecules of Biological Interest

We have now reviewed several theoretical and experimental data relevant to intermolecular forces in the biological milieu. However, it is not possible at the present moment to use these data to obtain reliable estimates for the interaction potential between biological molecules. It is therefore useful to describe some experimental data relevant to selected well-studied systems in order to convey an intuitive feeling of the forces that may be experienced by cell-surface macromolecules during cell-cell approach. Further, since affinity constants play a central role in the study of molecular interactions, we shall first discuss the significance of this important parameter.

1. Comments on the Physical Significance of Affinity Constants

The problems we shall now consider were discussed by Jencks in several reports.[94,97] Let us consider the association between two small molecules A and L maintained in solution at constant volume V and temperature T:

$$A + L \rightleftharpoons AL$$

When equilibrium is reached, the concentrations (A), (L), and (AL) of molecular species satisfy the well-known Guldberg-Wage formula:

$$(AL)/(A)(L) = K \exp(-\Delta F^\circ/RT) \tag{44}$$

where K is the affinity constant, R is the constant of perfect gases, and $-\Delta F^\circ$ is the standard free energy of the reaction. This is the free-energy change corresponding to the binding of 1 mol of free A to 1 mol of free B in a large reservoir where (A), (L), and (AL) are equal to 1 mol/ℓ, and temperature is T. Now, ΔF° may be split in three terms:

$$\Delta F^\circ = \Delta F^\circ_b + \Delta F^\circ_m + \Delta F^\circ_c \tag{45}$$

ΔF°_b is the binding free energy we considered in previous estimates. In vacuum, ΔF°_b would be temperature independent and might be calculated with quantum chemical techniques. In solution, ΔF°_b is a free energy likely to involve an entropic contribution since it is the net result of bond formation between A and B and rearrangement of solvent molecules (Equation 43). Thus, hydrophobic or electrostatic bonding involve a release of solvent molecules with concomitant energy increase.

ΔF°_m is the free-energy due to the loss of translational and rotational freedom caused by binding. It is well known from classical statistical mechanics that the rotational and translational free energy of a gas of rigid molecules of mass m, principal inertia moments I_A, I_B, and I_C, and molar volume V is[98]

$$F_m = -NkT \ln[(2\pi mkT/h^2)^{3/2} V/N] - NkT\ln[(\pi^{1/2}/\sigma) \cdot (8\pi^2 kT/h^2)^3 I_A I_B I_C] \tag{46}$$

where σ is a symmetry number that is 1 for an asymmetrical molecule and 2 for a molecule such as H_2O. The derivation is based upon Equations 3 and 4, assuming the intermolecular potential U is zero.

For example, the free energy calculated for a small molecule such as H_2O is (numerical values of structural parameters may be found in Reference 60);

$$F = -35.5 \text{ kJ/mol}$$

where the contributions of translation and rotation movements are 19.2 and 16.3 kJ/mol, respectively. Also, it is clear that F_m is only weakly dependent on molecular dimensions. Equation 46 met with some success in accounting from gas properties (vibration terms must be added[98]). However, the derivation of this formula (Equation 46) is not valid in condensed phases where the intermolecular potential is not negligible. The experimental value of melting enthalpy may provide an order of magnitude for F_m, if we assume that entropic term plays a major role in F_m and is low in solid phases. As an example, the entropy increase associated to the melting of water at 0 K is 22.3 J/mol/°K, corresponding to a free energy of 6.0 kJ/mol.[60]

Finally, $\Delta F°_c$ is the free energy associated to a possible conformational change of molecules A and L that might be required to allow bond formation. It is possible to obtain some information on the actual value of $\Delta F°_c$ and $\Delta F°_m$ in model systems, as was done by Jencks.[94] Indeed, let us assume that A is made of two groups, A_1 and A_2 ($A = A_1 - A_2$), and suppose we measure the affinity constants of the following three reactions:

$$A_1 - A_2 + L \leftrightharpoons A_1 - A_2 - L \quad (K_{12})$$

$$A_1 + L \leftrightharpoons A_1 - L \quad (K_1)$$

$$A_2 + L \leftrightharpoons A_2 - L \quad (K_2)$$

In a first approximation, we may assume that F_m is the same for all molecular species A_1, A_2, A_1–L, A_2–L, A_1–A_2–L (see Equation 46). In situations where no conformation change is required for bond formation between A_1–A_2 and L, Equations 44 and 45 yield:

$$\Delta F°_m = F_m = RT\ln(K_{12}/K_1K_2) \tag{47}$$

Now, if the formation of a A_1–L bond requires some molecular deformations when A_2 and L are bonded, the corresponding free energy term is related to F_m and affinity constants by:

$$\Delta F°_m + \Delta F°_c = RT\ln(K_{12}/K_1K_2) \tag{48}$$

Other formulas may be obtained if conformational changes are associated to the binding of isolated L and A_1 or A_2 molecules. $\Delta F°_c$ may thus be positive or negative. Indeed, on reviewing some experimental values for K_{12}, K_1, and K_2 in different model systems, Jencks found that $RT\ln(K_{12}/K_1K_2)$ might range between about 24 and -22 kJ/mol (see Reference 93 for more details). The important conclusion is that the binding energy between two macromolecules cannot be estimated by the mere addition of the contributions of different molecular interactions (i.e., hydrophobic, electrostatic, and hydrogen bonds). Also, a close matching of two molecular surfaces may result in high binding constants due to the contribution of $\Delta F°_m$. This is the multivalent bonding effect that is described later in the chapter. We shall now review some examples of molecular interactions in the biological milieu.

2. A Model for Specific Interactions: the Antigen-Antibody Reaction

Immunoglobulins are glycoproteins that are endowed with a particular capacity to bind selectively a variety of "foreign" substances called antigens. Binding sites are borne by specialized regions of the molecule called F_{ab} fragments (ab is for "antigen binding"). These

F_{ab} regions are made of a so-called ''light chain'' of about 25,000 mol wt and the half part of a ''heavy'' chain of 50,000 mol wt. The structural basis of the diversity of the specificity of immunoglobulin binding sites is the very particular occurrence of a high diversity in the primary structures of specialized parts of light and heavy chains called the ''complementarity determining regions''. These regions are spatially assembled into some sort of ''pocket'', with varying depth, by the remaining parts of chains that are called ''framework'' regions. The immunoglobulin (or antibody) molecule provided a unique model for the study of molecular associations. Experiments were done with antigens of varying size. Antibodies used were first heterogeneous antisera obtained by injecting animals with a given antigen. Later, homogenous immunoglobulins were found in the sera of patients suffering myeloma. These were said to be ''monoclonal'' since they were synthesized by a single clone of malignant lymphoid cells. Some monoclonal antibodies were found ''by chance'' to bind to small molecules such as dinitrophenol or phosphorylcholine with fairly low affinity (10^5 to 10^6 ℓ/mol) as compared to ''true'' antigen-antibody interactions (K ranged between 10^{-8} and 10^{-10} ℓ/M in many cases) and this provided a model for studying molecular interactions. More recently, cell-fusion techniques adapted by Kohler and Milstein allowed production of monoclonal antibodies of the desired specificity (see Reference 97 for a more detailed description of immunoglobulin structure).

The standard free energy of the binding of dinitrophenol by specific antibodies ranged between about -35 and -72 kJ/mol with an entropy component comprised between 90 and -120 J/mol/°K (see Reference 100).

The binding region was studied with X-ray diffraction techniques. The antibody site of myeloma proteins appeared as a grove or pocket of varying depth and a size of about 15 × 6 Å.[99,101] Precise localization of the antigen and antibody atomic sites suggested the possibility of at least 5 to 10 hydrogen, electrostatic, or apolar bonds between different atoms of the interacting molecules, as was found on a phosphorylcholine-binding myeloma protein[99] and a lysozyme-binding monoclonal antibody.[102] In the latter model, which is more representative of ''true'' antigen-antibody interactions, the binding region extended outside the ''crevice'' that had been considered as the antibody combining site in earlier studies. Further, that the establishment of a large antigen-antibody binding region may require some conformational changes of interacting molecules is consistent with the recent findings that the antibody combining sites of antigenic molecules may be preferentially located on regions where the atomic mobility is maximum.[103]

The importance of the electric charge was addressed in a study made on the binding of *p*-azophenyl trimethylammonium by specific polyclonal antibodies.[104] When the charged molecule was replaced with the isosteric uncharged group *p*-azo-*t*-butylbenzene, of similar shape and polarizability, the binding free energy was decreased by about 5 kJ/mol.

The importance of the molecular shapes in antigen-antibody reactions has been well known for many years. Pauling and colleagues studied the binding of *o*-, *m*-, and *p*-hydroxyphenylazo- groups by antibodies specific for each of these structures. Changing the position of the hydroxyl group resulted in modifications of the binding free energy that might be as high as 17 kJ/mol.[105] Using molecular models, the authors concluded that this energy might correspond to the deformation of the antibody site by about 1 Å.

A final interesting point is the ability of a molecule to engage in simultaneous bonds. Horwick and Karush[106] coated a bacteriophage with dinitrophenol (DNP) groups and compared the affinity of binding of monovalent F_{ab} fragments specific for DNP and complete divalent immunoglobulins of similar specificity. The binding constant was 100-fold less with F_{ab} than with intact molecules (the affinity constant for F_{ab}-DNP association was 10^7 M^{-1}). In a similar experiment made with F_{ab} and pentavalent immunoglobulin molecules, the intact molecules were bound to the DNP-coated surface with 10^6-fold higher efficiency than F_{ab} fragments. Similar findings were obtained with a divalent DNP-specific myeloma protein, yielding a 100 to 300 higher binding efficiency in multivalent conditions.[107]

3. Interactions Involving Carbohydrate Molecules

Molecular interactions involving carbohydrates are of particular relevance to cell adhesion due to the high amount of accessible polysaccharide chains on cell surfaces. We shall describe a few examples.

A protein purified from bovine nasal cartilage proteoglycan was found to bind hyaluronic acid (a ubiquitous polymer of glucuronyl-*N*-acetyl D-glucosamine). The binding could be efficiently inhibited with decasaccharides (i.e., five disaccharide units) obtained by hyaluronic-acid degradation, not with octasaccharides.[108]

In an interesting review on the binding behavior of carbohydrate chains, Rees[109] ascribed the lipid-binding ability of some carbohydrate chains to the presence of a cavity surrounded by methyl groups in the helix formed by these chains. Also, he ascribed the property of some sugars to bind divalent cations with an affinity constant of order 10 to 100 ℓ/mol to peculiar configurations or hydroxyl groups. The cation radius might critically influence binding since αD allopyranose bound Ca^{2+} with 30-fold higher affinity than Mg^{2+}.

Also, the autoassociation between proteoglycan chains was demonstrated and was found to depend on the composition of the polysaccharide chains.[110,111]

Finally, many authors reported associations between glycosaminoglycans and proteins involved in cell adhesion such as fibronectin[112] or N-CAM[113] (see also Chapter 9). Hence, polysaccharide chains may bind to other polysaccharides or proteins. Binding may involve repetitive low-affinity interactions[108] and is correlated to the detailed stereochemical configurations of interacting molecules.[109]

4. Lipid-Protein Interactions

Whereas membrane intrinsic proteins are well known to bind to surrounding lipids, several authors found that molecules such as concanavalin A,[114] immunoglobulins,[115,116] or fibronectin[117] were able to interact with phospholipid vesicles. However, the molecular significance of these findings remains ill understood.

D. Conclusion

A complete understanding of the mechanisms of molecular interactions requires further progress in several fields of physics and chemistry. Recent results relevant to small molecular models suggest that the geometrical properties of interacting molecules may play an important role in molecular associations. This means that physical-chemical parameters such as charge- or hydrogen-bonding ability cannot account for all of the binding behavior of cell-surface molecules.

However, due to the high diversity of these molecules, it is tempting to look for approximate means of modeling cell adhesion by introducing ''macroscopic'' parameters in order to account for a combination of various low-affinity low-specificity molecular interactions. This approach will be briefly reviewed in the last section of this chapter.

IV. FORCES BETWEEN MACROSCOPIC BODIES

Although the same basic phenomena are responsible for the interactions between molecules and macroscopic bodies, different theoretical and experimental tools were used to deal with both classes of systems. We shall sequentially review the data on the DLVO theory and hydration forces, and we shall only give some brief comments on the phenomenon of steric stabilization that is a particular aspect of the physical processes discussed in Chapter 2.

A. The DLVO Theory

The DLVO theory was named after Derjaguin and Landau[118] and Verwey and Overbeek,[119] who modeled the interaction between charged hydrophobic colloidal particles as a balance between electrodynamic attraction (the aforementioned r^{-6} potential) and electrostatic repulsion. We shall describe the basic theory, then we shall discuss its relevance to biological systems.

1. Basic Theory

a. Electrodynamic Attraction

Let us consider two semi-infinite media (1) and (2) with parallel plane surfaces separated by a distance d. Assuming pairwise additivity of electrodynamics interactions, the potential energy of interaction may be calculated by summation of the r^{-6} attraction between individual molecules. This was done by Hamaker,[120] who obtained the following formula:

$$W = -A/12\pi d^2; \quad A = \pi^2 c_{12} \rho_1 \rho_2 \qquad (49)$$

where W is the interaction potential per unit area, ρ_1 and ρ_2 are the number of molecules per unit volume in media 1 and 2, and the interaction between a type 1 and a type 2 molecule at distance r is $-c_{12}/r^6$. A is called the Hamaker constant. We quote for the sake of completeness a formula for the interaction between two spheres of radius R separated by a gap of width d[121]:

$$W = -AR/12d \qquad (50)$$

When interacting bodies are embedded in a solvent, "effective" Hamaker constants may be obtained by simple combinations of Hamaker constants between different media. However, the assumption of pairwise additivity of interactions may not be warranted, as discussed above. This problem was overcome by Lifshitz,[62] who was able to relate the interaction between material bodies to their dielectric properties by considering polarization fluctuations, as was mentioned. He found a d^{-2} law for the interaction between parallel semi-infinite media at small distance. Further, it was possible to relate the Hamaker constant to measurable spectroscopical properties of the interacting media. Gingell and Parsegian[122,123] were thus able to obtain an estimate ranging between 3.5×10^{-21} and 6.8×10^{-21} J for the Hamaker constant between two hydrocarbon solutions separated by water.

However, the Lifshitz theory is valid only within a range of values for d. First, it obviously fails when d is lower than a few diameters of the solvent molecules, as was made clear in the previous section. Secondly, the correlation between periodic fluctuations of the polarization of interacting media is less efficient in decreasing the free energy of the system when the period of these fluctuations is not much higher than d/c, where c is the velocity of light (i.e., about 3×10^8 m/sec). A theoretical treatment of this effect was first worked out by Casimir and Polder.[124] It was found that the correction to the Hamaker formula (Equation 49) became significant when d was higher than a few hundreds of angstrom units. The "retarded" potentials decay as d^{-3} instead of d^{-2} at such distances.

As reviewed by Israelachvili,[125] the Lifshitz theory was subjected to experimental check by direct measurement of the interaction between parallel glass or mica surfaces. For practical purposes, the surfaces were cylinders with perpendicular axes, which was equivalent to the interaction between spheres. The particular structure of mica allowed the preparation of surfaces that were optically smooth with an accuracy of order of 1 Å. Theoretical and experimental results were consistent when the distance was varied between 15 and 10,000 Å. An important conclusion of the author is that asperities on the order of 10 Å could decrease the adhesion between solid surfaces by several orders of magnitude.[125]

b. Electrostatic Repulsion

The second interaction involved in the DLVO theory is an electrostatic repulsion between charged surfaces. The basis of the simplest calculation is to use the linearized Poisson-Boltzmann equation to derive the potential between two parallel-charged plates. The force experienced by a plate per unit area is the sum of an electric force $\sigma\vec{E}$ (where σ is the charge density and $\vec{E}$ is the electric field on the surface) and an osmotic force resulting from the difference between the ion concentration near the surface and at infinity. If the surface charges are considered as fixed, the repulsive pressure between two plates of respective charge densities σ_A and σ_B is

$$P = (2/\epsilon)(\sigma_A^2 + \sigma_A\sigma_B[\exp(\kappa d) + \exp(-\kappa d) + \sigma_B^2]/[\exp(\kappa d) - \exp(-\kappa d)] \quad (51)$$

where κ is the reciprocal Debye-Huckel length and ϵ is the dielectric constant of the medium. When d is much higher than 1, Equation 51 yields:

$$P = (2/\epsilon)\sigma_A\sigma_B\exp(-\kappa d) \qquad (52)$$

More refined theoretical treatments of the force between planar-charged plates were worked out (see Reference 12b for a recent example and Reference 127 for a criticism of the theory for small d). However, the general conclusion that the interaction force is proportional to $\exp(-\kappa d)$ at distances higher than 10 to 20 Å remains valid. Since $1/\kappa$ is highly dependent on the ionic concentration (it is 100 and 8.1 Å when the ionic strength is 0.001 and 0.15 *M*, respectively), the range of repulsion is expected to be much higher at low ionic strength than in biological media.

2. Conclusion and Relevance to Biological Systems

When electrodynamic attraction and electrostatic repulsion are combined, the interaction potential between two parallel plates of thickness L separated by a gap of width d is obtained by using Equation 4 and integrating Equation 52, which yields:

$$W(d) = -A/12\pi[1/d^2 - 2/(d + L)^2 + 1/(d + 2L)^2] + 2\sigma^2\exp(-\kappa d)/\epsilon \qquad (53)$$

The general shape of W is as follows when "reasonable" values of A and σ are used: W first decreases due to the predominance of the attraction term. Then a minimum (called the secondary minimum) is found at a distance of 50 to 200 Å. W then increases and becomes positive. Finally, W decreases again when d is decreased to a few angstrom units, due to the d^{-2} term. The macroscopic theory is no longer valid for such small distances, but it is assumed that W reaches a "primary minimum" corresponding to molecular contact.

The general features of the DLVO theory were checked experimentally by studying the coagulation of colloid particles (see, e.g., Reference 128) or measuring the interaction between mica plates coated with charged groups.[125]

Since a reasonable agreement was obtained between theoretical and experimental results in several models, it was quite reasonable to try to apply the DLVO theory to biological surfaces, as was proposed by Curtis[129] and Weiss[130] in the early 1960s. However, the problem with this theory is that (1) it is possible to accommodate many different experimental results by a slight adjustment of parameters A and σ (Figure 3), (2) the macroscopic theory is questionable when intercellular distances are the size of intrinsic membrane molecules protruding out of the lipid bilayers, and (3) as was previously discussed,[48,131] the negative cell-surface charge distribution is not two-dimensional but is scattered in the cell coat. Since the Debye-Huckel screening is very strong at physiological ionic strength ($\exp[-\kappa d]$ is 1/500 when d is 50 Å), electrostatic repulsion may be much higher than estimated with the above theory. This point was discussed quantitatively, and it was found that the electrostatic repulsion between two charged surfaces separated by a 100-Å gap was 20,000-fold higher

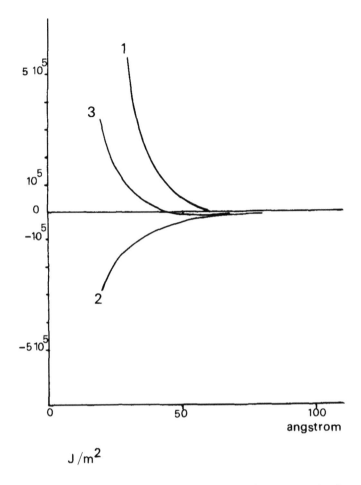

FIGURE 3. Dependence of the DLVO potential on Hamaker constant and surface charge density. The interaction potential per unit area (ordinate) between two charged plates of 40 Å width at various distances (abscissa) was calculated for various values of the Hamaker constant A and surface charge density σ. (1) A $= 10^{-21}$ J, $\sigma = 0.032$ C/m^2, (2) A $= 6 \times 10^{-21}$ J, $\sigma = 0.0048$ c/m^2, (3) A $= 3 \times 10^{-21}$ J, $\sigma = 0.016$.

if the surface density of charges was replaced with a uniform volume charge density filling a region of 100-Å thickness.[48] These considerations may account for the following experimental data: when Gingell and Todd measured the distance between glutaraldehyde-treated red cells and a charged oil-saline interface with interference reflection microscopy (see Chapter 9), they found a gap width of about 1000 Å when the ionic strength was very low (corresponding to a Debye-Huckel length of about 2000 Å).[132] Analyzing the balance of forces, Parsegian and Gingell concluded that the obtained value was consistent with a Hamaker constant comprised between 5×10^{-21} and 8×10^{-21} J, which was quite reasonable.[133] Further, when Goldsmith and colleagues used the traveling microtube technique (see Chapter 5 for a description of this method) to analyze the collisions between individual latex particles[134] or fixed erythrocytes,[135] they concluded that the DLVO theory could account for the interaction between latex particles in 50% glycerol at 1 mM KCl concentration. However, for higher ionic strength, the double-layer repulsion did not seem responsible for an impossibility for surfaces to approach closer than circa 100 Å. Similarly, in 10 mM NaCl, the erythrocyte surfaces could not approach closer than 40 nM, which seemed inconsistent with the DLVO theory. The general conclusion from these and other experiments[131] may be that the DLVO theory can account for the interactions between biological surfaces at very low ionic strength (say, 1 mM or less) when the Debye-Huckel length is higher than the size of cell-surface molecular asperities. However, at physiological ionic strength, the DLVO theory cannot account quantitatively for the interaction between cell membranes, since short-distance molecular interactions may occur before DLVO repulsion is significant. However, several reports suggest that electrostatic repulsion may prevent cell adhesion in physiological media.[131]

We shall now describe a nonelectrostatic short-range repulsion between extended surfaces embedded in aqueous medium.

B. Hydration Forces

As early as 1977, LeNeveu, Parsegian, and colleagues[136,137] measured with X-ray diffraction the separation between lecithin bilayers prepared in an excess of water and partially dehydrated by exposure to hypertonic solutions or high pressures. They concluded that the removal of water between bilayers was opposed by very important "hydration forces" with an exponential dependence on distance and a decay distance of about 2 Å. The net repulsive force per unit area was fitted with the following formula:

$$P = 7 \times 10^8 \exp(-d/2.56) \tag{54}$$

where d is the distance between bilayers expressed in angstroms and P is in N/m². Further studies on the interaction between mica surfaces revealed that when these were made hydrophobic by coating with hexadecylammonium, a short-range repulsive force was found:

$$F = 0.14 \, R \, \exp(-d/10) \tag{55}$$

where R is the radius of mica-coated cylinders (in meters), F is the force (in newtons), and d is the gap width (in angstroms). This force was not dependent on solute electrolytes. Further studies revealed that the repulsion between lipid bilayers was dependent on the nature of lipids,[139] and both theoretical considerations[139] and Monte-Carlo simulations[140] showed that the repulsion might be ascribed to an interaction between structured water layers associated to the surfaces, in accordance with aforementioned results concerning solute-solvent interactions. Hence, a repulsion might occur between hydrophilic structures (corresponding to the difficulty to remove hydration molecules) as well as hydrophobic structures (these were shown to be surrounded by structured water layers; also, oscillating potentials were sometimes obtained[140]). It is not known to what extent the cell glycocalyx can affect these interactions.

C. Steric Stabilization

As was pointed out by Napper,[141] electrostatic repulsion is poorly efficient in preventing molecular contact between colloidal particles in solutions of high ionic strength such as biological media. An alternative way of stabilizing colloidal suspension is to coat the particles with hydrophilic polymers. These data are relevant to cell adhesion since the approach of two cell membranes may be opposed by a force generated by the local increase of polysaccharide concentration (when cell coats begin to interpenetrate) and macromolecule squeezing between bilayers. As a direct experimental proof of this possibility, Klein[142] reported that collagen-coated mica surfaces exerted measurable mutual repulsion when the thickness of the gap between them decreased to 600 to 700 nm,[142] and rough estimates suggested that steric stabilization might efficiently decrease cellular interactions.[48] However, the interaction between polymer-coated surfaces is quite a complex phenomenon, and the present state of knowledge in this field is described in Chapter 3.

V. CONCLUSION

Several theoretical and experimental results concerning intermolecular forces were reviewed in order to provide a realistic, if approximate, feeling for the forces likely to be experienced by the cell surface during the cell-cell or cell-substrate approach. However, it is clearly impossible at the present time to derive a reliable force-distance law for real systems. It is hoped that the methods we described will lead to progress during the coming years.

REFERENCES

1. **Moore, W. J.**, *Chimie Physique*, Dunod, Paris, 1965, 608.
2. **March, J.**, *Advanced Organic Chemistry*, John Wiley & Sons, New York, 1985, 23.
3. **Schuster, P.**, The fine structure of the hydrogen bond, in *Intermolecular Interactions: From Diatomics to Biopolymers*, Pullman, B., Ed., John Wiley & Sons, New York, 1978, 363.
4. **Yon, J.**, *Structure et Dynamique Conformationnelle des Protéines*, Hermann, Paris, 1969, 57.
5. **Maitland, G. C., Rigby, M., Brian Smith, E., and Wakeham, W. A.**, *Intermolecular Forces — Their Origin and Determination*, Clarendon Press, Oxford, 1981.
6. **Singer, S. J. and Nicolson, G. L.**, The fluid mosaic model of the structure of cell membranes: cell membranes are viewed as two-dimensional solutions of oriented globular proteins and lipids, *Science*, 175, 720, 1972.
7. **Gomperts, B.**, *The Plasma Membrane — Models for Structure and Function*, Academic Press, London, 1977, 1.
8. **Alberts, B., Bray, D., Lewis, J., Raff, M., Roberts, K., and Watson, J. D.**, *Molecular Biology of the Cell*, Garland Press, New York, 1983, 256.
9. **Levine, Y. K.**, Physical studies of membrane structure, *Prog. Biophys. Mol. Biol.*, 24, 1, 1972.
10. **Kates, M. and Kuksis, A., Eds.**, *Membrane Fluidity*, Humana Press, Clifton, N.J., 1980.
11. **Wu, E.**, Lateral diffusion in phospholipid multibilayers measured by fluorescence recovery after photobleaching, *Biochemistry*, 16, 3936, 1977.
12. **Spiegel, S., Kassis, S., Wilchek, M., and Fishman, P. H.**, Direct visualization of redistribution and capping of fluorescent gangliosides on lymphocytes, *J. Cell Biol.*, 99, 1575, 1984.
13. **Karnovsky, M. J., Kleinfeld, A. M., Hoover, R. L., and Klausner, R. D.**, The concept of lipid domains in membranes, *J. Cell Biol.*, 94, 1, 1982.
14. **Steck, T. L. and Fox, C. F.**, Membrane proteins, in *Membrane Molecular Biology*, Sinauer Associates, Stamford, Conn., 1972, 58.
15. **Evans, E. A. and Skalak, R.**, *Mechanics and Thermodynamics of Biomembranes*, CRC Press, Boca Raton, Fla., 1980, 172.
16. **LeBouteiller, P. P., Mishal, Z., Lemonnier, F. A., and Kourilsky, F. M.**, Quantification by flow cytometry of HLA class I molecules at the surface of murine cells transformed by cloned HLA genes, *J. Immunol. Methods*, 61, 301, 1983.
17. **Sieckmann, D. G., Scher, I., and Paul, W. E.**, B lymphocyte activation by anti-immunoglobulin antibodies, in *Physical Chemical Aspects of Cell Surface Events in Cellular Regulation*, DeLisi, C. and Blumenthal, R., Eds., Elsevier, New York, 1979, 325.
18. **Davies, D. R. and Metzger, H.**, Structural basis of antibody function, *Annu. Rev. Immunol.*, 1, 87, 1983.
19. **Germain, R. N. and Malissen, B.**, Analysis of the expression and function of class II major histocompatibility complex-encoded molecules by DNA-mediated gene transfer, *Annu. Rev. Immunol.*, 4, 281, 1986.
20. **Abbas, A. K., Ault, K. A., Karnovsky, M. J., and Unanue, E. R.**, Non-random distribution of surface immunoglobulin on murine B lymphocytes, *J. Immunol.*, 114, 1197, 1975.
21. **Abbas, A. K., Dorf, M. E., Karnovsky, M. J., and Unanue, E. R.**, *J. Immunol.*, 116, 371, 1976.
22. **Sabatini, D. D., Guepp, E. B., Rodriguez, E. J., Dolan, W. J., Roblins, E. S., Papadopoulos, S., Nanov, I. E., and Rindler, M. J.**, Biogenesis of epithelial cell polarity, in *Spatial Organization of Eukaryotic Cells*, McIntosh, J. R., Ed., Alan R. Liss, New York, 1983, 416.
23. **Frye, L. D. and Edidin, M.**, The rapid intermixing of cell surface antigens after formation of mouse-human heterokaryons, *J. Cell Sci.*, 7, 319, 1970.
24. **Schlessinger, J., Koppel, D. E., Axelrod, D., Jacobson, K., Webb, W. W., and Elson, E. L.**, *Proc. Natl. Acad. Sci. U.S.A.*, 73, 2409, 1976.

25. **Dragsten, P., Henkart, P., Blumenthal, R., Weinstein, J., and Schlessinger, J.,** Lateral diffusion of surface immunoglobulin, Thy-1 antigen and a lipid probe in lymphocyte plasma membrane, *Proc. Natl. Acad. Sci. U.S.A.,* 76, 5163, 1979.

26. **Smith, B. A., Clark, W. R., and McConnell, H. M.,** Anisotropic molecular motion on cell surfaces, *Proc. Natl. Acad. Sci. U.S.A.,* 76, 5641, 1979.

27. **Loor, F.,** Structure and dynamics of the lymphocyte surface, in *B and T Cells in Immune Recognition,* Loor, F. and Roelants, G. E., Eds., John Wiley & Sons, New York, 1977, 153.

28. **Henis, Y. I. and Elson, E. L.,** Inhibition of the mobility of mouse lymphocyte surface immunoglobulin by locally bound concanavalin A, *Proc. Natl. Acad. Sci. U.S.A.,* 78, 1072, 1981.

29. **Flanagan, J. and Koch, G. L. E.,** Cross-linked surface Ig attaches to actin, *Nature (London),* 273, 278, 1978.

30. **Poo, M. M., Poo, W. J. H., and Lam, J. W.,** Lateral electrophoresis and diffusion of concanavalin A receptors in the membrane of embryonic muscle cells, *J. Cell Biol.,* 76, 483, 1978.

31. **Leblond, C. P. and Bennett, G.,** Elaboration and turnover of cell coat glycoproteins, in *The Cell Surface in Development,* Moscona, A. A., Ed., John Wiley & Sons, New York, 1974, 29.

32. **Winzler, R. J.,** Carbohydrates in cell surfaces, *Int. Rev. Cytol.,* 29, 77, 1970.

33. **Montesano, R., Mossaz, A., Ryser, J. E., Orci, L., and Vassalli, P.,** Leukocyte interleukins induce cultured endothelial cells to produce a highly organized glycosaminoglycan-rich pericellular matrix, *J. Cell Biol.,* 99, 1706, 1984.

34. **Roden, L.,** Structure and metabolism of connective tissue proteoglycans, in *The Biochemistry of Glycoproteins and Proteoglycans,* Lennarz, W. J., Ed., Plenum Press, New York, 1980, 267.

35. **Stryer, L.,** *Biochemistry,* W.H. Freeman, San Francisco, 1981, 200.

36. **Roblin, R., Albert, S. O., Gelb, N. A., and Black, P. H.,** Cell surface changes correlated with density dependent growth inhibition — glycosaminoglycan metabolism in 3T3, SV3T3 and ConA-selected revertant cells, *Biochemistry,* 14, 347, 1975.

37. **Engel, J., Oedermatt, E., and Engel, A.,** Shapes, domain organizations and flexibility of laminin and fibronectin, two multifunctional proteins of the extracellular matrix, *J. Mol. Biol.,* 150, 97, 1981.

38. **Forstner, G. G.,** Incorporation of (1-¹⁴C) glucosamine by rat intestinal microvillus membrane, *Biochim. Biophys. Acta,* 150, 736, 1968.

39. **Rittenhouse, H. G., Rittenhouse, J. W., and Takimoto, L.,** Characterization of the cell coat of Ehrlich ascite tumor cells, *Biochemistry,* 17, 829, 1978.

40. **Kilarski, W.,** Some ultrastructural features of the cell surface after SV40 transformation, *Cancer Res.,* 35, 1797, 1975.

41. **Woods, A., Hook, M., Kjellen, L., Smith, C. G., and Rees, D. A.,** Relationship of heparan sulfate proteoglycans to the cytoskeleton and extracellular matrix of cultures fibroblasts, *J. Cell Biol.,* 99, 1743, 1984.

42. **Hay, E. E.,** Cell and extracellular matrix: their organization and mutual dependence, in *Spatial Organization of Eukaryotic Cells,* MacIntosh, J. R., Ed., Alan R. Liss, New York, 1983, 509.

43. **Cook, G. M. W.,** Techniques for the analysis of membrane carbohydrates, in *Biochemical Analysis of Membranes,* Maddy, A. H., Ed., Chapman & Hall, London, 1976, 283.

44. **Gingell, D. and Vince, S.,** Long range forces and adhesion: an analysis of cell-substratum study, in *Cell Adhesion and Motility,* Curtis, A. S. G. and Pitts, J. D., Eds., Cambridge University Press, London, 1980, 1.

45. **Capo, C., Garrouste, F., Benoliel, A. M., Bongrand, P., and Depieds, R.,** Nonspecific binding by macrophages: evaluation of the influence of medium-range electrostatic repulsion and short-range hydrophobic interaction, *Immunol. Commun.,* 10, 35, 1981.

46. **Mehrishi, J. N.,** Molecular aspects of the mammalian cell surface, *Prog. Biophys. Mol. Biol.,* 25, 3, 1972.

47. **Jones, I. S.,** A theory of electrophoresis of large colloidal particles with adsorbed polyelectrolytes, *J. Colloid Interface Sci.,* 68, 451, 1979.

48. **Bongrand, P., Capo, C., and Depieds, R.,** Physics of cell adhesion, *Prog. Surf. Sci.,* 12, 217, 1982.

49. **Eylar, E. H., Madoff, M. A., Brody, O. V., and Oncley, J. L.,** The contribution of sialic acid to the surface charge of the erythrocyte, *J. Biol. Chem.,* 237, 1992, 1962.

50. **Brush, S. G.,** *Statistical Physics and the Atomic Theory of Matter,* Princeton University Press, Princeton, N.J., 1983, 204.

51. **Hirschefelder, J. O., Curtiss, C. F., and Bird, R. B.,** *Molecular Theory of Gases and Liquids,* John Wiley & Sons, New York, 1954.

52. **Margenau, H. and Kestner, N. R.,** *Theory of Intermolecular Forces,* Pergamon Press, Oxford, 1969.

53. **Hansen, J. P. and McDonald, I. R.,** *Theory of Simple Liquids,* Academic Press, New York, 1976.

54. **Jorg nsen, W. L., Chandrasekhar, J., Madura, J. D., Impey, R. W., and Klein, M. L.,** Comparison of s: ple potential functions for simulating liquid water, *J. Chem. Phys.,* 79, 926, 1983.

55. **Metropolis, M., Rosenbluth, M. N., Teller, A. H., and Teller, E.,** Equation of state calculation by fast computing machines, *J. Chem. Phys.,* 21, 1087, 1953.
56. **Binder, K. and Stauffer, D.,** A simple introduction to Monte Carlo simulation and some specialized topics, in *Applications of the Monte Carlo Method in Statistical Physics,* Binder, K., Ed., Springer-Verlag, Berlin, 1983, 1.
57. **Alder, B. J. and Wainwright, T. E.,** Phase transition for a hard sphere system, *J. Chem. Phys.,* 27, 1208, 1957.
58. **Stillinger, F. H. and Rahman, A.,** Improved simulation of liquid water by molecular dynamics, *J. Chem. Phys.,* 60, 1545, 1974.
59. **Barker, J. A. and Henderson, D.,** What is "liquid"? Understanding the states of matter, *Rev. Mod. Phys.,* 48, 587, 1976.
60. **Eisenberg, D. and Kauzmann, W.,** *The Structure and Properties of Water,* Oxford University Press, Oxford, 1969.
61. **Cohen-Tannoudji, C., Diu, B., and Laloe, F.,** *Mécanique Quantique,* Hermann, Paris, 1973.
62. **Lifshitz, E. M.,** The theory of molecular attractive forces between solids, *Sov. Phys. J. Exp. Theor. Phys. U.S.S.R.,* 2, 73, 1956.
63. **London, F.,** Zur theorie und Systematik der Molekularkrafte, *Z. Phys.,* 63, 245, 1930.
64. **Ketalaar, J. A. A.,** *Liaisons et Propriétés Chimiques,* Dunod, Paris, 1960, 105.
65. **Clementi, E. and Habitz, P.,** A new two-body water-water potential, *J. Phys. Chem.,* 87, 2815, 1983.
66. **Bohm, H. J., Ahlrichs, R., Scharf, P., and Schiffer, H.,** Intermolecular potentials for CH_4, CHF_3, CH_3Cl, CH_2Cl_2, CH_3CN and CO_2, *J. Chem. Phys.,* 81, 1389, 1984.
67. **Eyring, H. S., Walter, J., and Kimball, G. E.,** *Quantum Chemistry,* John Wiley & Sons, New York, 1944.
68. **Coulson, C. A.,** *Valence,* Clarendon Press, Oxford, 1952.
69. **Murrell, J. N., Kettle, S. F. A., and Tedder, J. M.,** *Valence Theory,* John Wiley & Sons, New York, 1965.
70. **Szabo, A. and Ostlund, N. S.,** *Modern Quantum Chemistry — Introduction to Advanced Electronic Structure Theory,* Macmillan, New York, 1982.
71. **Bohm, H. J. and Ahlrichs, R.,** A study of short-range repulsions, *J. Chem. Phys.,* 77, 2028, 1982.
72. **Matsuoka, O., Clementi, E., and Yoshimine, M.,** CI study of the water dimer potential surface, *J. Chem. Phys.,* 64, 1351, 1976.
73. **Claverie, P.,** Elaboration of approximate formulas for the interactions between large molecules: application in organic chemistry, in *Intermolecular Interactions: From Diatomics to Biopolymers,* Pullman, B., Ed., John Wiley & Sons, New York, 1978, 69.
74. **Pauling, L.,** *The Nature of the Chemical Bond,* Cornell University Press, Ithaca, N.Y., 1972, 244.
75. **Okazaki, S., Nakanishi, K., and Touhara, H.,** Computer experiments on aqueous solutions. I. Monte Carlo calculations on the hydration of methanol in an infinitely dilute aqueous solution with a new water-methanol pair potential, *J. Chem. Phys.,* 78, 454, 1983.
76. **Ben Naim, A.,** Thermodynamics of dilute aqueous solutions of non-polar solutes, in *Water and Aqueous Solutions,* Horne, R. A., Ed., John Wiley & Sons, New York, 1972, 425.
77. **Chothia, C.,** Principles that determine the structure of proteins, *Ann. Rev. Biochem.,* 53, 537, 1984.
78. **Richards, F. M.,** Areas, volumes, packing and protein structure, *Ann. Rev. Biophys. Bioeng.,* 6, 151, 1977.
79. **Pangali, C., Rao, M., and Berne, B. J.,** A Monte Carlo simulation of the hydrophobic interaction, *J. Chem. Phys.,* 71, 2975, 1979.
80. **Pratt, L. R. and Chandler, D.,** Theory of the hydrophobic effect, *J. Chem. Phys.,* 67, 3683, 1977.
81. **Nakanishi, K., Ikari, K., Okazaki, S., and Touhara, H.,** Computer experiments on aqueous solutions. III. Monte Carlo calculations on the hydration of tertiary butyl alcohol in an infinitely dilute aqueous solution with a new water-butanol pair potential, *J. Chem. Phys.,* 80, 1656, 1984.
82. **Lee, C. Y., McCammon, J. A., and Rossky, P.,** The structure of water at an extended hydrophobic surface, *J. Chem. Phys.,* 80, 4448, 1984.
83. **Zichi, D. A. and Rossky, P.,** The equilibrium solvation structure for the solvent-separated hydrophobic bond, *J. Chem. Phys.,* 83, 797, 1985.
84. **Weast, R. C. and Astle, M. J., Eds.,** *Handbook of Chemistry and Physics,* CRC Press, Boca Raton, Fla., 1983, E51.
85. **Kirkwood, J. H.,** The dielectric polarization of polar liquids, *J. Chem. Phys.,* 7, 911, 1939.
86. **Hoye, J. S. and Stell, G.,** Dielectric theory for polar molecules with fluctuating polarizability, *J. Chem. Phys.,* 73, 461, 1980.
87. **Levesque, D., Weis, J. J., and Patey, G. N.,** Charged hard spheres in dipolar hard sphere solvents, *J. Chem. Phys.,* 72, 1887, 1980.
88. **Hirata, F., Rossky, P. J., and Montgomery Pettitt, B.,** The interionic potential of mean force in a molecular polar solvent from an extended RISM equation, *J. Chem. Phys.,* 78, 4133, 1983.

89. **Patey, G. N. and Valleau, J.,** A Monte Carlo method for obtaining the interionic potential of mean force in ionic solutions, *J. Chem. Phys.,* 63, 2331, 1975.

90. **Pollock, E. L., Alder, B. J., and Pratt, L. R.,** Relation between the local field at large distances from a charge or dipole and the dielectric constant, *Proc. Natl. Acad. Sci. U.S.A.,* 77, 49, 1980.

91. **Bockris, J. O'M. and Reddy, A. K. N.,** *Modern Electrochemistry,* Vol. 2, Plenum Press, New York, 1970.

92. **Rasaiah, J. C. and Friedman, H. L.,** Integral equation methods in the computation of equilibrium properties of ionic solutions, *J. Chem. Phys.,* 48, 2742, 1968.

93. **Waisman, E. and Lebowitz, J. L.,** Mean spherical model integral equation for charged hard spheres, *J. Chem. Phys.,* 56, 3086, 1972.

94. **Jencks, W. P.,** On the attribution and additivity of binding energies, *Proc. Natl. Acad. Sci. U.S.A.,* 78, 4046, 1981.

95. **Fersht, A. R., Shi, J. P., Knill-Jones, J., Lowe, D. M., Wilkinson, A. J., Blow, D. M., Brick, P., Carter, P., Wayer, M. N. Y., and Winter, G.,** *Nature (London),* 314, 235, 1985.

96. **Kauzmann, W.,** Some factors in the interpretation of protein denaturation, *Adv. Protein Chem.,* 14, 1, 1959.

97. **Page, M. I. and Jencks, W. P.,** Entropic contributions to rate acceleration in enzymic and intramolecular reactions and the chelate effect, *Proc. Natl. Acad. Sci. U.S.A.,* 68, 1678, 1971.

98. **Hill, T. L.,** *Introduction to Statistical Thermodynamics,* Addison-Wesley, Reading, Mass., 1960, 161.

99. **Turner, M. W.,** Structure and functions of immunoglobulins, in *Immunochemistry: An Advanced Textbook,* Glynn, L. E. and Steward, M. W., Eds., John Wiley & Sons, New York, 1977, 1.

100. **Kabat, E. A.,** *Structural Concepts in Immunology and Immunochemistry,* Holt, Rinehart & Winston, New York, 1968, 79.

101. **Padlan, E. A., Davies, D. R., Rudikoff, S., and Potter, M.,** Structural basis for the specificity of phosphorylcholine-binding immunoglobulins, *Immunochemistry,* 13, 945, 1976.

102. **Amit, A. G., Maruizza, R. A., Phillips, S. E. V., and Poljak, R. J.,** Three-dimensional structure of an antigen-antibody complex at 6 Å, *Nature (London),* 313, 156, 1985.

103. **Tainer, J. A., Getzoff, E. D., Paterson, Y., Jobson, A., and Lerner, R. A.,** *Annu. Rev. Immunol.,* 3, 501, 1985.

104. **Karush, F.,** Immunologic specificity and molecular structure, *Adv. Immunol.,* 2, 1, 1962.

105. **Pauling, L. and Pressman, D.,** The serological properties of simple substances. IX. Hapten inhibition of precipitation of antisera homologous to the *o-, m-* and *p-*azophenylarsonic acid groups, *J. Am. Chem. Soc.,* 67, 1003, 1945.

106. **Horwick, C. L. and Karush, F.,** Antibody affinity. III. The role of multivalence, *Immunochemistry,* 9, 325, 1972.

107. **Bystryn, J. C., Siskind, G. W., and Uhr, J. W.,** Binding of antigen by immunocytes. I. Effect of ligand valence on binding affinity of MOPC 315 cells for DNP conjugates, *J. Exp. Med.,* 137, 1973.

108. **Hascall, V. C. and Heinegard, D.,** Aggregation of cartilage proteoglycans. II. Oligosaccharide competitors of the proteoglycan-hyaluronic acid interaction, *J. Biol. Chem.,* 249, 4242, 1974.

109. **Rees, D. A.,** Stereochemistry and binding behaviour of carbohydrate chains, *MTP Int. Rev.,* 1, 1975.

110. **Fransson, L. A.,** Interaction between dermatan sulphate chains. I. Affinity chromatography of copolymeric galactosaminoglycans on dermatan-sulphate-substituted agarose, *Biochim. Biophys. Acta,* 437, 106, 1976.

111. **Fransson, L. A., Nieduszynski, I. A., Phelps, C. F., and Sheehan, J. K.,** Interactions between dermatan sulphate chains. III. Light scattering and viscometric studies of self association, *Biochim. Biophys. Acta,* 586, 179, 1979.

112. **Latena, J., Ansbacher, R. A., and Culp, L. A.,** Glycosaminoglycans that bind cold insoluble globulin in cell-substratum adhesion sites of murine fibroblasts, *Proc. Natl. Acad. Sci. U.S.A.,* 77, 6662, 1980.

113. **Cole, G. J., Loewy, A., and Glaser, L.,** Neuronal cell-cell adhesion depends on interactions of N-CAM with heparin-like molecules, *Nature (London),* 320, 445, 1986.

114. **van der Bosch, J. and McConnell, H. M.,** Fusion of dipalmitoyl phosphatidylcholine vesicle membranes induced by concanavalin A, *Proc. Natl. Acad. Sci. U.S.A.,* 72, 4409, 1975.

115. **Weissmann, G., Brand, A., and Franklin, E. C.,** *J. Clin. Invest.,* 53, 536, 1974.

116. **Vandenbranden, M., de Ckoen, J. L., Jeener, R., Kanared, L., and Ruyschaert, J. M.,** Interaction of γ-immunoglobulins of different hydrophobicities, *Mol. Immunol.,* 18, 621, 1981.

117. **Rossi, J. D. and Wallace, B. A.,** Binding of fibronectin to phospholipid vesicles, *J. Biol. Chem.,* 258, 3327, 1983.

118. **Derjaguin, B. V. and Landau, L. D.,** Theory of the stability of strongly charged sols and of the adhesion of strongly charged particles in solutions of electrolytes, *Acta Physicochim. URSS,* 14, 633, 1941.

119. **Verwey, E. J. W. and Overbeek, J. Th. G.,** *Theory of the Stability of Lyophobic Colloids,* Elsevier, Amsterdam, 1948.

120. **Hamaker, H. C.,** The London-van der Waals attraction between spherical particles, *Physica,* 4, 1058, 1937.

121. **Israelachvili, J. N.,** Van der Waals forces in biological systems, *Q. Rev. Biophys.,* 6, 341, 1974.
122. **Gingell, D. and Parsegian, V. A.,** Computation of van der Waals interactions in aqueous systems using reflectivity data, *J. Theor. Biol.,* 36, 41, 1972.
123. **Parsegian, V. A.,** Long range physical forces in the biological milieu, *Ann. Rev. Biophys. Bioeng.,* 2, 221, 1973.
124. **Casimir, H. B. G. and Polder, D.,** The influence of retardation of the London-van der Waals forces, *Phys. Rev.,* 73, 360, 1948.
125. **Israelachvili, J. N.,** Forces between surfaces in liquids, *Adv. Colloid Interface Sci.,* 16, 31, 1982.
126. **Lozada-Cassou, M.,** The force between two planar electrical double layers, *J. Chem. Phys.,* 80, 3344, 1984.
127. **Spitzer, J. J.,** *A reinterpretation of the hydration forces near charged surfaces, Nature (London),* 310, 396, 1984.
128. **Shenkel, J. H. and Kitchener, J. A.,** A test of the Derjaguin-Verwey-Overbeek theory with a colloidal suspension, *Trans. Faraday Soc.,* 56, 161, 1960.
129. **Curtis, A. S. G.,** Cell contacts: some physical considerations, *Am. Nat.,* 94, 37, 1960.
130. **Weiss, L.,** The adhesion of cells, *Int. Rev. Cytol.,* 9, 187, 1960.
131. **Bongrand, P. and Bell, G. I.,** Cell-cell adhesion — parameters and possible mechanisms, in *Cell Surface Dynamics,* Perelson, A. S., DeLisi, C., and Wiegel, F. W., Eds., Marcel Dekker, New York, 1984, 459.
132. **Gingell, D. and Todd, I.,** Red blood cell adhesion. II. Interferometric examination of the interaction with hydrocarbon oil and glass, *J. Cell Sci.,* 41, 135, 1980.
133. **Parsegian, V. A. and Gingell, D.,** Red blood cell adhesion. III. Analysis of forces, *J. Cell Sci.,* 41, 151, 1980.
134. **Takamura, K., Goldsmith, H. L., and Mason, S. G.,** The microrheology of colloidal dispersions. XII. Trajectories of orthokinetic pair collisions of latex spheres in a simple electrolyte, *J. Colloid Interface Sci.,* 82, 175, 1981.
135. **Goldsmith, H. L., Lichtarge, O., and Tessier-Lavigne, M.,** Some model experiments in hemodynamics. VI. Two-body collisions between blood cells, *Biorheology,* 18, 531, 1981.
136. **LeNeveu, D. M., Rand, R. D., Parsegian, V. A., and Gingell, D.,** Measurement and modification of forces between lecithin bilayers, *Biophys. J.,* 18, 209, 1977.
137. **Parsegian, V. A., Fuller, N., and Rand, R. P.,** Measured force of deformation of lecithin bilayers, *Proc. Natl. Acad. Sci. U.S.A.,* 76, 2750, 1979.
138. **Israelachvili, J. and Pashley, R.,** The hydrophobic interaction is long range decaying exponentially with distance, *Nature (London),* 300, 341, 1982.
139. **Gruen, D. W. R., Marcelja, S., and Parsegian, V. A.,** Water structure near the membrane surface, in *Cell Surface Dynamics,* Perelson, A. S., DeLisi, C., and Wiegel, F. W., Eds., Marcel Dekker, New York, 1984, 59.
140. **Snook, I. K. and Van Megen, W.,** Solvation forces in simple dense fluids, *J. Chem. Phys.,* 72, 2907, 1980.
141. **Napper, D. H.,** Steric stabilization, *J. Colloid Interface Sci.,* 58, 390, 1977.
142. **Klein, J.,** Forces between mica surfaces bearing adsorbed macromolecules in liquid media, *J. Chem. Soc. Faraday Trans.,* 1, 79, 1983.

Chapter 2

MODEL POLYMERS AT INTERFACES

P. G. de Gennes

TABLE OF CONTENTS

I. INTRODUCTION

A solid (or fluid) surface can be deeply modified by polymer adsorption. India ink, for instance, is a water dispersion of carbon black particles stabilized by gum arabic, a polyuronic macromolecule. Without the gum, the carbon particles attract each other by van der Waals forces, and flocculate. With the gum, the particles repel each other and make a stable sol.

The cell surface is covered by a rich set of glycoproteins and glycolipids. The resulting cell-cell interactions have at least one part which is nonspecific, and which can hopefully be understood in terms of general physical concepts: our knowledge of colloid stabilization (or destabilization) by polymers must be transposed to the biosystems. During recent years, certain model colloids have evolved with simple, well-controlled solid surfaces and polymer adsorbates. Our aim in the present text is to present these systems, the new tools (experimental and theoretical) which have been applied to them, and the resulting concepts.

A. Choice of Polymer/Solvent Systems

We mentioned earlier the 4000-year-old invention of India ink, based on a stabilization of pigment by gum arabic. This is clearly a very bad example. The carbon black particles have a complicated surface, and the polymer is a complex polysaccharide, with at least three modes of interaction (charge effects, van der Waals, and hydrogen bonds). Most of the current physicochemical research on colloid stabilization has concentrated on simpler polymers. Recent references covering this sector are the book by Napper[1] and a review by Cohen-Stuart et al.[2] The model systems are defined below.

1. Linear Chains as Opposed to Branched Polymers

Branching introduces two complications: (1) the statistics of the branch points is often poorly controlled, and (2) the basic features of adsorbed (or grafted) layers, made of branched polymers, are still relatively unknown (we come back to this point in Section V).

2. Neutral Chains

The opposite case, of long polyelectrolytes in water solutions, is still poorly understood. Even in bulk solution the nature of the equilibrium phase (without salt) is still open to doubt.[3] Thus, in most of this review, we shall omit charge effects, but we mention some related observations in Section V.

3. Homopolymers

When a chain is a statistical copolymer $A_x B_{1-x}$ the adsorption properties of the component monomers A and B are often very different, and this generates many complications.

4. Flexible Chains

This choice is made first because adsorption of flexible chains leads to diffuse adsorbed layers, which play an important role in colloid science and also in cell adhesion. The adsorption of rigid chains is often very abrupt, with all the chains being trapped at the surface (and also sometimes denatured), and the resulting layer being dense and hard.

5. Solvent Conditions

When a chain is in a good solvent, monomer/solvent contacts are favored with respect to monomer/monomer contacts; there is an effective repulsion between monomers. In the opposite case, where the monomers begin to stick together, the chains tend to precipitate. As we shall see in Section II.C, this is an important source of adhesion, but often impractical, because insoluble chains are difficult to deliver at the right surface. In most of this text we focus on good solvent conditions.

In practice, the typical model chain for studies in organic, nonpolar solvents is polystyrene (PS). Polydimethylsiloxane (PDMS) is also of interest, and has one advantage: it remains fluid at room temperature, even in concentrated solutions. This avoids certain glassy features which occur in adsorption layers of PS (see Section IV). If we want to work in water, the main example of an available neutral, flexible chain is polyoxyethylene (POE).

B. Adsorption vs. Depletion

The polymer of interest will be adsorbed (or attached terminally) to a surface which can be solid (silica, polystyrene latex, etc.), liquid (emulsion droplets), or even a gas (the free surface).

It is essential to distinguish between the two following regimes:

1. If the surface prefers the polymer to the solvent, we have adsorption (Figure 1). In most practical cases, the free energy gained when a monomer is brought from the bulk solution to the wall is relatively large (i.e., is comparable to the thermal energy kT). This is called strong adsorption.
2. If the surface prefers the solvent to the polymer, the chains avoid the vicinity of the surface. A polymer solution then shows a depletion layer near the surface (Figure 2).[4,5]

In practice, the most important regime corresponds to "semidilute" solutions (where the coils overlap strongly, but the polymer volume fraction Φ is still small). The depletion layer then has a characteristic thickness $\xi(\Phi)$ which is a decreasing function of Φ, and is discussed in more detail later. Direct measurements of the depletion-layer thickness have been performed recently.[6]

The difference between adsorption and depletion is illustrated in Figure 3, where we show the surface tension γ of polymer solutions as a function of the concentration Φ. In the adsorption case, there is an immediate drop of $\gamma(\Phi)$ at $\Phi = 0$, due to the buildup of the adsorbed layer, which occurs even in very dilute solutions. In the depletion case, $\gamma(\Phi)$ increases very slowly with Φ. At low Φ, the polymer never touches the surface.

Let us quote one example: many polymers like PS adsorb from organic solvents on a glass surface, but if the glass is silanated (i.e., covered with short aliphatic tails, terminally attached by an Si-O bond to the glass), we often obtain a depletion layer. In the present review we are mainly concerned with polymers *bound* to a surface. We shall also consider

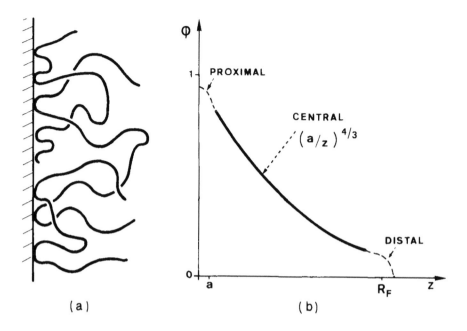

FIGURE 1.   Multichain adsorption from a good solvent: (a) qualitative aspect of the diffuse layer; (b) concentration profile $\Phi(z)$. Note the three regions: proximal (very sensitive to the details of the interactions), central (self similar), distal (controlled by a few large loops and tails).

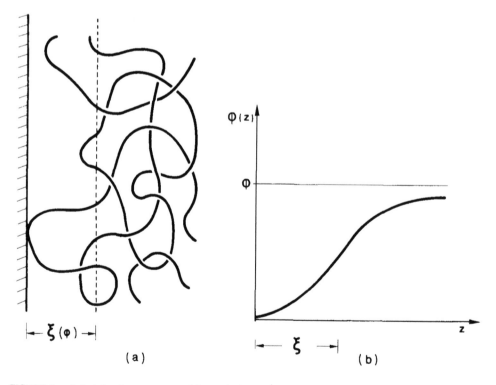

FIGURE 2.   A depletion layer near a repulsive wall. The thickness of the layer is the correlation length $\xi(\Phi)$. (a) Spatial aspect; (b) concentration profile.

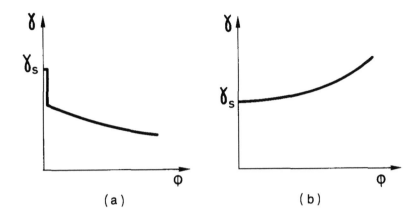

FIGURE 3. Surface tension γ of a polymer solution (volume fraction Φ). (a) Adsorptive case: as soon as some polymer is added, the tension drops abruptly; (b) depletion case: the interface remains nearly pure at low Φ, and γ is not very much affected. A rule of thumb to know if a particular polymer/solvent pair falls into case (a) or (b) amounts to comparing $\gamma(0)$ (pure solvent) and $\gamma(1)$ (pure polymer). If $\gamma(0) > \gamma(1)$ we expect to find case (a).

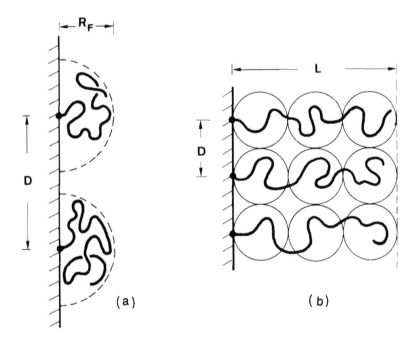

FIGURE 4. Two types of grafted surfaces: (1) low grafting density. The distance between heads D is larger than the coil size R_F. This is the "mushroom" regime. (b) high grafting density $(D < R_F) =$ "brush".

the depletion case. Depletion layers are important for certain processes where colloidal suspensions (or red blood cells) are induced to flocculate by addition of free polymer.

C. Three Modes of Fixation

Adsorption — This is one of the simplest methods available for colloid protection.[1,2]

Grafting — In situations where the polymer does not adsorb spontaneously, one can sometimes chemically attach the end of the chain to the surface (Figure 4). In good solvent

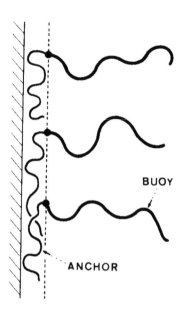

FIGURE 5.   Terminally attached chains
via block copolymers. The "anchor" A
precipitates against the wall, while the
"buoy" B protrudes toward the solution.

conditions, the chains repel each other, and the grafting reaction terminates when the chains
are disposed like adjacent "mushrooms" (Figure 4a). In some favorable cases, one can
push the density of grafted points to higher values; the result is the "brush" of Figure 4b,
with a relatively thick region of roughly constant concentration.

Use of block copolymers[1] — If a chain is made of one insoluble part A (the "anchor")
and one soluble part B (the "buoy"), it will often happen that A precipitates near the surface,
while B builds up an external brush (Figure 5). For instance, in water, A may be an aliphatic
chain, while B is made with POE or polyacrylamides. This method is extremely useful, but
the formation process of the adsorbed layer is delicate and poorly understood.

D. Recent Advances in Experiments and Theory

Originally the main observable in a polymer layer was the total coverage Γ (number of
monomers per square centimeter of surface). Currently we are learning more about thickness
of the layer by two techniques, hydrodynamics and ellipsometry.

1. Hydrodynamic Measurements

For instance, a spherical grain of radius R, moving with a velocity V, suffers a friction
force $6\pi\eta RV$ (where η is the solvent viscosity). However, if the grain is coated with polymer,
the force becomes $6\pi\eta(R + e_H)V$ where e_H is a certain hydrodynamic thickness.

2. Ellipsometry

A flat, bare surface usually reflects light of all polarizations. However, at one special
incidence (the Brewster angle), with light polarized in the plane of incidence, the reflection
coefficient vanishes. This cancellation is suppressed as soon as the surface is covered by an
adsorbed layer.[7] The residual reflectance at the Brewster angle is mainly dependent on the
total coverage Γ, i.e., the number of monomer units per square centimeters in the layer,
but there is a correction term, which does depend on the detailed concentration profile c(z)
in the layer. A complete study[8] shows that what is measured is the first moment $\bar{z}$ of the
profile. The ellipsometric thickness e_ℓ is thus:

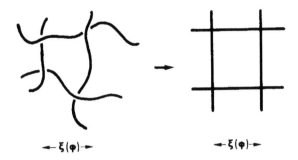

FIGURE 6. A polymer solution (volume fraction Φ) idealized as a "grid" with the same mesh size $\xi(\Phi)$.

$$e_\ell = 2\bar{z} = 2\Gamma^{-1} \int_0^\infty c(z) \, z \, dz$$

where z is the distance to the wall. (The factor 2 can be understood from the trivial case of a uniform layer $c(z) = c_o \ (z < \ell)$ where $\bar{z} = \ell/2$.

The hydrodynamical measurements are simple and can be performed on colloidal particles. The ellipsometric method requires a neat, flat surface, and is thus more restricted. We shall see later that they supplement each other rather nicely.

More sophisticated techniques, using evanescent waves[6] or neutron scattering[9,10] are just beginning to be used.

3. Forces between Surfaces

Thanks to the work of Israelashvili,[11] it is now possible to measure the forces between two mica plates separated by very small, controlled distances h ($10 \ \text{Å} < h < 10^4 \text{Å}$). This technique is extremely well adapted to the case where each plate is covered with a polymer layer. The pioneer in this field was Klein.[12-15] We shall discuss some of his results in Section III.

4. Theoretical Tools

For many years the structure of polymer layers was derived by some form of self-consisted field theory, where each chain feels an average potential, with a short range partly due to the wall, and a long-range repulsive part proportional to the concentration profile $c(z)$. This existed in two versions: one analytic, compact but restricted to the main features due to Jones and Richmond;[16] the other, more numerical but very detailed, due to Scheutens and Fleer.[17] It then became progressively recognized, however, that the mean-field approach is not reliable for these layers, which are close to being two-dimensional systems (the mean field in two dimensions being notoriously incorrect). A completely different scheme, incorporating the correlations in a rigorous fashion, has been constructed from general scaling laws.[18] In particular, this showed that adsorbed layers have a remarkable property called *self similarity*, discussed in Section II.B. The scaling analysis is not precise (it does not predict the numerical coefficients in any formula), but it is simple, and does provide an improved insight. Most of our discussion is based on it.

5. Scaling Analysis

The starting point of the scaling analysis is the current understanding of polymer solutions with overlapping coils (the so-called "semidilute" solutions).[19] An idealized representation of such a solution is shown in Figure 6. The most characteristic feature is the mesh size ξ which is a decreasing function of Φ:

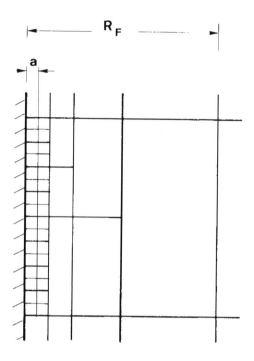

FIGURE 7.   An adsorbed polymer layer represented as
a ''self-similar grid''. At any distance z from the wall,
the local mesh size is equal to z.

$$\xi = a\Phi^{-3/4} \tag{1}$$

where a is a monomer size ($\sim$ 3Å). ξ is typically in the range 0 to 100 Å. Another important
parameter is the number g of monomers in one subunit of size ξ. The scaling structure of
g is

$$g = \Phi^{-5/4} \tag{2}$$

Knowing relations 1 and 2, one can predict most properties of a bulk solution.[19] For instance,
the osmotic pressure $\pi(\Phi)$ is simply

$$\pi = kT/\xi^3 \tag{3}$$

We shall see in the following sections that the structure of many adsorbed, or grafted, layers
in good solvents can also be described very compactly in terms of ξ and g.

II. STRUCTURE OF ONE LAYER

A. Adsorbed Chains
1. The ''Self-Similar Grid''
   Both the hydrodynamic and ellipsometric measurements show that the thickness of an
adsorbed layer is large (hundreds or thousands of angstroms). This is indeed what is expected
from the theory of an adsorbed layer at equilibrium. The basic clue is provided by Figure
7, showing what we call the ''self-similar grid'' structure of the layer. At any distance z
from the wall, the local mesh size is equal to z. This statement gives us the structure of the
profile[18,19] if

$$\xi[\Phi(z)] = z \qquad (4)$$

Using Equation 1, we find

$$\Phi(z) = (a/z)^{4/3} \qquad (5)$$

Thus, the profile decreases very slowly. Of course, the self-similar construction shown in Figure 7 must break down at very small z, and also at very large z. It is thus important to state what the two limits (z_{min} and z_{max}) are.

For the (usual) case of strong adsorption (as defined in Section I) the lower limit is a monomer size:

$$z_{min} = a \qquad (6)$$

For the (usual) case of an adsorbed layer facing nearly pure solvent, the upper limit is the size of one coil in bulk solution, which we call R_F (the Flory radius):

$$z_{max} = R_F = aN^{3/5} \qquad (7)$$

where N is the number of monomers per chain. Note that the relation between R_F and N is identical to the relation between g (Equation 2) and ξ (Equation 1): $\xi = ag^{3/5}$. Because N is often large (10^3 to 10^4), the interval (z_{min}, z_{max}) is large and the self-similar features of Figure 7 are indeed meaningful.

2. Relation to Experimental Data

The total coverage Γ predicted from Equation 5 is

$$\Gamma = \int_a^\infty \frac{\Phi(z)dz}{a^3} \cong \frac{1}{a^2} \qquad (8)$$

where a^3 is a monomer volume. We see that Γ corresponds roughly to one full monolayer of dense polymer, but this should not mislead us: there are long loops and tails in the structure, extending up to the higher cutoff R_F. A detailed hydrodynamic analysis shows that one single loop (or tail) is enough to perturb the flow of solvent near the wall very significantly.[18,20,21] The conclusion is that the hydrodynamic thickness e_H should be comparable to the coil size R_H:

$$e_H \sim aN^{3/5} \qquad (9)$$

Equation 9 is supported by certain data on polystyrene latices covered by POE (in water),[22] but some other experiments (on the same system!) give a higher exponent ($e_H \sim N^{0.8}$).[23]

We have seen that the ellipsometric thickness e_ℓ should be given:

$$e_\ell \sim \Gamma^{-1} \int_a^{RF} \frac{\Phi(z)}{a^3} zdz$$

This integral is also very sensitive to the cutoff R_F:

$$e_\ell \sim a^{1/3} R_F^{2/3} \sim aN^{2/5} \qquad (10)$$

Equation 10 is in excellent agreement with data by Kawaguchi and Jarahashi[24] which were

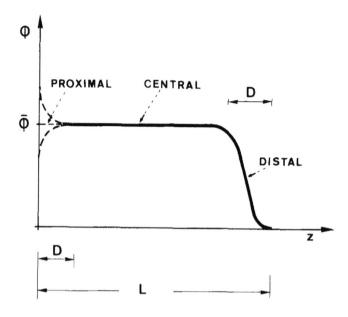

FIGURE 8.   The concentration profile inside a brush. In the central region the profile is flat. In the proximal region the profile depends on the details of the monomer-wall concentration.

taken before any theoretical prediction. Note the remarkable difference between e_H (Equation 9) and e_ℓ (Equation 10). With these very diffuse profiles, each experimental method may measure a different thickness.

B. Terminally Attached Chains

We restrict our attention to chains which are *not adsorbed* on the surface. (In the opposite case, with adsorption, we return to Figure 7, and grafting is not very relevant.)

The two main regimes have been shown in Figure 4a and b. If the grafting level is low, we have separate "mushrooms", each with a size $\sim R_F$. If the grafting density is high, we have a "brush". The scaling structure of the brush in good solvents has been deciphered first by Alexander.[25] He points out that the crucial parameter is the average distance D between attachment points. The brush builds up a region of roughly uniform concentration $\overline{\Phi}$ (Figure 8) and *the mesh size in the brush is equal to D.*

$$\xi(\overline{\Phi}) = D \tag{11}$$

The chains in the brush are stretched out. As shown on Figure 4b, they can be described as a linear sequence of subunits each of size D and number of monomers g_D (Equation 2):

$$g_D = \Phi^{-5/4} = (D/a)^{5/3}$$

The overall thickness of the brush is then

$$L = \frac{N}{g_D} D = Na\left(\frac{a}{D}\right)^{2/3} \tag{12}$$

Unfortunately, at present we do not have very clear-cut data on these "brushes". The brush regime requires a high density of attached chains ($D \ll R_F$), and this is not achieved easily in grafting. It may be reached, however, with block copolymers.

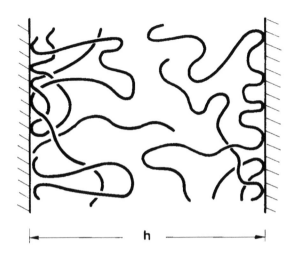

FIGURE 9. Two adsorbed layers brought into contact, but before any bridging.

C. More General Situations

All our presentation of adsorbed layers, or of brushes, assumed a good solvent and no additives.

1. If we decrease the quality of the solvent, the structures tend to become more compact. For adsorbed layers, this regime has been computed in some detail by Klein and Pincus.[26]
2. If we add some free polymer (chemically identical to the grafted chains) to a brush, the structure also tends to shrink, and many different regimes can occur.[27] One basic source of shrinkage is a ''screening effect'': the mobile polymers tend to screen out the repulsive interactions between monomers from the brush.

III. INTERACTIONS BETWEEN TWO PARALLEL PLATES

A. Plates Coated by Adsorbed Polymer

The basic experiment is idealized in Figure 9. In stage 1, two mica plates are incubated with polymer solutions at large distances h. In a second stage, h is reduced to some small, prescribed value (monitored by Fabry Pérot interferometry) such that the two diffuse layers overlap (h $<$ 2R$_F$). Then the forces are measured.

There have been some disputes concerning even the *sign* of the force. The most detailed experiments (by Klein and co-workers[12-15]) have been performed with POE (of molecular weights in the range 10^5 to 3×10^5 daltons) in water at room temperature (i.e., in a rather good solvent). They show that in slow (hopefully reversible) conditions the force between the plates is *repulsive* at all measurable distances. What are the theoretical predictions? It turns out that the conditions of the experiment have to be stated in detail.[18]

1. If the equilibrium was complete, allowing for some polymer to flow out of the gap when the two plates are squeezed together, one expects attraction; adjustable layers should lead to adhesion.
2. In practice, the polymer moves only very slowly along the mica, and a more plausible condition is to impose constant coverage (fixed Γ) on both plates. Indeed, certain optical measurements in the gap (which can be done using another set of Fabry Pérot fringes) suggest that this condition is satisfied for the Klein experiments. At fixed Γ

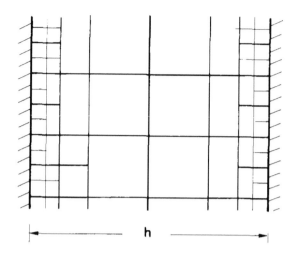

FIGURE 10.   The self-similar grid for two adsorbed layers in contact. The maximum mesh is of order h/2, where h is the gap.

one expects the force to be repulsive and to be given by the osmotic pressure at the midpoint (Figure 10). Since the mesh size at this point is h/2, we expect from Equation 3 that the force (per unit area) is

$$F \sim (constant)\ kT/(h/2)^3 \sim kT/h^3 \tag{13}$$

or that the interaction energy U (per square centimeter) scales like

$$U \sim kT/h^2 \quad (a \ll h \ll R_F) \tag{14}$$

This is compatible with the Klein data, but the exponent remains to be established in more detail.

3.   An interesting complication is brought in by *bridging* (Figure 11). After some time of contact, the chains from one plate may establish bridges with the other plate. The final equilibrium number n_b of bridges per square centimeter may be directly understood from the self-similar picture of Figure 10. The spacing between possible bridges is on the order of h/2; thus, the scaling structure of n_b must be

$$n_b \sim 1/h^2 \quad (h < R_F) \tag{15}$$

Equation 14 can be translated into a prediction for the number of bridges between two spherical grains, of radius R_{grain} separated by a gap $h < R_F \ll R_{grain}$. The total number is then

$$n_{total} \sim hR_{grain}\ n_b \sim \frac{R_{grain}}{h} \tag{16}$$

How can we detect the bridges? If we pull the grains apart very slowly (remaining constantly in thermal equilibrium), the bridge should retract without any great effect on the forces. However, most practical velocities of separation are fast compared to the bridging kinetics, and then the bridges should act as elastic springs, keeping the grains together. Scaling arguments predict a bridging energy $\sim kT/h^2$, very similar to

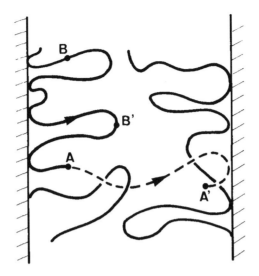

FIGURE 11. The bridging process: one chain end moves from A to A′, while the other end moves from B to B′.

Equation 14, but with a reversed sign. Thus, adhesive properties may tell us something about the bridges. We shall come back to the kinetic aspects in Section IV.

4. A (possibly related) striking experimental effect is obtained under conditions of "starvation".[14] Here, the plates are incubated at large h (always in a good solvent), but only for a short time. The coverage Γ is lower than its equilibrium value. Then, when the two plates are brought together, a strong attractive force shows up at intermediate distances. This is probably due to a fast bridging process.

B. Plates with Terminally Attached Chains

Experiments using the Israelashvili machine with block copolymers (anchor + buoy) are currently underway.[28] However, the exact state of these chains, their surface density, etc., still must be described. At present, we can only give the theoretical predictions, corresponding to the "mushroom" and "brush" regimes of Figure 4a and b.

1. Interactions between Brushes

These interactions have been computed by many methods, in particular the self-consistent theory[29] and (more recently) scaling arguments.[18] The two brushes come into contact when $h = h_c = 2L$, where L (Equation 12) is linear in N. At $h < h_c$ the two brushes are squeezed against each other, but they do not interdigitate (Figure 12).

The polymer concentration inside each brush increases $\Phi = \Phi h_c/h$. This gives two contributions to the force: (1) the osmotic pressure inside each brush increases, and (2) the elastic restoring forces (tending to thin out the brush) decrease.

In the single brush problem, the osmotic and elastic terms balanced each other exactly at the equilibrium thickness (L). In the compressed brush (thickness h/2), they do not, and the result is a force per square centimeter.

$$F \cong \frac{kT}{D^3} \left[\left(\frac{h_c}{h}\right)^{9/4} - \left(\frac{h}{h_c}\right)^{3/4} \right] \quad (h < h_c) \tag{17}$$

Here, D is always the average distance between attachment points. The first term in the bracket is the osmotic term; the second one is the elastic term. At strong compressions the osmotic term should dominate completely.

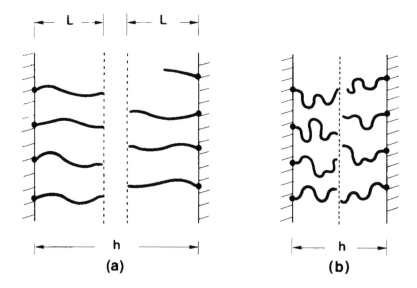

FIGURE 12.   Two brushes (a) before contact $h > h_c$ (b) under compression $h < h_c$. The two brushes do not intermix very much.

2. Interactions between Dilute Mushrooms

Here, the interaction shows up first at a critical gap thickness $h = h_c \sim R_F$. Thus, h_c is now a weaker function of N ($h_c \sim N^{3/5}$). At $h << h_c$ the chain is squeezed in a thin gap, and scaling arguments predict a confinement energy per chain.[19]

$$U_1 \cong kT\left(\frac{R_F}{h}\right)^{5/3} = NkT\left(\frac{a}{h}\right)^{5/3} \qquad (18)$$

The energy per square centimeter of plates is U_1/D^2 and the corresponding force is

$$F \cong \frac{kT}{D^2 h_c}\left(\frac{h_c}{h}\right)^{8/3} \qquad (h < h_c) \qquad (19)$$

The recent data by Hadzioannou and co-workers[28] on copolymers A = polyvinylpyridine B = polystyrene give overall forces F which increase more weakly than these predictions (Equations 17 or 19) at small h. However, many complications could arise (e.g., van der Waals attractions between the dense A parts), and further studies are needed to check whether D is independent of h in these experiments.

C. Effects of Free Polymer

The only case which has been discussed up to now corresponds to an additive which is chemically identical to the polymer already present in the layer.

1. Adsorbed Chains Plus Semidilute Solutions

The theoretical concentration profile is qualitatively sketched in Figure 13. We still have two self-similar adsorbed layers, but they extend out only to a thickness $\xi(\phi_b)$, where ξ is the characteristic mesh size of the solution (concentration ϕ_b). This leads to two regimes:

1.   If the thickness of the gap h is larger than $\sim 2\xi(\phi_b)$ the two adsorbed layers do not couple, and we expect negligible forces between the two plates (in thermal equilibrium).

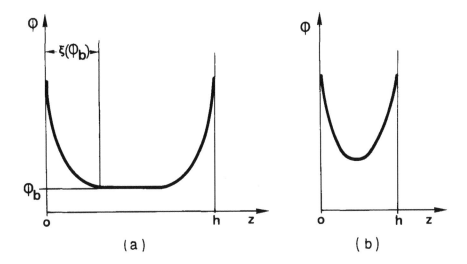

FIGURE 13.   Concentration profile for two adsorbed layers in the presence of mobile polymer (concentration Φ_b): (a) the bulk correlation length $\xi(\Phi_b)$ is small, or the gap large [h $\gtrsim 2\xi(\Phi_b)$]; (b) reverse conditions [h $< 2\xi(\Phi_b)$].

2.   If h is smaller than $2\xi(\phi_b)$, the two diffuse layers do interact, and we should recover the forces described in Section III.A.

At the moment, we have no direct check on these predictions. It may well be that in practice, various time-dependent effects complicate the behavior; exchanges between solution and adsorbed layers may be very slow. We shall come back to this point in Section IV.

2. Brushes Plus Semidilute Solutions

This case is not far from a very practical problem in colloid stability, well described in the book of Napper.[1] Let us start with grains protected by terminally attached chains, giving a stable sol. Then add some mobile chains; it is often found that at a threshold concentration Φ_1 the sol flocculates. Then add more polymer: at a second concentration Φ_2 the floc redissolves!

The theory of these effects has been worked out[30,31] and is summarized in Figure 14. One essential parameter is the ratio of the solution concentration Φ to the brush concentration $\bar{\Phi}$.

When $\Phi << \bar{\Phi}$ the brush acts as a strongly repulsive region for the mobile chains. This creates a depletion layer (of thickness $\xi[\Phi]$) just out of the brush. When h is decreased, and a good overlap occurs between the depletion layers from both sides, we expect an *attraction* between the plates. The origin of this attraction is shown in Figure 15. When this force leads to a grain/grain attraction energy stronger than kT, the grains flocculate; this corresponds to the first threshold Φ.

When Φ increases and becomes comparable to the brush concentration $\bar{\Phi}$, many things change:

1.   The depletion layer thickness $\xi(\Phi)$ develops on the order of D, and is thus hidden by the intrinsic width of the brush surface (which is also on the order of D).
2.   A mobile chain sees a strong repulsive potential in the brush region, but sees also a strong repulsive potential in the outer solution. Thus, the potential is essentially constant in all the liquid space and we lose the depletion layer. Then we should return to a repulsive regime between brushes. The details of this repulsion are complex: the brush

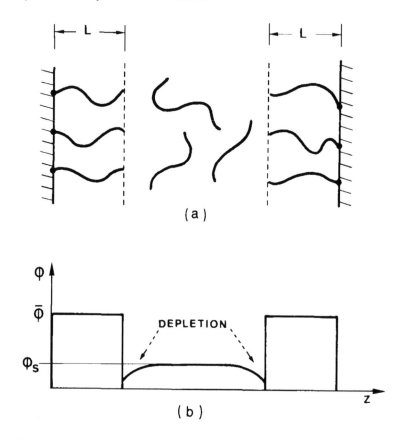

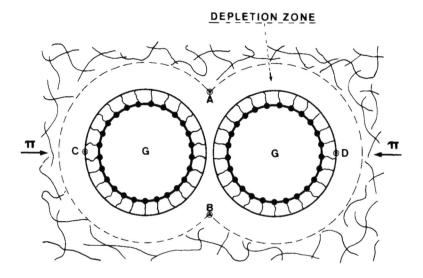

FIGURE 14.   Effect of mobile polymer (concentration Φ_x) on the interactions between two surfaces protected by "brushes" of internal concentration $\bar{\Phi} \gg \Phi_x$. The mobile chains are depleted near the brushes. (a) Spatial aspect; (b) concentration profile.

FIGURE 15.   Origin of the grain/grain attraction in the presence of depletion layers: in regions A and B there is no osmotic pressure from the polymer, while in regions C and D the osmotic pressure from the outer solution pushes the grains together.

begins to collapse under the osmotic pressure of the mobile chains, etc. However, it is natural to expect stabilization at $\Phi = \Phi_2 \sim \bar{\Phi}$.

IV. DYNAMICAL PROBLEMS

A. Basic Results on Exchange between Adsorbed Layers and Solutions

Can a molecule leave an adsorbed layer, and go into solution, or the reverse? Two groups of observations are important here.

1. Stability under Washing

If we prepare an adsorbed layer of long chains (large N) by slow incubation, and then wash it with pure solvent, we usually find that the layer does not redissolve at all. The classical interpretation of this fact was the following:

A finite fraction f of the monomers is in direct contact with the wall. The fraction f is measured by various techniques, and in particular by nuclear resonance, the spectrum of the f fraction being solid-like and broad, while the spectrum of the complement $1 - f$ are liquid-like and narrow.[2] Typically, $f \sim 0.1$ to 0.8. The binding energy of the chain to the wall is on the order of fNU_b where U_b (the adsorption energy) is on the order of kT for our regime of strong adsorption. This binding energy is then reduced by the repulsive monomer/monomer interactions in the good solvent. This repulsive energy is itself a finite fraction of ψ of the above estimate ψfNU_b with $\psi \sim 1/2$ to $1/3$ depending on the detailed model. The net result is a binding energy $f(1 - \psi)NU_b \sim 0.1\,NU_b$. The Boltzmann factor for removing a chain would then be on the order of

$$\tau \sim \exp(-0.1\,NU_b/kT) \sim \exp(-N/10) \qquad (20)$$

With N values above 10^3 this Boltzmann factor is extremely small, and this could explain the stability under washing. We shall show, however, that this argument is utterly wrong, as suggested by the experiments quoted below.

2. Allowed Exchanges at Finite Bulk Concentrations

It has been known for some time that an adsorbed layer, when faced with bulk solution, can exchange some molecules with the solution. The first hint came from situations where the adsorbed chains are relatively short, while the solution chains are longer.[32] Then the long chains displace the short ones.

A more detailed proof was obtained in 1985 by isotopic exchange measurements performed in Strasbourg.[33] Here one starts with radioactively labeled chains covering a set of grains, and achieving a coverage Γ^*. Then the system is washed with a solution of chemically identical, unlabeled chains (concentration Φ_b). During periods $\lesssim 1$ day, there is a net loss of radioactivity, ruled by a kinetic equation of the form

$$\frac{d\Gamma^*}{dt} = -k\Gamma^*\phi_b \qquad (21)$$

As pointed out in Reference 33, this resembles a second-order process: (bound chain)* + free chain → (free chain)* + bound chain.

However, we know that for these model systems, there is no reason for a (1,1) association. Thus, Equation 21 raises an interesting question. The current answer is described in Section IV.B. Note that Equation 21 is compatible with the aforementioned stability of adsorbed layers with respect to pure solvent. When $\Phi_b \to 0$ the rate of Equation 21 vanishes.

B. Tentative Interpretation[34]

1. Absence of Very High Energy Barriers

To extract a chain from the adsorbed layer we need not overcome the huge barrier described by Equation 20. Physically, this may again be understood, thanks to the self-similar representation of Figure 7. When one particular coil wants to move (from the surface outward) in the self-similar grid, it meets a dense repulsive grid only in the early steps; later, the grid is quite open. The final result is that the barrier factor is not given by Equation 20, but is a much weaker function of N:

$$\tau = N^{-x} \quad (x \sim 0.3) \tag{22}$$

With $N \sim 10^3$, $\tau \sim 0.1$ exchanges can occur rather freely.

2. The Saturation Condition

An important feature of the adsorbed layer is the strong repulsive interaction between adjacent chains. A single chain is very firmly adsorbed; adding a second chain, we find a little less binding energy. Ultimately, adding 1,2 . . . p chains, we can fill the adsorbed layer up to an equilibrium coverage Γ. However, if we insist on adding more chains, their free energy becomes higher than the bulk solution level, and they "spill out" fast. Thus, there is a strong restoring force, tending to maintain Γ constant. When this feature is injected into the adsorption and desorption rates, one finds that chains can escape from an equilibrated adsorbed layer only if some chains from the solution immediately come and maintain Γ nearly constant. Thus, in the Strasbourg experiments, it is natural that the rate for escape $(- \, d\Gamma^*/dt)$ be proportional to the outer concentration Φ_b, as observed in Equation 21. This condition of nearly constant Γ is what we call the *saturation condition*. It is somewhat similar to the condition of electrical neutrality which is met in an ion exchanger. The exchanger can liberate its own mobile ions (for instance, Na^+) only if some other positive ions are available in the ambient bath.

To summarize: we understand that the energy barriers are weak, and also that the saturation condition enforces rate equations which resemble second-order kinetics (Equation 21). There remain many delicate points, however. To predict the factor k in Equation 21, we need a detailed description of the crawling motions of chains in the barrier region. This crawling is complicated by another effect, present with many polymer systems, and described below.

C. The Glassy Layer

An important feature of adsorbed layers has been discovered recently by Cohen-Stuart.[35] They studied adsorbed layers of conventional polymers (POE, PVA, etc.) at the *free surface* of a solvent. Performing delicate rheological measurements on this thin layer, they found that it shows a nonzero yield strength. The adsorbed layer is not fluid, but behaves like a (weak) solid crust.

The most plausible interpretation of this effect[36] is based on the observation that these polymers, in thin bulk phase at room temperature, are in a *glassy state*. In solution, they have a lower glass transition temperature $T_g(\Phi)$, but we know from Equation 5 that the first few monomer layers near the surface are quite dense. In this region we are probably dealing with a glass. This slows down all creeping motions of the chains.

Future experimental work on the dynamics should probably focus first on selected polymer systems which remain liquid even at room temperature in their bulk phase: elastomers or polydimethylsiloxane. However, even with these systems, there is a danger that the adsorbing surface (when it is solid) still favors a certain "freezing" of the motions in the first layer.

D. Kinetics of Bridging: Conjecture

Let us return to Figure 11 and consider two overlapping layers, brought at a distance h

$< 2R_F$ at time 0. Qualitatively, one may think of bridging in terms of exchange of polymer between one layer (e.g., the left layer) and an equivalent solution (replacing the layer on the righthand side). The concentration of the equivalent solution would then be the concentration at the midpoint.

This scheme allows one to describe bridging in terms of rate equations similar to Equation 21. Again, there is no high barrier due to polymer repulsions, and the whole line of thought can be transposed.[34] However, the same difficulty also remains: the magnitude of the rate constants will often be controlled by the glassy features of the inner sublayers. There is one possibly important complication, if we want to compare the rate constants at different values of the gap h: when h is decreased, the concentration in the inner sublayers raises slightly; this in turn increases the glassy features. This process is probably important, but as of now it is not at all controlled, experimentally or theoretically.

V. FROM MODEL SYSTEMS TO BIOSYSTEMS: A LIST OF OPEN QUESTIONS

There remains a huge gap between the simple interface/polymer problems which we discussed, and the surface of a living cell, covered with complex proteoglycans and hyaluronates. In this section we shall try to do three things: summarize our main ideas from model systems, construct a list of open problems, and search for possible lines of research.

A. Attractions and Repulsions in Model Systems

In Section III we found that even in this relatively simple case, the sign of the long-range interactions depended critically on certain conditions.

1. If the system is *not adjustable* (constant coverage Γ) we expect, and find, purely repulsive forces.
2. If the system is adjustable by migration of molecules in and out of the interplate region, we can find strong attractions. This "adjustable" case is deeply related to certain recent models of cellular adhesion.[37]
3. Another important feature is the presence or absence of *bridging*. When bridging has taken place, and if the bridges can be considered as *permanent* on the time scale of the experiments, they behave as elastic springs coupling the two plates. If we force the plates to separate fast, we expect a yield stress, but the scaling structure of this yield stress is yet unknown. Experiments on this are needed.

A remark on *poor solvents*: if the chains are close to precipitation, two polymer-covered plates show an attraction at intermediate range. This regime has been studied in detail by Klein[14] with PS in cyclohexane, and there is a good theoretical picture for it.[26] However, it is probably not very relevant to the cell-cell problems.

B. Charge Effects

Colloidal grains (which are usually negative in water) can often be flocculated by cationic polyelectrolytes for obvious reasons. A more surprising feature is that anionic polyelectrolytes can also behave as flocculants if the solution contains divalent cations (which link the solid and the polymer), and if the salt content is not too low (good screening).

With the Israelashvili machine, Klein has studied the long-range forces between mica plates in water with polylysine adsorbed on the plates.[15] He finds a fascinatingly complex set of interactions, suggestive of a competition between two states of the adsorbed layer (a "dilute" state and a "compact" state?).

C. Special Features of Branched Systems

Proteoglycans are branched structures; we clearly need a certain understanding of adsorption or grafting statistics with ramified objects. Very little is known at present.

Branched structures are more dense than linear chains. Recent experiments by Bouchaud and co-workers[38] on condensation products (polymethanes) obtained with bi- and trifunctional units, just below a gel point, and selected later by gel permeation chromatography, show that a branched molecule of N subunits in good solvent conditions has a size

$$R \sim aN^{1/2} \tag{23}$$

This is to be compared to the law for linear chains $R \sim aN^{3/5}$. The branched structure is more compact, and this should have many consequences.

Stochastically branched chains, attached chemically at one point on a surface, will probably achieve only the "mushroom" type of coverage, and not overlap with each other.

What is the concentration profile in an adsorbed layer of (homodisperse) branched chains? We can guess the answer, at least for the case of stochastic branching, by a sequence of scaling arguments similar to Equations 1, 2, 4, and 5.

We consider first a semidilute solution of branched objects (concentration Φ) and define a correlation length ξ and a typical number of monomers g in the correlation volume ξ^3:

$$g = \frac{\Phi}{a^3} \xi^3 \tag{24}$$

The relation between g and ξ is the transposition of Equation 23, namely:

$$\xi = ag^{1/2} \tag{25}$$

Solving Equations 23 and 24 we get:

$$\xi = a\Phi^{-1} \tag{26}$$

Now we are equipped to construct the adsorption profile $\Phi(z)$ assuming self similarity, and imposing the condition 4, we arrive at:

$$\Phi \cong a/z \tag{27}$$

and all the ellipsometric and hydrodynamic consequences can be worked out. We need experiments on these lines.

D. Effects of Polymer Additives

We encountered many different effects, which we try to sort out here.

1. Colloids protected by "brushes" with a sharp concentration profile (Figure 8) can be destabilized by low concentrations of additives (mobile chains) which are chemically identical to the "brush" (see Section III.C). If we are dealing with a weaker protection (the "mushroom" regime), the process is far less clear.
2. Similar effects are expected if the additive is chemically different from the brush, but does not like to mix with it. All depletion effects are maintained.
3. If the additive does mix with the brush (i.e., a significant, negative enthalpy of mixing is present) we also expect an additive force, but for completely different reasons. To our knowledge, this process has not yet been studied systematically.

4. If we adsorb chains on bare surfaces, we can sometimes reach a significant level of bridging. If the bridges have a long lifetime, they will show up as an attractive coupling between the surfaces.

5. If we add mobile chains to surfaces which are already covered by an *adsorption* layer, most of the effects quoted above (1 to 4) can probably occur. A typical biophysical example is the aggregation of red blood cells by dextran;[39] further complications, related to charge effects in the adsorbed layer, have also been invoked for this case.[40]

ACKNOWLEDGMENTS

The views on polymers at interfaces which have been presented here represent the result of a very cooperative (and slow) buildup. Alex Silberberg, Jacob Klein, L. Terminassian, R. Varoqui, and recently, M. Cohen-Stuart were our main experimental advisers. S. Alexander, F. Brochard, J. F. Joanny, L. Leibler, and P. Pincus constructed a large part of the theory. More recent (and future) experiments connected to this theme and stimulating for us theorists were set up by H. Hervet, J. M. de Meglio, R. Ober, F. Rondelez, C. Taupin, and very recently by L. Auvray.

REFERENCES

1. **Napper, D.**, *Polymeric Stabilisation of Colloidal Dispersions*, Academic Press, New York, 1983.
2. **Cohen-Stuart, M., Cosgrove, T., and Vincent, B.**, Experimental aspects of polymer adsorption at solid-solution interface, *Adv. Colloid Interface Sci.*, in press.
3. **Hayter, T., Jannink, G., Brochard, F., and de Gennes, P. G.**, Correlations and dynamics of polyelectrolyte solutions, *J. Phys. (Paris), Lett.*, 41, 451, 1980.
4. **Oosawa, F. and Asakura, S.**, Surface tension of high polymer solutions, *J. Chem. Phys.*, 22, 1255, 1954.
5. **Joanny, J. F., Leibler, L., and de Gennes, P. G.**, Effects of polymer solutions on colloid stability, *J. Polym. Sci. (Phys.)*, 17, 1073, 1979.
6. **Allain, C., Ausserre, D., and Rondelez, F.**, Direct optical observation of interfacial depletion layers in polymer solutions, *Phys. Rev. Lett.*, 49, 1694, 1982.
7. **Azzam, R. and Bashara, N.**, *Ellipsometry and Polarised Light*, North-Holland, Amsterdam, 1977.
8. **Charmet, J. C. and de Gennes, P. G.**, Ellipsometric formulas for an inhomogeneous layer, *J. Opt. Soc. Am.*, 73, 1777, 1983.
9. **Barnett, K., Cosgrove, T., Crowley, T., Tadros, J., and Vincent, B.**, in *The Effects of Polymers on Dispersion Stability*, Tadros, J., Ed., Academic Press, New York, 1982, 183.
10. **Auvray, L.**, *C.R. Acad. Sci. (Paris)*, in press.
11. **Israelashvili, J. and Adamans, G.**, Measurement of forces between two mica surfaces in aqueous electrolyte solutions in the range 0—100 nm, *Trans. Faraday Soc. I* 74, 975, 1978.
12. **Klein, J. and Luckham, P.**, Forces between two polyethylene oxide layers immersed in a good aqueous solvent, *Nature (London)*, 300, 429, 1982.
13. **Klein, J. and Luckham, P.**, Long-range attractive forces between two mica surfaces in an aqueous polymer solution, *Nature (London)*, 308, 836, 1984.
14. **Klein, J.**, Forces between two polymer layers adsorbed at a solid-liquid interface in a poor solvent, *Adv. Colloid Interface Sci.*, 16, 101, 1982.
15. **Luckham, P. and Klein, J.**, Forces between mica surfaces bearing adsorbed polyetiolyte, poly-L-lysine, in aqueous media, *Trans. Faraday Soc. I*, 80, 865, 1984.
16. **Jones, J. S. and Richmond, P.**, Effects of excluded volume on the conformation of adsorbed polymers, *Trans. Faraday Soc. II*, 73, 1962, 1977.
17. **Scheutens, J. and Fleer, G.**, in *The Effects of Polymers on Dispersion Properties*, Tadros, J., Ed., Academic Press, New York, 1982, 45.
18. **de Gennes, P. G.**, Polymer solutions near an interface. I. Adsorption and depletion layers, *Macromolecules*, 14, 1637, 1981; **de Gennes, P. G.**, Polymers at an interface. II. Interaction between two plates carrying adsorbed polymer layers, *Macromolecules*, 15, 492, 1982.
19. **de Gennes, P. G.**, *Scaling Concepts in Polymer Physics*, Cornell University Press, Ithaca, N.Y., 1985.

20. **de Gennes, P. G.,** Mobilité electrophorétique de grains avec polymères adsorbés, *C.R. Acad. Sci. (Paris),* 297(II), 883, 1983.
21. **Cohen-Stuart, M., Waajen, F., Cosgrove, T., Vincent, B., and Crowley, T.,** Hydrodynamic thickness of adsorbed polymer layers, *Macromolecules,* 17, 1825, 1984.
22. **Kato, T., Nakamura, K., Kawaguchi, K., and Takahashi, A.,** Quasi-elastic light scattering measurements of polystyrene-latice and confirmation of poly(oxyethylene) adsorbed in the latice, *Polym. J.,* 13, 1037, 1981.
23. **Cosgrove, T., Crowley, T., Cohen-Stuart, M., and Vincent, B.,** in Polymer adsorption and dispersion stability, *ACS Symp.,* 240, 147, 1984.
24. **Kawaguchi, M. and Jarahashi, A.,** Adsorption of polystyrene onto a metal surface in good solvent conditions, *Macromolecules,* 16, 631, 1983; **Kawaguchi, M. and Jarahashi, A.,** Molecular weight dependence of the thickness of the polystyrene layer adsorbed onto a metal surface in good solvent conditions, *Macromolecules,* 16, 1465, 1983.
25. **Alexander, S.,** Adsorption of chain molecules with a polar head. A scaling description, *J. Phys. (Paris),* 38, 983, 1977.
26. **Klein, J. and Pincus, P.,** Interaction between surfaces with adsorbed polymers: poor solvents, *Macromolecules,* 15, 1129, 1982.
27. **de Gennes, P. G.,** Conformations of polymers attached to an interface, *Macromolecules,* 13, 1069, 1980.
28. **Hadzioannou, G., Patel, S., Granick, S., and Tirrell, M.,** Forces between surfaces of block copolymers adsorbed on mica, *J. Am. Chem. Soc.,* 108, 2869, 1986.
29. **Dolan, A. and Edwards, S. F.,** Theory of the stabilization of colloids by adsorbed polymer, *Proc. R. Soc. London Ser. A,* 337, 509, 1974; **Dolan, A. and Edwards, S. F.,** The effect of excluded volume on polymer dispersant action, *Proc. R. Soc. London Ser. A,* 343, 427, 1975.
30. **Gast, A., Russel, W., and Hall, C. K.,** *J. Colloid Interface Sci.,* in press.
31. **Vincent, B.,** Abstr. no. 106, 5th Conf. Surface Colloid Sci., Potsdam, N.Y., 1986.
32. **Thies, C.,** The adsorption of polystyrene-polymethylmethacrylate. Mixtures at a solid-liquid interface, *J. Phys. Chem.,* 70, 3783, 1966; **Cohen-Stuart, M., Scheutens, J., and Fleer, G.,** Polydispersity effects and the interpretation of polymer adsorption isotherms, *J. Polymer Sci. (Phys.),* 18, 559, 1980; **Furusawa, K., Yamashita, K., and Konno, K.,** Adsorption of monodisperte polystyrene onto porous glass. I. Preference — adsorption and displacement of high-molecular weight species, *J. Colloid Interface Sci.,* 86, 35, 1982.
33. **Pferrerkorn, E., Carroy, A., and Varoqui, R.,** Dynamic behavior of flexible polymers at a solid/liquid interface, *J. Polymer Sci. (Phys.),* 23, 1997, 1985.
34. **de Gennes, P. G.,** Pénétration d'une chaine dans une couche adsorbée: échanges solution/adsorbant et pontage de grains colloïdaux, *C.R. Acad. Sci. (Paris),* 301(II), 1399, 1985.
35. **Cohen-Stuart, M.,** *Colloids Surfaces,* 17, 91, 1986.
36. **Kremer, K.,** *J. Phys.,* 47, 1269, 1986.
37. **Bell, G., Dembo, M., and Bongrand, P.,** Conception between nonspecific repulsion and specific bounding, *Biophys. J.,* 45, 1051, 1984.
38. **Bouchaud, E., Delsanti, M., Adam, M., Daoud, M., and Durand, D.,** *J. Physiol. (Paris),* in press.
39. **Snabre, P., Grossman, G., and Mills, P.,** Effects of dextran polydispersity on red blood cell aggregation, *Colloid Polym. Sci.,* 263, 478, 1985.
40. **Snabre, P. and Mills, P.,** Effect of dextran polymer on glycocalyx structure and cell electrophoretic mobility, *Colloid Polym. Sci.,* 263, 494, 1985.

Chapter 3

SURFACE PHYSICS AND CELL ADHESION

Pierre Bongrand, Christian Capo, Jean-Louis Mege, and Anne-Marie Benoliel

TABLE OF CONTENTS

I. INTRODUCTION

When one attempts to use the aforementioned data on intermolecular forces to estimate the interaction between two cell surfaces approaching each other, many *ad hoc* assumptions are required to obtain numerical results. Indeed, when short intercellular distances are considered (say, when the polar headgroups of lipid bilayers are less than a few hundred angstrom units apart) the precise shape, flexibility, and charge distribution of membrane molecules may heavily influence intercellular forces. Further, even if the details of membrane surface structures were known, accounting for all of them would lead to intractable equations.

It is therefore instructive to consider approximate procedures that were developed and tested in order to measure the properties of actual complex surfaces and predict the outcome of an interaction between two different bodies. In the present chapter, we review some data relevant to the physical chemistry of surfaces. It must be borne in mind that no new kind of basic interaction is involved, but only a different and semiempirical way of dealing with complex systems. Selected basic principles and results are described, together with some applications to cell adhesion. The mathematically inclined reader has to be warned that the methods we deal with are highly empirical, and the basic question is to know whether they work, not whether they can be rigorously justified, which is certainly not the case at the present time.

II. BASIC DEFINITIONS

The surface free energy γ of a material (solid or liquid) body may be defined as the free energy required to create a plane free surface of unit area. The surrounding medium is thought of as a dilute phase that is not supposed to interact with the surface (see a later section on the problem of adsorption).

It is well known that the free energy of a surface is dependent on its curvature.[1] However, this dependence may be neglected when the curvature radii of the surface are much larger than molecular dimensions, which may be considered as a legitimate assumption as far as cell membranes are concerned. Indeed, the radius of curvature of a typical microvillus is about 0.1 μm.[2]

In most cases, the quantities to be considered are interfacial free energies. The interfacial free energy γ_{12} between phases 1 and 2 is the free energy required to create an interface of unit area between these phases. It must be emphasized that the mere use of γ_{12} as a material constant involves the important assumption that the structure of the interface between materials 1 and 2 is only dependent on their physicochemical properties. If we use a parameter γ_{12} to describe the interaction between two cell surfaces, we implicitly assume that some equilibrium interfacial structure can be reached. This may be tenable if cells are not subjected to large external forces, and experimental conditions allow equilibrium configurations to be reached, which may require several hours.

An important parameter is the work of adhesion W_{12} between bodies 1 and 2. This is the

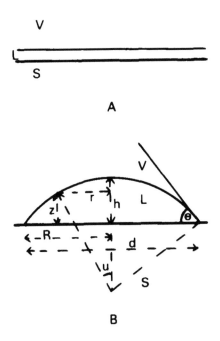

FIGURE 1. Interaction of a liquid droplet with a solid surface. V: vapor, L: liquid, S: solid. (A) Complete wetting; (B) partial wetting. (θ is the contact angle h [droplet height], R and d are the radius and diameter of the liquid/solid contact disc).

work done by the system when two free surfaces of unit area of 1 and 2 are brought into contact. Simple thermodynamical arguments show that the work of adhesion is related to surface and interfacial energies by the following equation:

$$W_{12} = \gamma_1 + \gamma_2 - \gamma_{12} \tag{1}$$

When both phases are identical, Equation 1 reads:

$$W_{11} = 2\gamma_1 \tag{2}$$

Now, when two bodies (1 and 2) embedded in a material medium (3) are brought into contact, the work done per unit of contact area is

$$W^3_{12} = \gamma_{13} + \gamma_{23} - \gamma_{12} \tag{3}$$

If W^3_{12} is positive, bodies 1 and 2 will be expected to stick together when they are embedded in medium 3. Clearly, what we may ask from surface physics is to provide a way of deriving W^3_{12} from a set of experimental parameters measured on the surfaces of two cells (1 and 2) and the aqueous medium (3). In this respect, an instructive model is the interaction between a liquid droplet and a solid surface resembling a cell membrane, since this may involve molecular interactions and configurational changes that are likely to occur during cell-cell adhesion. Further, the former process may be easier to study.

When one drops a small amount of liquid L on a plane solid surface S, two possibilities may occur (Figure 1):

1. The liquid spreads on the solid. It is said to wet the solid (Figure 1A).
2. The liquid-solid attraction is too weak to overcome the liquid intermolecular forces. This is the partial wetting case (Figure 1B). The liquid-solid contact area is surrounded by a three-phase line 1, and the angle θ between the solid surface and a plane tangent to the droplet free surface on any point of 1 is called the "contact angle".

If the droplet is small enough that gravitational forces are negligible, its upper surface is a portion of a sphere. It may be shown[3] that this occurs when $h(d/2)g\rho/\gamma$ is much smaller than 1, where ρ is the liquid density, g is 9.81 m/sec² and γ is the liquid surface energy. In practice, this condition is fulfilled when the droplet volume is lower than a few cubic millimeters.[3] θ may then be easily deduced from the values of the free energies of the liquid (γ_L) and solid (γ_S) surfaces and the liquid-solid interface (γ_{LS}) by minimizing the total free energy of the system.[4] The basic equation was discovered by Thomas Young in 1805 and reads:

$$\gamma_L \cos\theta = \gamma_S - \gamma_{SL} \tag{4}$$

Several points of caution must be noted:

1. If the liquid and solid phases are in equilibrium with the saturating vapor of the liquid V at pressure p, γ_L and γ_S should be replaced with γ_{LV} and γ_{SV}. Some gas adsorption on the solid is expected to occur, resulting in the formation of a film. Using Gibbs-Duhem procedure, it may be shown that:[5]

$$\gamma_{SV} = \gamma_S - \pi = \gamma_S - RT \int_0^p \Gamma \, d\ln p \tag{5}$$

   where π is the film pressure and Γ is the number of adsorbed gas molecules per solid unit area at gas pressure p.
2. In many practical cases (see Section III) contact angle systems are out of equilibrium. This is made obvious when a contact angle is measured (1) immediately after depositing a droplet, which yields an advancing contact angle (θ_a), and (2) after reducing the droplet size, e.g., by pipette aspiration, which yields a receding contact angle (θ_r); θ_a is usually higher than θ_r. There are many possible explanations for this hysteresis effect, as discussed in Section III.

III. GENERAL METHODS FOR STUDYING INTERFACIAL ENERGIES

Many methods were devised to study interfacial energies, and we shall describe only a few basic ones. For further details, we refer to general textbooks on interface science.[6]

The important point is that only liquid-gas or liquid-liquid interfacial energies can be accurately and unambiguously measured.

A. Measurement of the Surface Energy of Liquids

The surface energy of liquids is sometimes considered as a force, tending to reduce the free liquid surface. This is called surface tension and is expressed in Newton/m instead of Joule/m². The concept of surface tension will not be used in this chapter.

We shall describe three common ways of measuring liquid surface energy (Figure 2).

1. The Capillary Rise Method

The height h of liquid rise into a thin capillary tube of radius r is measured. Assuming that the meniscus is spherical (which requires that r be small) the surface tension γ may be obtained by minimizing the system free energy:

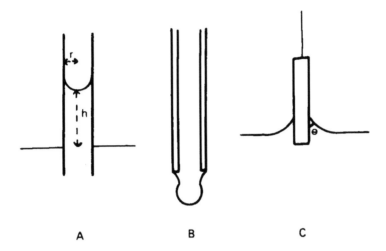

FIGURE 2. Methods of measuring the surface energy of liquids. (A) Capillary rise method; (B) drop weight method; (C) Wilhelmy slide method.

$$\gamma = \Delta\rho \cdot g \cdot h \cdot r \cdot/(2 \cos\theta) \qquad (6)$$

where θ is the contact angle (Figure 2A), $\Delta\rho$ is the difference between the liquid and gas phase densities, and g is the gravity acceleration. Numerical corrections may be used to account for the deviation of the meniscus sphericity.[6] Also, if the contact angle is not zero, its experimental value may be poorly reproducible due to aforementioned hysteresis effects.

2. The Drop Weight Method

This consists of measuring the weight of a sufficient number of liquid drops formed with a tube or pipette of known geometry. The drop weight w is related to the tube radius r and liquid surface tension γ through the following formula:

$$\gamma = w/(2\pi rf) \qquad (7)$$

where f is a numerical factor that depends on the drop shape and is often approximated as 1, although the error involved may be 20% or more.[6]

3. The Wilhelmy Slide Method

This consists of immersing a thin plate into the liquid, then raising it and measuring the traction F exerted by the liquid with an electrobalance (Figure 2C). The surface tension is given by:

$$\gamma = F/(p \cos\theta) \qquad (8)$$

where p is the length of the three-phase line (i.e., about twice the plate length).

Surface energy determinations require careful removal of surface active molecules from the bulk liquid, since these might get adsorbed on interfaces and substantially lower interfacial energies (Equation 5).

B. Surface Energy of Solids

As emphasized above, it is much more difficult to study the solid-liquid or solid-vapor interface than the surface of liquids. However, this point deserves a detailed description since it is of particular relevance to cell adhesion.

A direct way of assaying solid-liquid interaction would be to measure the heat produced by mixing a liquid with a solid of known surface area, using a fine powder to ensure that this area is sufficient to yield measurable effects. However, microcalorimetric measurements require specialized equipment. Further, evaluating the solid area is not an easy task. Hence, more indirect but easier techniques were devised to probe interfaces involving solid bodies. We shall describe some methods that were applied to biological surfaces with particular emphasis on contact angle and phase partition techniques.

1. Contact Angle Measurements
a. Experimental Procedure

The basic method consists of depositing a liquid drop on a solid surface exposed to a gas phase or a liquid medium.

Producing suitable droplets is easy, provided devices of small enough aperture are used. Several authors successfully used micrometer syringes[7] or glass micropipettes[3,7] obtained with microelectrode pullers currently used by electrophysiologists and enlarged to a tip diameter of 50 to 100 μm by simple manipulation.[3]

Contact angles are then measured. Van Oss and colleagues,[8] who pioneered the use of contact angle measurements in biological systems, used a telescope with cross hairs they attached to a goniometer. Gerson[7] adapted a stereomicroscope and a prism to project the image of a drop onto a ground-glass screen, where he performed angle measurements. However, assembling these devices may prove difficult in a biological laboratory. A simpler method consists of producing drops out of a suspension of latex beads (about 1 μm diameter). Examination with a standard microscope equipped with a micrometer-bearing eyepiece allows simultaneous determination of the drop diameter d (with the micrometer) and apparent height h′. (This may be achieved by sequentially focusing on the lowest and highest visible beads and reading the vertical displacement on the micrometer screw. What is measured is not the actual height since the drop acts as a lens). The contact angle θ is related to d and h′ by the simple formula:

$$2 \, h'/d = (1 - \cos\theta)/(\sin\theta + (n - 1) \sin\theta\cos\theta) \tag{9}$$

where n is the refractive index within the drop.

The only limitation of this method is that the drop has to be deposited on a transparent substrate.

b. Derivation of Young's Equation

(This section may be skipped by the reader uninterested in mathematical details.)

Let us consider an axisymmetric liquid droplet of volume V deposited on a solid substrate. Let γ_L, γ_S, and γ_{SL} be the interfacial energies of the liquid-vapor, solid-vapor, and solid-liquid interfaces. Define z(r) as the distance between the droplet upper and lower surface at distance r ($0 \leqslant r \leqslant R$) of the axis (Figure 1B; R is the radius of the solid-liquid contact disc). Supposing R is fixed, we may find z(r) by minimizing the total free energy of the system at constant volume. Introducing a Lagrange multiplier λ,[9] this is achieved by looking for a function z that makes minimum:

$$I = \int_0^R \gamma_L \cdot 2\pi r(1 + z'^2)^{1/2} \, dr + \rho g \int_0^R \pi r z^2 dr + \lambda \int_0^R 2\pi r z dr \tag{10}$$

(z′ is for dz/dr, ρ is the liquid density, and g is 9.81 msec⁻²). When z is varied by δz, we obtain for the variation δI, after integrating by parts:

$$\delta I = 2\pi\gamma_L(rz'\delta z/(1 + z'^2)^{1/2})_0^R + 2\int_0^R \{\delta z(-\gamma_L(z' + z'')/(1 + z'^2)^{1/2} + \dots$$

$$\gamma_L rz'^2 z''/(1 + z'^2)^{3/2} + \rho gzr + r\lambda\}dr \tag{11}$$

where z'' is d^2z/dr^2. Since $z(R)$ is zero, making the coefficient of δz equal to zero yields the Laplace equation:

$$(\lambda + 2\rho gz)/\gamma_L = z'/[r(1 + z'^2)^{1/2}] + z''/(1 + z'^2)^{3/2} \tag{12}$$

Assuming that the drop is small enough that the gravitational term is negligible, we find that the droplet upper surface is spherical by writing (Figure 1B):

$$z' = tgu; \quad r = \gamma_L \cdot \sin u/(2\lambda)$$

Now, let z_R be the function z corresponding to a solid liquid interface of radius R; R may be determined by noticing that the total surface energy J is minimum:

$$J = \int_0^R \gamma_L \cdot 2\pi r(1 + z'^2)^{1/2} dr + \pi R^2(\gamma_{SL} - \gamma_S)$$

Using equations 11 and 12 and noticing that the droplet volume $\int_0^R 2\pi rz_R dr$ is constant when R is varied, we obtain:

$$0 = dJ/dR = 2\pi R\gamma_L(1 + z'^2)^{1/2} + 2\pi\gamma_L Rz'/(1 + z'^2)^{1/2} dz_R/dR(R)$$

$$+ 2\pi R(\gamma_{SL} - \gamma_S) \tag{13}$$

Since $Z_R(R)$ is always zero, $dz_R/dR(R) = -z'_R(R)$; hence, Equation 13 yields:

$$(\gamma_S - \gamma_{SL})/\gamma_L = z'^2/(1 + z'^2)^{1/2} - (1 + z'^2)^{1/2} = \cos\theta \tag{14}$$

which is Young's equation.

c. Heterogeneous or Rough Surfaces

Two special cases are of practical interest. First, when a solid surface is made of patches of materials of contact angle θ_1 and θ_2 (with a given liquid L) and partial areas s_1 and s_2, replacing γ_S and γ_{SL} with $s_1\gamma_{S1} + s_2\gamma_{S2}$ and $s_1\gamma_{S1L} + s_2\gamma_{S2L}$, respectively, Equation 14 yields:

$$\cos\theta = s_1\cos\theta_1 + s_2\cos\theta_2 \tag{15}$$

This equation, first reported by Cassie,[10] relies on the assumption that the three-phase line remains circular. The validity of this hypothesis was discussed by Neuman and Good,[4] who concluded from numerical simulations that the Cassie equation was valid when the scale of surface heterogeneities was lower than 0.1 μm, when θ_1 and θ_2 were 30° and 40°, respectively. A much more general discussion is presented in the recent review by de Gennes.[11]

Second, when a surface is rough, the actual surface area may differ from the apparent area. Wenzel[12] proposed the following equation (as a logical consequence of Equation 14):

$$\cos\theta' = r(\gamma_S - \gamma_{SL})/\gamma_L \tag{16}$$

where r is the ratio between the actual (or microscopic) area and the apparent (or geometric) area. This parameter was called the "roughness factor". θ' is the "apparent" contact angle.

d. Significance of Experimental Contact Angle Values

Contact angle determinations are difficult to perform due to the possibility of several artifacts. In addition to the sensitivity of interfacial energies to minute amounts of surface active impurities, a major problem is the common occurrence of hysteresis. When the volume of a liquid droplet is decreased, the contact angle θ may decrease to a value θ_r (or receding contact angle) substantially lower than that measured on a freshly deposited droplet (i.e., the advancing contact angle, θ_a). Clearly, any value between θ_a and θ_r may be found. For example, if the height of the tip of the pipette used to deposit the droplet is increased, θ may be decreased. Hence, both θ_a and θ_r should be determined when a surface is studied. As written by Good[13] 10 years ago, "no journal should print any paper which does not identify angles, e.g., as to whether attempts were made to observe θ_a and θ_r".

Several mechanisms may be responsible for hysteresis. We shall describe them with some details, since these phenomena may play a role during biological adhesion.

i. Rough or Heterogeneous Surfaces

When a liquid droplet is deposited on a heterogeneous surface, the shape of the three-phase line may differ from a circle. Joanny and de Gennes[14] recently described a model for contact angle hysteresis. They showed that the liquid-vapor interfacial term might be represented as an elastic restoring force opposing the deformations of the triple line. Using a quantitative expression for this force, they showed that the three-phase line became wiggly in response to substrate heterogeneities. Considering the case where the amplitude of the fluctuations of $\gamma_{SL} - \gamma_S$ were lower than $\gamma_{LV} \cdot \theta^2$ (θ is the contact angle that was assumed to be small but finite), they showed that the size of the triple line fluctuations was about $(L \cdot x)^{1/2}$, where x was the correlation length of the substrate fluctuations and L the droplet size. Hence, fluctuations on the order of 1 μm might be expected for a droplet of a few microliters and defects of correlation length about 1 nm. Such fluctuations should be detected by microscopic observation of the triple line (see Section V.A).

Another finding of these authors was that the equation of equilibrium of the triple line yielded more than one solution, suggesting the possibility of hysteresis (see also Reference 15 for a discussion of contact angles on heterogeneous surfaces).

A basic assumption of these models is that the substrate does not exhibit any structural change during droplet spreading. We shall now describe some experimental arguments suggesting that such changes are likely to occur when a solid surface is coated with flexible polymer molecules, such as are found on cell surfaces.

ii. Hysteresis as a Consequence of Conformational Changes on the Substrate

Nearly 40 years ago, Cassie[10] suggested that contact-angle hysteresis might be due to the existence of several states of adsorption of molecules from the liquid phase on the substrate for a given value of the chemical potential (e.g., a state of extensive adsorption with low entropy and low potential energy and a state of lower adsorption with higher energy and entropy). Indeed, much experimental evidence supports the possibility of substrate structural changes during solid-liquid interaction. Here are some examples.

Zisman[5] obtained monolayers of hydrocarbon chains by the so-called "retraction method". This consisted of pulling a glass or metal plate out of a solution of an amphipathic substance such as heptadecylamine in oil or water. When he measured the advancing and receding contact angles of water on these surfaces, he found equal values (about 90°) when the monolayer was retracted out of an aqueous hydrocarbon solution, whereas in other cases θ_a was higher than θ_r by about 10°. He concluded that water could penetrate closely packed

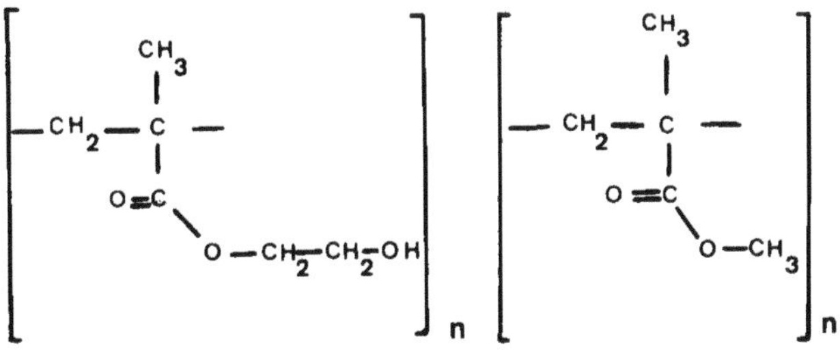

POLY HYDROXYETHYLMETHACRYLATE   POLYMETHYLMETHACRYLATE

FIGURE 3.   Molecular structure of polyhema and polymma.

hydrocarbon chains, and these could undergo configurational changes from "anhydrous" to "hydrated" conformation.

Holly and Refojo[16] measured the advancing and receding contact angle of water on poly(2-hydroxyethylmethacrylate), an antiadhesive hydrogel used to make contact lenses (Figure 3). They found very high hysteresis (for example, θ_a and θ_r were 84° and 8.8° on a gel containing 42.9% water and maintained in equilibrium with a water-saturated atmosphere). These authors suggested that hydrophobic regions of the polymer became oriented outward when this was exposed to air. Indeed, the polymer surface was optically smooth, suggesting that hysteresis was not due to heterogeneity. Further, the advancing contact angle was not decreased, and indeed sometimes increased, as the water content of the gel increased, which was expected to facilitate conformational changes by increasing the mobility of the polymer segments. Hence, the advancing contact angle was concluded to be a poor indicator of the hydrophobicity of a surface.

Other authors reported direct experimental evidence for the occurrence of conformational changes of the molecules bound to an interface. Ratner and colleagues[18] used the ESCA (electron microscopy for chemical analysis) technique to study the surface composition of silicone rubber surfaces radiation-grafted with polymers such as poly-2-hydroxyethylmethacrylate or polyacrylamide. They found substantial differences between the surface stoichiometry of different atomic species when hydrated and dehydrated samples were compared, suggesting the occurrence of important structural changes upon dehydration.

Finally, many studies made on hydration/dehydration of biological substances (such as textile fibers or proteins) demonstrated some hysteresis when the water content of tested samples was plotted vs. relative water pressure. These data are consistent with the occurrence of conformational changes between molecular states of varying hydration.[19,20]

Hence, it seems convincingly established that polymer-coated surfaces may undergo definite conformational changes when interacting with water.[17,21] The advancing and receding contact angle of water on these surfaces may respectively probe the "hydrophobicity" of the dehydrated and hydrated conformations.

Other causes of hysteresis, such as penetrability of the solid by liquid molecules or changes of the liquid molecular structure near the interface,[22] will not be described here since their actual importance in the systems that are of interest for us is not well documented. Also, penetration of the solid by the liquid may be considered a special case of conformational change.

We shall now review briefly other methods of studying the surface of solid particles.

2. Phase Partition Methods

A quite versatile way of assaying the properties of a population of small particles is to introduce them in a tube containing two immiscible phases, shaking vigorously, and looking at the particle distribution when phases settled again. An example is shown in Figure 4.

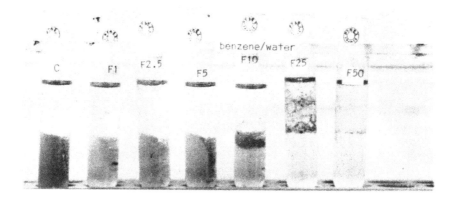

FIGURE 4.   Phase partition techniques. Sheep erythrocytes were incubated with various concentrations (1/1000 to 50/1000) of formaldehyde. Then they were deposited in glass tubes containing 1.5 mℓ of physiological saline (bottom phase) and 1.5 mℓ of benzene (top phase). The tubes were vigorously agitated. Phases reappeared about 30 sec later. Tube C (control) contained untreated erythrocytes. (From Capo, C. et al., *Immunology*, 36, 501, 1979. With permission.)

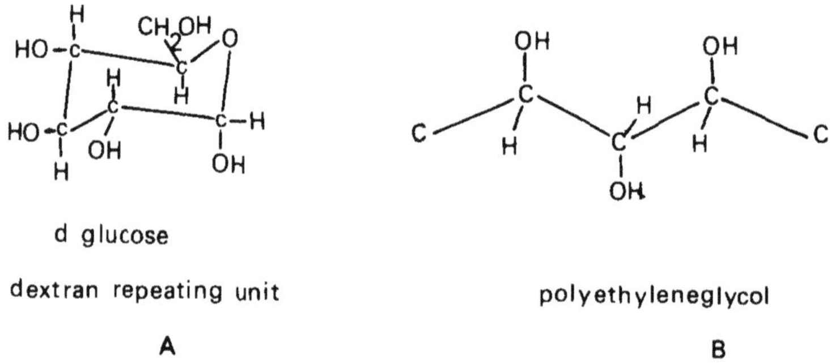

d glucose

dextran repeating unit   polyethyleneglycol

A   **B**

FIGURE 5.   Molecular structure of dextran and polyethyleneglycol.

Using a benzene-water two-phase system, normal sheep erythrocytes went into the bottom aqueous phase whereas formaldehyde-treated red cells gathered into the interface or went into the upper hydrophobic phase.[23]

However, this system does not discriminate between different cell types with only slight surface differences. In this case, two-phase polymer systems may be more useful.[24,25]

Since phase-partition methods are easily accessible and were applied to biological systems, some details may be useful. First, it must be noted that the appearance of two separate phases on mixing two polymers is a quite common phenomenon, called "incompatibility".[26] This may be explained as follows: when the molecular weight of molecular systems is increased, the entropy increase tending to favor the mixing of different molecules becomes less and less important relatively to the interaction energies. Hence, even a minimal preference of monomers for self association suffices to trigger phase separation. Further, such preference may be expected to be a common occurrence, considering the properties of intermolecular forces (Chapter 1). Indeed, Dobry and Boyer-Kawenoki[26] reported that even identity between the main chain or substituents alone was not sufficient to ensure compatibility between different synthetic polymers.

Second, it appears that the most popular polymer mixture in biological research is dextran-polyethyleneglycol. The basic structure of these molecules is shown in Figure 5. Clearly, it is difficult to guess what is assayed by this system on inspection of structural formulas. Albertsson[24] provided some experimental evidence suggesting that dextran was less hydrophobic than polyethyleneglycol: when dextran was substituted with increasing amounts of hydrophobic hydroxypropyl groups, it became less and less compatible with dextran, and more and more compatible with polyethyleneglycol. More hydrophilic particles are thus expected to partition into the dextran-rich (bottom) phase.

Third, it appears that in addition to the nature of polymers, the ionic composition of the solution influences partition. This may be due to the existence of a potential difference between phases, which may influence the partition of charged particles.[24,27] Walter[27] estimated that the potential difference between dextran- and polyethyleneglycol-rich phases was a few millivolts (with the upper phase positive) if phosphate buffer was used to make solutions isotonic, whereas no potential difference was found with sodium chloride. These phenomena might be ascribed to ion-polymer interactions.

It must be pointed out that a few millivolts may strongly influence the partition behavior of large particles. Recall that, according to Boltzmann's law, the partition factor for a particle of charge q across an interface with a potential difference V is about $\exp(q \cdot V/25)$, where V is expressed in millivolts and q in electronic charges (i.e., 1.6×10^{-19} C).

Indeed, the partition of different erythrocyte species between phosphate/NaCl dextran- and polyethyleneglycol-rich solutions was strongly correlated to their electrophoretic mobility.[27]

A fourth remark is that, in contrast with usual chemical equilibriums, the partition constant is the ratio between the absolute quantities of particles in the upper and lower phases, not their concentrations.[24,27] The following explanation was proposed by Albertsson:[24] during the shaking step, a microemulsion is formed with the appearance of a very small droplets of a phase in the other one; hence, if a given droplet is too small to contain more than one particle, this will be partitioned independently from the others, and the number of particles found in a given phase will be equal to the probability for a particle to go to the corresponding phase in an isolated droplet system.

In conclusion, determining the partition of cells between two immiscible polymer phases may seem an appealing way of studying their surface properties, since the measurements may be performed without damaging cells. Further, even small partition differences may be detected by repeating many times the separation process. This may be achieved with an automatic apparatus (a countercurrent distribution system) that may be used to separate different cell subpopulations.[24,28]

3. Other Methods for Studying the Surface of Solid Particles

We shall briefly describe two more methods that were used to study the surface of solid particles of biological interest.

a. The Solidification Front Method

The principle is as follows: a particle suspension is subjected to controlled freezing by establishing a temperature gradient under microscopic observation[29] (e.g., the suspension is deposited in a groove made in a suitable cell, and one extremity is cooled). The solidification front is examined to determine whether particles are rejected or engulfed. If the solidification is slow enough, it may be expected that particles will be rejected if $\gamma_{PS} - \gamma_{PL}$ is positive (where γ_{PS} and γ_{PL} are the particle-solid and particle-liquid interfacial energies).

Omenyi and Neumann[29] found that biphenyl and naphthalene readily engulfed hydrophobic teflon- or silicon-coated glass particles (the contact angle of water on these surfaces was 104° and 106°, respectively) whereas the more hydrophilic nylon and acetal particles were rejected (the contact angle of water on nylon and acetal was found to be 62° and 61°, respectively).

The problem remains of finding a quantitative interpretation for these results, which requires the use of empirical formulas to interpret γ_{PS} and γ_{PL}. This point is discussed in Section IV.

b. The Droplet Sedimentation Method

An interesting method for assaying the surface properties of small particles was recently described by Omenyi and colleagues.[30] The principle consists of depositing into a carefully thermostated container deuterium oxide, then an aqueous suspension of particles in a water/dimethylsulfoxide mixture. The system was considered as unstable when small droplets from the upper suspension entered the bottom D_2O phase within less than 3 to 4 min.

The authors plotted the maximum particle concentration compatible with stability vs. the proportion of dimethylsulfoxide. They found a maximum that was interpreted as the value for which particles experienced minimal van der Waals forces, which was supposed to occur when particles and the surrounding medium had similar surface energy (see Section IV for the origin of this view). Studying glutaraldehyde-treated erythrocytes as particles, these authors obtained a maximum stability when the interfacial energy of the water/dimethylsulfoxide solution was 65 erg/cm² (i.e., 65 mJ/m²).

In conclusion, we described several methods allowing direct measurement of the surface energy of liquids. Other methods were described for probing the solid-vapor and solid-liquid interfaces. Although these methods yielded some qualitative information on these systems, quantitative interpretation of the experimental data (e.g., contact angles) requires some theoretical tools. In the following section, we shall review some data pertinent to the molecular significance of interfacial energies. Then we shall give some details on the search for combining rules between interfacial energies.

IV. MOLECULAR SIGNIFICANCE OF INTERFACIAL ENERGIES

A. Theoretical Estimates of Interfacial Energies

Although it is not possible to derive surface energies from intermolecular potentials (except in some simple cases) it may be useful to try to get some intuitive understanding of the relationship existing between both quantities. For a more complete discussion of the significance and statistical mechanical estimate of surface free energies, we refer to the monograph of Rowlinson and Widom.[31]

The process of creating a unit free area of some medium may be split in two sequential steps:

1. Instantaneous breaking of intermolecular bonds without any rearrangement of molecular distribution functions within half bodies (i.e., separation). This process is expected to result in net energy increase.
2. Molecular rearrangement near the free area (relaxation) with concomitant free-energy decrease. The energy and entropy variations may be positive or negative.

In order to further explore the significance of these steps, we shall consider two systems of increasing complexity.

1. System of Apolar Spherical Molecules

Due to its simplicity, this model was extensively studied with a variety of theoretical models.[32] Let us consider a homogeneous isotropic fluid with a mean density of n molecules per unit volume. Let g be the radial distribution function (the molecular density at a distance r from a given molecule is $n \cdot g(r) \cdot dr$). Denote u(r) as the interaction energy between two molecules at distance r.

a. Step 1

Defining an axis Oz, we consider the increase per unit area W of the potential energy of the molecules located in the half space $z > 0$ when all molecules with a negative coordinate

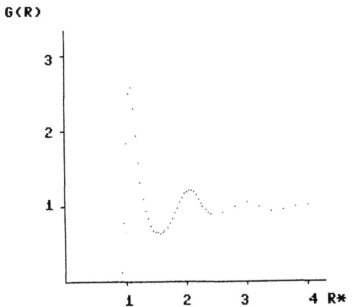

FIGURE 6. Radial distribution function in a Lennard-Jones fluid. The potential is $U(r) = 4\epsilon[(\sigma/r)^{12} - (\sigma/r)^6]$. The fluid density is $0.85/\sigma^3$, which corresponds to liquid argon, and R^* is r/σ. The temperature T is $1.273 \cdot \epsilon/k$. The curve was drawn using numerical data from Reference 33.

z are removed, without allowing any redistribution of the remaining molecules. Assuming that the interaction energy between molecules is pairwise additive, we may write:

$$W = -\int_0^\infty n\,dz \int_{r=z}^\infty u(r)ng(r) \cdot 2\pi r(r - z)dr \qquad (17)$$

which yields, after simple algebraic manipulations:

$$W = -\pi n^2 \int_0^\infty u(r) \cdot g(r) \cdot r^3 dr \qquad (18)$$

In order to get a clearer understanding of the significance of the above formula, we need some numerical estimates of u and g. Since we are interested in qualitative figures, we may choose a simple Lennard-Jones potential for u (Chapter 1):

$$u(r) = 4\epsilon[(\sigma/r)^{12} - (\sigma/r)^6]$$

We may then make use of the extensive molecular dynamics simulations of Verlet[33] to obtain an estimate for g(r). A typical radial distribution function, corresponding to a liquid state, is shown in Figure 6 (see legend for notations).

Clearly, it is possible to define the nearest neighbors of a given molecule as the molecules located at a distance comprised between 0 and 1.56σ (corresponding to the first minimum of g). Numerical integration shows that there are about 12 molecules in the first shell which is the value obtained in an array of closely packed spheres.

Now, it is of interest to ask what is the contribution of the breaking of nearest-neighbor interactions to the energy increase W. For this purpose, we may calculate the energy W(r) obtained by replacing the intermolecular potential u with the truncated one:

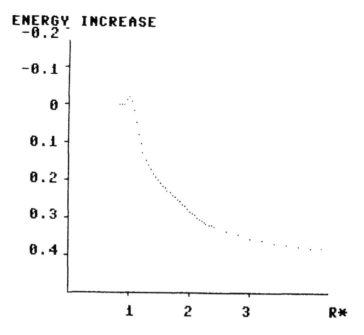

FIGURE 7.   Intermolecular potential in a Lennard-Jones fluid. A truncated Lennard-Jones potential was used to evaluate the intercellular potential. The ordinate represents the calculated potential when the Lennard-Jones potential is replaced with zero when r/σ is higher than the abscissa.

$$u_{R^*} = \begin{cases} u(r) & (r < R^*) \\ 0 & (r > R^*) \end{cases} \tag{19}$$

The "truncated" integral

$$I_{R^*} = \int_0^\infty g(r)u_{R^*}(r)r^3dr$$

is shown in Figure 7 as a function of R*. It is found that nearest-neighbor interactions account for about half the value of W.

b. Step 2

It remains for us to account for the effect of the difference between the molecular distribution near the inferface and within the bulk fluid. Unfortunately, accurate experimental data on the density profile near an interface are scanty,[30] and we have to refer to computer simulation. Miyazaki and Barker[34] used a Monte Carlo simulation to study the surface tension of a Lennard Jones fluid, under conditions close to the triple point for argon (the reduced density $n{\cdot}\sigma^3$ was 0.85, and the reduced temperature was 0.7; see the Figure 6 legend for notations). These authors estimated that relaxation resulted in about a 15% decrease of the total surface energy, due to both energy and entropy increase.

The main effect of relaxation was to smooth the density profile of the fluid near the interface, yielding a region of intermediate density. According to computer simulations, the thickness of this transition zone may be a few molecular diameters[30,35] under conditions far enough from the critical point.

However, the above results may not be very useful in understanding what occurs near the cell surface due to the complexity of membrane-coating macromolecules. It is therefore of interest to consider more complex systems. Water is a good model — although still much too simple — due to its relevance to biological problems.

2. The Water Surface

Many authors studied the structure of liquid water in view of its own theoretical interest as well as its importance in understanding biological systems.[36-39] The main peculiarity of this medium is its high structure due to the existence of strong hydrogen bonds between molecules. A first point in trying to understand water structure is to estimate the interaction potential between water molecules. Although sophisticated *ab initio* calculations were performed for this purpose,[40] simple empirical formulas were found to yield satisfactory results when used to perform computer simulations.[39] The results are as follows: *ab initio* determination of the minimal interaction energy yields estimates of about −5 kcal/mol.[40,41] Further, computer simulation led to the view that in liquid water each molecule was bound to a mean number of 3.5 to 4 nearest neighbors by a possibly distorted hydrogen bond, the energy of which varied around 4 kcal/mol.[39,42] These findings are consistent with the experimental value of the enthalpy of vaporization (i.e., 10.5 kcal/mol at 298 K[36]), and support the view that hydrogen bonding plays a dominant role in the determination of water structure. Hence, it is likely that more than 50% of the binding energy is accounted for by nearest-neighbor interactions in water (recall that the definition of nearest neighbors in liquid water involves some arbitrariness, allowing only crude considerations). In a first approximation, assuming that the potential energy of a water molecule is accounted for by four hydrogen bonds of 5 kcal/mol, each involving four nearest neighbors located 2.8 Å apart, Equation 17 yields for the energy required to create an interface of unit area without allowing any relaxation:

$$\gamma = 0.163 \text{ J/m}^2 = 163 \text{ erg/cm}^2$$

(Compare to the experimental value of 0.073 J/m^2 at room temperature.) It is well known that the interfacial excess entropy of liquid water is much lower than that of nonpolar liquids. Both computer simulation[43] and theoretical estimates[44] suggest that this is due to a preferential orientation of the molecules located near the interface, resulting in a simultaneous decrease of potential energy and entropy with a net free-energy decrease.

Two main conclusions may be drawn from the above data: (1) nearest-neighbor interactions may play a dominant role in the interaction between two media, and (2) surface energies are dependent on the flexibility and freedom of involved molecules, since reorientations and conformational changes may result in substantial free-energy variations.

Also, it appears that the complexity of surface phenomena makes it an elusive task to try to derive theoretical formulas for calculating interfacial energies. Hence, it is of interest to look for empirical rules providing some crude estimates for these energies.

B. Search for Combining Rules to Estimate Interfacial Energies

1. Liquid-Liquid Interfaces

Since our aim is to derive the interaction between two biological surfaces from individual material properties, it is of obvious interest to ask whether this can be done with liquids, since in this field proposed theories can be directly checked with experimental data.

The simplest approximation is probably to assume that the work of adhesion between two media (1 and 2) is the product of characteristic parameters of these bodies. The so-called "geometric mean" approximation yields:

$$W_{12} = 2(\gamma_1 \cdot \gamma_2)^{1/2}$$

$$\gamma_{12} = \gamma_1 + \gamma_2 - 2(\gamma_1\gamma_2)^{1/2} \tag{20}$$

Table 1
TEST OF COMBINING RULES FOR INTERFACIAL ENERGIES

Liquid 1	Liquid 2	γ_1	γ_2	γ_{12}	$\gamma_1 + \gamma_2 - \gamma_{12}$	$2(\gamma_1\gamma_2)^{1/2}$	γ_2^d	γ_2^p
Water	*n*-Octane	72.8	21.8	50.8	43.8	79.7	21.8	0.0
	n-Octanol	72.8	27.5	8.5	91.8	89.5	—	—
	CCl$_4$	72.8	27.0	45.0	54.8	88.7	26.8	0.2
	Benzene	72.8	28.9	35.0	96.6	86.2	—	—
	Perfluorodibutylether	72.8	12.2	51.9	33.1	59.6	12.2	0.0
	Perfluorotributylamine	72.8	16.8	25.6	64.0	69.9	11.8	5.0
Heptane	Perfluorodibutyether	20.4	12.2	3.6	29.0	31.6	10.3	1.3
	Perfluorotributylamine	20.4	16.8	1.6	35.6	37.0	15.5	1.3

Note: The experimental values of the surface tension of different liquids[45] were used to determine the work of adhesion ($\gamma_1 + \gamma_2 - \gamma_{12}$) and compare it to the geometric mean $2(\gamma_1\gamma_2)^{1/2}$. Also these values were used to estimate γ_2^d and γ_2^p for liquid 2, assuming that γ^p is zero for octane or heptane and using Equation 21.

This approximation is not unreasonable in view of the different theoretical formulas obtained for intermolecular forces (Chapter 1), and it may be expected to hold when a single kind of interaction plays a dominant role. We shall not describe theoretical attempts to validate this or related formulas since they rely on too many approximations and assumptions to yield safe conclusions. Only comparison with a sufficient variety of experimental data can provide a reasonable check of these procedures.

It may be noticed that Girifalco and Good[45] defined the "interaction parameter" ϕ as $(\gamma_{12} - \gamma_1 - \gamma_2)/2(\gamma_1\gamma_2)^{1/2}$. Empirically, ϕ is found to range between 0.5 and 1.15[6] and is generally highest when media 1 and 2 are fairly "similar".

A very interesting consequence of Equation 20 is the following if two surfaces (1 and 2) embedded in a medium (3) are brought in close contact, the variation of the free energy of the system is

$$dF = \gamma_{12} - \gamma_{13} - \gamma_{23}$$

$$dF = 2(\gamma_1\gamma_3)^{1/2} + 2(\gamma_2\gamma_3)^{1/2} - 2(\gamma_2\gamma_1)^{1/2} - 2\gamma_3 \tag{21}$$

It is a well-known result of elementary polynoma theory that dF is positive when γ_3 ranges between γ_1 and γ_2. If this happens, surfaces 1 and 2 are not expected to stick together within medium 3.

However, when Equation 20 is checked with experimental data (Table 1), marked discrepancies are found. As an example, water is found to interact much more strongly with octanol than with carbon tetrachloride, although both liquids exhibit similar surface energy. A further step to account for the selectivity of intermolecular forces is to distinguish between various kinds of interactions by expressing the work of adhesion as a sum of several components (for reviews, see, e.g., References 46 and 47). A common approximation is

$$W_{12} = 2(\gamma_1^d\gamma_2^d)^{1/2} + 2(\gamma_1^p\gamma_2^p)^{1/2}$$

$$\gamma_i = \gamma_i^d + \gamma_i^p \tag{22}$$

where parameters γ_i^d and γ_i^p account for the capacity of a given substance to be involved in dispersive and polar interactions, respectively. Equation 22 may be further refined by defining more than two interaction terms (e.g., by distinguishing between dipole-dipole interactions and hydrogen bonding[48]) or replacing geometric means by more sophisticated formulas such as the harmonic mean[49]:

$$W_{12} = 4\,\gamma_1^d\gamma_2^d/(\gamma_1^d + \gamma_2^d) + 4\,\gamma_1^p\gamma_2^p/(\gamma_1^p + \gamma_2^p) \tag{23}$$

Unfortunately, any further sophistication may reduce the predictive value of these formulas if the number of adjustable parameters is increased.

The validity of Equation 21 may be checked as follows: γ^d and γ^p may be derived from the experimental value of interfacial energies, assuming that γ^p is zero for heptane or octane (this reasonable hypothesis is commonly used in this field). As shown in Table 1, γ^d and γ^p could be calculated for five fluids (unsolvable equations were obtained with octanol and dipropylamine, suggesting a particularly strong interaction of these liquids with water). However, when the theoretical values of γ^p and γ^d for perfluorodibutylether and perfluorotributylamine were determined twice with two sets of experimental values (using water or heptane as test fluids), the agreement between calculated parameters was rather poor. The problem with simple liquids is that it is difficult to find a sufficient variety of immiscible liquids to test thoroughly (Equation 21). More complex molecules were used by Wu,[50] who studied molten polymers. The advantage of this model is that these liquid are usually immiscible, allowing experimental determination of interfacial energies in many different systems. Also, they may be more similar to cell surface molecules than simpler liquids.

Studying different polymers such as polymethylmethacrylate, polystyrene, or polyvinylacetate, Wu calculated γ^d and γ^p by means of Equation 22. Using molten polyethylene as a test fluid with 0^p, he took advantage of the obtained parameters to predict the energy of 12 interfaces, and met with reasonable agreement between theoretical and experimental values (the mean relative error was about 20%).

It is concluded that Equations 21 and 22 may yield rather consistent values for parameters ^{d} and ^{p}, as well as crude estimates for liquid/liquid interfacial energies *in a reasonably wide spectrum of experimental situations*. However, these are expected to fail in such situations as the following: a polar molecule such as dimethylsulfoxide contains no available proton for hydrogen bonding (this is a so-called dipolar aprotic fluid[51]); however, it is likely to form hydrogen bonds with hydrogen donors. Hence, higher γ^p should be used to calculate the work of adhesion of this substance with a hydrogen bond donor, rather than to estimate its surface energy. The problem is that splitting γ^p into different parameters would result in a need for too many adjustable parameters to be of practical use in many situations. Perhaps when surfaces not too different (such as biological membranes) and test fluids are used, and when sufficiently complex molecules are used so that all kinds of interaction might occur, Equation 22 may hold. This possibility must be subjected to experimental checking.

2. Liquid-Solid Interfaces
As will be shown, two main problems arise when solid/liquid interfaces are studied.

1. Since interfacial energies involving solids cannot be measured directly, only indirect checking of empirical equations can be achieved by relying on the Young equation. It must therefore be borne in mind that solid-liquid and solid-vapor interfacial energies are not measurable parameters in usual situations, which makes their use somewhat dangerous.
2. Considering the experimental data that were accumulated, it seems that some predictive schemes may be of limited value in describing the behavior of a homogenous series of surfaces of liquids. Forgetting these limitations may result in serious errors and misinterpretations.

We shall now describe some selected experimental data. The approach pioneered by Zisman[5] consisted of measuring the contact angle of a series of liquids on a solid surface. Plotting $\cos\theta$ vs. γ_{LV}, he often obtained smooth lines. Extrapolating to $\theta = 0$ (or $\cos\theta = 1$), he obtained a value of γ_{LV}. He called the critical surface tension γ_C. This parameter is interesting in that a liquid was predicted to wet a solid if its surface energy was lower than γ_C.

However, it rapidly appeared that this approach was essentially valid only in particular systems. Studying nine representative polymer surfaces with polar and nonpolar fluids, Dann[52] concluded that dispersive components could be used to predict contact angles by means of the so-called Good-Girifalco-Fowkes-Young equation that read (neglecting adsorption of solids):

$$\cos\theta = (-\gamma_L + 2 - \gamma_s^d \gamma_L^d)^{1/2}/\gamma_L \tag{24}$$

This equation was valid only if apolar solid and liquids or combination of a polar and an apolar medium were assayed. In this case, it might be derived from Equations 14 and 20. Clearly, if a solid is probed with a series of apolar liquids, the critical surface tension γ_C is thus expected to be equal to the dispersive component γ_s^d (Equation 22). However, biological adhesion involves interactions between polar media. The critical surface tension of biological surfaces is expected to be dependent on the series of test fluids used for its determination.

Several authors used extensive compilations of experimental results to derive empirical relationships which they called "equations of state" between γ_L, γ_{SL}, and γ_S. These equations might be used to predict the contact angle of a liquid of known interfacial energy on a solid surface after probing this surface with a single liquid. The problem remains to assess the range of validity of these formulas.

An interesting study of the contact angle of different liquids on polymer surfaces (cellulose, polyvinyl alcohol, or polymethylmethacrylate) may be relevant to biological adhesion.[55] The authors preferred replacing the vapor phase V with a second liquid, since they noticed that a water droplet positioned on a dry polymer such as cellulose gradually penetrated into the film, with a steady decrease in the contact angle (see Section III.B.1). This instability disappeared when a droplet was made out of a polar liquid (water, ethyleneglycol, formamide, or glycerol) in a hydrocarbon such as hexane. Deriving works of adhesion with the Young equation, the authors concluded that the geometric-mean expression was appropriate for representing the dispersive interaction. However, polar interactions seemed better represented by evaluating the number of electron donor/acceptor interactions and multiplying this value by a parameter that was estimated to range between 2 to 4 kcal/mol. Some experimental results relevant to polymer-liquid interaction are listed in Table 2.

C. Conclusion

We described selected experimental data relevant to interfacial energies. Some possibly useful conclusions are listed below:

1. Polymer-coated surfaces may display different conformations depending on previous treatments and the surrounding medium.
2. Measuring the advancing and receding contact angle of water on a polymer-coated surface may allow probing of the interaction energy between water and the surface for two different molecular conformations of the polymer molecules.
3. The aforementioned contact angle determinations may prove difficult due to the instability of the tested systems. Replacing the vapor phase with a liquid such as octane may improve the reproducibility of measurements.

Table 2
CONTACT ANGLE MEASUREMENTS ON POLYMER
SURFACES

Polymer	Phase L (droplet)	Phase V	Contact angle[a]	Ref.
Cellulose (dried)	Water	Air	11.8	55
		Octane	44	55
Polyhydroxymethylmethacrylate	Air	Water	165	56
(Gel — 40% H$_2$O)	Octane	Water	164	56
Polymethylmethacrylate	Air	Water	121	56
(Gel — 3% H$_2$O)	Octane	Water	91	56
Polyethylene	Water	Air	85	57
Polystyrene	Water	Air	77	57

Note: Different polymer surfaces were assayed for contact angle with various fluids.

[a] Advancing contact angle.

4. The interaction between two media involves dispersive and polar interactions. Electron donor/acceptor interactions might play a dominant role in these systems.
5. Some empirical equations and concepts were evolved to deduce works of adhesion from interfacial and surface energies. These equations were found to hold with some limited range of validity, and it is not impossible that biological membranes be amenable to such treatment. However, too few experimental data are available at the present time to assess this hypothesis.

Now we shall review some experimental data in order to know whether the above concepts may be helpful in understanding cell adhesion.

V. USE OF THE CONCEPTS AND METHODS OF SURFACE PHYSICS TO STUDY CELL ADHESION

Three main points will be described: the measurement of cell surface properties, applications to cell-substrate interactions, and relevance to phagocytosis.

A. Measurement of Cell Surface Energy
The first series of quantitative studies on surface energy is probably due to van Oss and Gillman[58] (see Reference 8 for a review). These authors measured the advancing contact angle of water on dried monolayers of various cells or bacteria. This technique was used by other authors in later years[59-61] and was found to yield rather consistent results (see Table 3). However, several problems are raised when one tries to understand the meaning of experimental data:

1. Drying probably alters the conformation of cell surface molecules. Hence, what is probed by advancing contact angle measurements is probably the maximum surface hydrophobicity (indeed, receding contact angles are zero in this system,[60] and increasing the rigidity of surface membrane molecules may decrease the advancing contact angle[60]).
2. As was reported by van Oss, cell contact angles are independent of the substrate when the cell density is high enough. A systematic study made on different cell types confirmed that a plateau value was obtained when more than 30% of the substrate area appeared coated with cells on microscopic examination[60]. As shown in Figure 8, the

Table 3
SURFACE PROPERTIES OF DIFFERENT CELL TYPES

	Contact angle		c^a
Cell type	Water Dried monolayer	Fluorocarbon Aqueous medium	μN/m
Granulocyte			
Human	18° [8]		0.84[25]
Rat	13° [8]		
Monocyte/macrophage			
P388D1 line	25.7° [60]		
Rat macrophage	22.8° [60]	147° [64]	
Porcine macrophage			0.93[25]
Lymphocyte			
Rat thymocyte	36° [60]		
Mouse spleen			2.3[25]
Erythrocyte			
Human	15° [8]		0.6[25]
Sheep	15.7° [8], 12.8[60]	164° [64]	

Note: Cells were deposited on different substrates and assayed for contact angles
(1) after air drying, with water droplets, and (2) without drying, by depositing
dense fluorocarbon droplets within aqueous NaCl solution.

[a] Critical interfacial tension of spreading (measured with dextran/
polyethyleneglycol).

three-phase line is quite smooth when sufficient cell density (say, more than 50%
substrate coverage) is achieved.

3.   An important point is that monolayers must be mechanically stable; cells that are not
detached near the three-phase line must be checked microscopically. Further, obtaining
a stable monolayer may be a problem with some cell species, especially erythrocytes.

4.   It may be checked that the contact-angle value is only weakly dependent on the relative
water pressure achieved during cell drying[60] and the duration of the drying process[8,60]
(if this is more than about 30 min).

5.   As previously emphasized, contact angles might depend on cell surface roughness as
well as intrinsic surface energy (Equation 16). However, this may not be a problem
since drying resulted in marked cell stretching on the substrate, as demonstrated with
scanning electron microscopy.[60]

Finally, the main criticism of this technique was the possibility that the molecular structure
of cell surfaces be drastically altered on drying. Two alternative procedures were proposed
to overcome this difficulty.

Gerson and Akit[62] used a dextran/polyethyleneglycol two-phase system to study both the
partition of different cell lines and the contact angle of dextran-rich droplets deposited on
cell layers immersed in polyethyleneglycol-rich phase. Using different polymer concentrations, he obtained a "critical interfacial tension for spreading" with an extrapolation procedure analogous to Zisman's (that yielded a critical surface tension). The interesting point
of this technique was that cells were not damaged by this treatment and membrane molecules
might be expected to display their native conformation. Another point of interest is that the
order of magnitude of the critical interfacial tension for spreading was about 10^{-6}J/m^2, which
was at least 1000-fold lower than "usual" interfacial energies shown in Table 1. Such low
values could be measured since the interfacial tension between immiscible phases was very

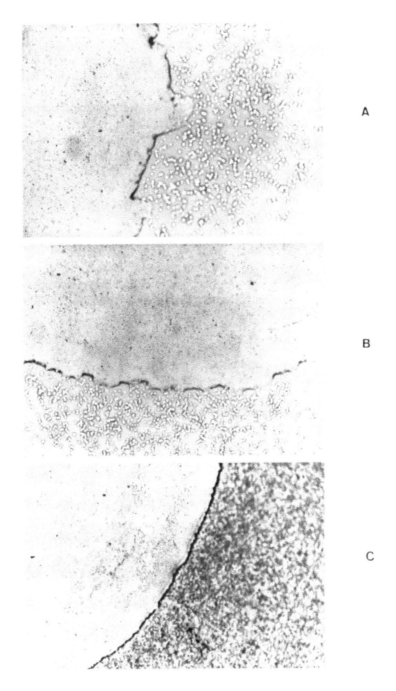

FIGURE 8.   Contact angles on heterogeneous substrates. Different amounts of thy-mocytes were deposited on a substrate. The fraction of the substrate area coated with cells was (A) 0.06, (B) 0.09, and (C) 0.58. Cells were then air dried before depositions of a droplet of latex suspension of about 2 $\mu\ell$. Details of the three-phase line are shown.

low. However, the molecular significance of the measured parameter remains poorly understood.

Some experimental values of the critical surface tension are shown in Table 3.

In an interesting study, McIver and Schurch[63] used a similar system to measure the contact angle of different surfaces. They assayed a hydrophobic fluorocarbon surface (1) bare, (2) coated with phospholipids, and (3) coated with phospholipids and glycophorin (a highly glycosylated integral protein of the erythrocyte membrane). The measured contact angles were 180°, 150°, and 95°, respectively, as compared to 100° for erythrocytes. When a hydrophobic test fluid (dibutylphthalate) was used to probe the same surfaces immersed in a 0.15% NaCl solution, the respective contact angles were about 5°, 90°, and 175°. Hence, cell surface glycoproteins may play a dominant role in the determination of contact angle values.

Another method that was developed by Gerson consisted of depositing droplets of a dense fluorocarbon on cell monolayers immersed in saline solution. Clearly, the more hydrophobic a surface, the lower the contact angle. A problem with this technique is that Θ is not very different from 180°, which may impair the accuracy of measurements. Also, it is difficult to rule out any conformational change of cell surface molecules on contact with an hydrophobic medium. Some experimental contact angles values are shown in Table 3.

The following conclusions are suggested by the above data:

1. Whereas phospholipid bilayers are fairly hydrophobic surfaces, cell membrane macromolecules are probably responsible for the higher hydrophilicity of these particles, which makes them more stable than lipid vesicles. Cells must therefore be viewed as coated with an anti-adhesive layer. A similar procedure may be used to stabilize colloid particles by coating them with hydrophilic polymers, which was called "steric stabilization".[65]

2. Under some circumstances, the cell surface may undergo conformational changes in order to constitute a more hydrophobic structure, as happens on dehydration. It is not known whether this phenomenon plays a physiological role, e.g., in stabilizing cell adhesion or promoting cell fusion. These conformational changes probably play a role in contact-angle hysteresis.

3. There are indeed actual differences between the contact angle properties and phase partition behavior of different cell types (Table 3). It is not clear whether these differences are responsible for behavioral differences between these cells. However, these findings may allow efficient separation of different cell subpopulations.[28]

B. Relationship between Surface Energies and Adhesion

As will be shown below, adhesion in serum-containing media is complicated by the occurrence of many competitive interactions and ligand-receptor associations (see also Chapter 9). It may thus be useful to describe first some experimental data obtained in protein-free conditions, which may be simpler to understand.

1. Adhesion in Protein-Free Solutions

As shown by Baier (quoted in Reference 66), exposing any solid substrate to blood results in immediate (within a few seconds) adsorption of a biopolymeric material. Using the technique of "multiple attenuated internal reflexion", Baier and Weiss[67] found that an internal reflexion prism exposed to fetal calf serum became coated with a film that was essentially made of a glycoprotein that could not be extracted by water or a saline solution (which was ascribed to adsorption-induced denaturation). Further studies done on the adsorption of proteins on various substrates showed that this adsorption occurred on nearly all tested surfaces, and resulted in some protein denaturation as shown by circular dichroism measurements[68] or a combination of ellipsometry and radiotracer detection.[69] Interestingly, fibronectin (a protein involved in fibroblast adhesion to surrounding tissues) but not immunoglobulin displayed conformation differences when adsorbed on hydrophobic as compared to hydrophilic substrates.[69]

Similarly, when Klebe and colleagues[70] studied the adhesion of Chinese hamster ovary cells to different polymers in protein-free conditions, they found that cells adhered to all substrates, except poly(hydroxyethylmethacrylate) and agarose (a gel-forming branched polysaccharide[71]). However, it may be added that:

1. Macrophages are known to stick to a variety of foreign surfaces, but these cells displayed different behavior when they were bound to glass or to a more hydrophobic polystyrene surface[72] since the latter cells displayed markedly decreased ability to bind hydrophobic glutaraldehyde-treated erythrocytes.
2. Whereas lymphocytes and macrophages adhered to "bare" plastic dishes in protein-free solutions, only the latter cells substantially adhered to polystyrene latex beads under similar physicochemical conditions.[74] This point may be of importance when one tries to understand why cells do not extensively adhere to one another in protein-free suspensions. This problem arises even if it is noted that "nonspecific" intercellular binding is often observed when intercellular contacts are promoted.[74]

Neumann and colleagues[75] studied the adhesion of human leukocytes and platelets to different polymers of varying surface energies in saline solutions containing different concentrations of dimethylsulfoxide. Further, they measured the liquid surface energy and they estimated the surface energies of the substrates as described in Section IV.B.2. Interestingly, when they plotted the fraction of adherent neutrophils vs. the estimated free energy of the substrate, they found that adhesion was independent of the substrate in a 7.5% dimethylsulfoxide solution (the surface energy γ_{LV} of which was 69 erg/cm²). When γ_{LV} was higher than 69 erg/cm², adhesion was a decreasing function of the substrate surface energy, and when γ_{LV} was lower than 69 erg/cm², the reverse behavior was observed. Since the estimated surface energy of substrates ranged between 16.4 and 66.7 erg/cm² (as estimated with the standard Zisman procedure) and was thus lower than the surface energy of tested liquids, the results were consistent with the view that 7.5% dimethylsulfoxide solution displayed maximum similarity with the leukocyte surface, since adhesion is expected to occur when the cell surface displays maximum similarity with the substrate relative to the medium, in accordance with previous remarks (see Section IV.B and Reference 76 for a similar conclusion in a study made on bacterial adhesion to polymer surfaces).

The results obtained with platelets were complicated by the interesting finding that these cells (in contrast with neutrophils) continuously released surface active material that lowered the solution surface energy.

There remains now a study of the influence of soluble substances on cell-substrate adhesion. In this respect, it is of interest to note that the surface energy of a 1% bovine albumin solution was found to be 51.3 erg/cm².

2. Adhesion in Biological Media

It is well known that the presence of serum may impair cell-substrate adhesion. In a very interesting study, Curtis and Forrester[77] clarified this important point by showing that (1) the binding of different proteins such as fibronectin, β_2 macroglobulin, or albumin to normal or hydroxylated polystyrene surfaces was correlated to the adhesion of BHK (baby hamster kidney) cells and leukocytes to these surfaces in protein-free media, and (2) hydroxylated polystyrene bound less β_2-macroglobulin and albumin, but more fibronectin than normal polystyrene, and cells were found to be more adhesive to the former substrate in the presence of serum. Also, the low adhesion of fibronectin to normal polystyrene was ascribed to a competition with "anti-adhesive molecules", in accordance with previous findings by Grinnell and Feld.[78]

It must be added that these findings are relevant to initial adhesion since cultured cells are known to release adhesion molecules that are thought to play a role in cell-substrate interaction.

Other experimental data may give some hint as to the molecular significance of these findings. Margolis and colleagues[79] studied the adhesion of platelets to lipid substrates obtained by drying chloroform/methanol phospholipid solution on glass substrates. Adhesion occurred (in presence of serum) when the lipid phase was gel-like, not "liquid crystalline". Further, glutaraldehyde cross-linking of the latter lipid molecules induced both a solid state and platelet adhesion. Also, Maroudas[80] compared the capacity of different substrates to bind cells and suggested that fibroblasts might be unable to bind to "yielding hydrophilic" or "rigid hydrophobic" surfaces, thus proposing the important concept that the mechanical properties of the substrate might play a role in adhesion, in addition to hydrophobicity/hydrophilicity. This finding is in line with the importance of substrate roughness in cell adhesion.[81,82]

The above data support the following conclusions:

1. The order of magnitude of surface energy effects is usually far higher than is required to ensure cell-cell adhesion (see Chapter 7).
2. The finding that a cell will adhere to virtually any substrate in a protein-free medium is consistent with the general rule that two similar surfaces that "resemble" each other more than water will tend to adhere in aqueous solutions.
3. Serum contains anti-adhesive proteins (such as albumin) that may act by coating potentially adhesive substrates and lowering the medium surface energy. Also, some "adhesive" proteins display both nonspecific substrate binding ability and affinity for specific cell surface receptors (e.g., fibronectin; see Chapter 9). Hence, the selectivity of adhesion in protein-containing solutions may be largely due to competition for adsorption between different proteins.
4. In fact, what is to be explained is the reason why adhesion does not occur in all situations where the work of adhesion is expected to be positive (e.g., why leukocytes do not usually adhere to each other). Perhaps some parameters such as cell surface roughness or molecular flexibility may play a role in this respect.

C. Surface Energy and Phagocytosis

It has long been known that phagocytic cells could engulf inert particles such as colloidal carbon, and it was suggested in the beginning of this century that the ingestion process might be driven by interfacial energy differences.[83] It was further proposed that serum-mediated enhancement of the phagocytosis of different bacterial species, a phenomenon called "opsonization", was correlated to a modification of the microorganism physical properties due to its coating by different proteins.[84] This physicochemical view of the phagocytic process might be supported by experimental studies made on immiscible liquid droplets by Torza and Mason.[85] When two droplets made of media 1 and 2 and immersed in medium 3, encountered each other, droplet 1 was engulfed by droplet 2 when the following inequalities were observed:

$$\gamma_{23} - \gamma_{12} - \gamma_{13} < 0$$

$$\gamma_{12} - \gamma_{13} - \gamma_{23} < 0$$

$$\gamma_{13} - \gamma_{21} - \gamma_{32} > 0$$

It was only in 1972 that van Oss and colleagues[58] reported a quantitative test of the physicochemical concept of phagocytosis. Using a variety of phagocytic cells and bacterial species, they measured the contact angle of water on dried cell or bacterium monolayers, and measured the phagocytosis of these bacteria by phagocytic cells in protein-deprived media. They reported a close correlation between particle hydrophobicity and phagocytosis since a given bacterial species was ingested only when it had a higher contact angle than the phagocyte. Further, when hydrophilic encapsulated bacteria were coated with specific antibodies, their contact angle was increased together with phagocytosis efficiency.[86]

Similar conclusions were reached by Stendahl and colleagues,[87] who assayed rough mutants of *Salmonella typhimurium* for susceptibility to phagocytosis (in a medium containing 0.1% bovine albumin as proteins). These authors also assayed partition in a dextran/polyethyleneglycol two-phase system and found that the mutants that resisted phagocytosis appeared to partition in the dextran-rich bottom phase. Further, they studied the effect of specific antibodies on their model. They reported that IgG antibodies made *Salmonella* more susceptible to phagocytosis and more hydrophobic, whereas IgA antibodies had opposite effects on both partition and phagocytosis.[89] However, the correlation between partition behavior and susceptibility to phagocytosis was not absolute, especially when "smooth" bacteria with a polysaccharide coating were assayed.

Similar findings were reported in a quite different experimental model. Rabinovitch[89] reported that aldehyde-treated erythrocytes were readily ingested by phagocytic cells, and further studies showed that aldehyde treatment resulted in marked increase of the hydrophobicity of these particles, as demonstrated with a two-phase benzene/water or butanol/water partition system[23] or contact angle measurements.[64] Phagocytosis was found to occur in both the presence and absence of serum proteins.

Additional data shed some light on the effect of serum proteins on phagocytosis: whereas polystyrene latex beads are readily ingested by polymorphonuclear leukocytes in protein-free buffer,[90] preincubating them with less than 100 µg/mℓ albumin resulted in significant protein adsorption and concomitant decrease of phagocytosis. However, latex beads are readily ingested by phagocytic cells in serum-containing media, which is a convenient way of identifying phagocytes.

An interesting series of experiments allows some better understanding of these results. Absolom and colleagues[91] measured the adsorption of different serum proteins on several bacterial species together with the effect of this adsorption on phagocytosis. They found that (1) the extent of adsorption was positively correlated with the contact angle of water on dried bacterium monolayers, (2) immunoglobulin G (monoclonal or polyclonal) displayed five- to tenfold higher adsorption than albumin, and (3) immunoglobulin adsorption enhanced particle phagocytosis.

Now, in view of these and other findings, we may propose the following interpretation:

1. Phagocytic cells are endowed with membrane receptors with a specificity for immunoglobulins,[92] complement,[93] or sugars commonly found on a variety of cell surfaces.[94,95] Several of these receptors were characterized antigenically and biochemically.[96-98]

2. Mononuclear phagocytes (such as monocytes and macrophages) may[99] secrete complement molecules that will be adsorbed on some particles (and especially on so-called "complement activators", such as many bacterial species).

3. Phagocytes are also endowed with a so-called "nonspecific recognition system" that allows binding and ingestion of "hydrophobic" particles such as some bacteria, latex spherules, and aldehyde-treated erythrocytes, and different binding mechanisms display different behavior (e.g., as to sensitivity to metabolic inhibitors[100]).

4. A possible explanation for the correlation between contact angle and phagocytosis

increase after immunoglobulin adsorption, as well as greater efficiency of immuno-globulin adsorption on hydrophobic particles as compared to albumin adsorption, would be that nonspecific phagocytosis is a very primitive cell function (indeed, phagocytosis is a universal cell function, found in "primitive" species such as amoeba[101]). Im-munoglobulin and complement might have evolved from primitive "sticking" non-specific recognition molecules, followed by the appearance of "specific" immunoglobulin or complement receptors on cells. This might explain the correlation found between the hydrophobicity, opsonizing activity, and nonspecific adsorbability of different serum proteins, and the correlation between susceptibility to phagocytosis in serum-deprived medium and immunoglobulin adsorption in presence of serum pro-teins that was found in a series of different bacterial species. Clearly, the synergistic effect of different mechanisms of particle phagocytosis makes it very difficult to obtain a quantitative view of the actual importance of individual recognition systems.

VI. GENERAL CONCLUSION

Although the experiments we described may have raised more problems than they solved, it may be useful to list some conclusions that may guide future work:

1. The order of magnitude of nonspecific surface forces is sufficient to mediate strong adhesion or repulsion between particles of the size of living cells. (Compare the order of magnitude of surface energies and surface affinity estimates by Evans.)
2. Under protein-free conditions, the general features of the dependence of cell adhesion on the medium and substrate surface properties may be accounted for by simple approximations (Section IV.B).
3. It is not clear why intercellular adhesion may not occur in a protein-free medium, under conditions where the work of adhesion is expected to be positive. Perhaps the geometrical and mechanical properties of the cell surface as well as the flexibility of membrane molecules play a role in this respect.
4. In usual serum-containing media, adhesion may be dominated by competition for adsorption between different soluble macromolecules. Specific ligand-receptor inter-actions probably play an important role in many situations of biological importance.

Probably a more thorough understanding of molecule behavior (Chapters 1 and 2) will be required to better understand the experimental data described in this chapter, which should help in understanding the mechanisms of cell adhesion.

REFERENCES

1. **Hirschefelder, J. O., Curtiss, C. F., and Bird, R. B.,** *Molecular Theory of Gases and Liquids,* John Wiley & Sons, New York, 1964, 336.
2. **Alberts, B., Bray, D., Lewis, J., Raff, M., Roberts, K., and Watson, J. D.,** *Molecular Biology of the Cell,* Garland Press, New York, 1983, 582.
3. **Farnarier, C., Capo, C., Balloy, V., Benoliel, A. M., and Bongrand, P.,** Simple microscopical meas-urement of contact angles on transparent substrates, *J. Colloid Interface Sci.,* 99, 164, 1984.
4. **Neumann, A. W. and Good, R. J.,** Thermodynamics of contact angles. I. Heterogeneous solid surfaces, *J. Colloid Interface Sci.,* 38, 341, 1972.
5. **Zisman, W. A.,** Relation of the equilibrium contact angle to liquid and solid constitution, *ACS Adv. Chem. Ser.,* 43, 1, 1976.

6. **Adamson, A. W.**, *Physical Chemistry of Surfaces*, John Wiley & Sons, New York, 1976, 340.
7. **Gerson, D. F.**, Methods in surface physics for immunology, in *Immunological Methods*, Vol. 2, Lefkovits, I. and Pernis, B., Eds., Academic Press, New York, 1981, 105.
8. **van Oss, C. J., Gillman, C. F., and Neumann, A. W.**, *Phagocytic Engulfment and Cell Adhesiveness as Cellular Surface Phenomena*, Marcel Dekker, New York, 1975, 11.
9. **Sommerfeld, A.**, *Lectures on Theoretical Physics*, Vol. 1, Academic Press, New York, 1964, 68.
10. **Cassie, A. B. D.**, Contact angles, *Discuss. Faraday Soc.*, 3, 11, 1948.
11. **De Gennes, P. G.**, Wetting: statics and dynamics, *Rev. Mod. Phys.*, 57, 827, 1985.
12. **Wenzel, R.**, Resistance of solid surfaces to wetting by water, *Ind. Eng. Chem.*, 28, 988, 1936.
13. **Good, R. J.**, Surface free energy of solids and liquids: thermodynamics, molecular forces and structure, *J. Colloid Interface Sci.*, 59, 398, 1977.
14. **Joanny, J. F. and de Gennes, P. G.**, A model for contact angle hysteresis, *J. Chem. Phys.*, 81, 552, 1984.
15. **Pomeau, Y. and Vannimenus, J.**, Contact angle on heterogeneous surfaces: weak heterogeneities, *J. Colloid Interface Sci.*, 104, 477, 1985.
16. **Holly, F. J. and Refojo, M. F.**, Wettability of hydrogels. I. Poly(2-hydroxyethylmethacrylate), *J. Biomed. Mater. Res.*, 9, 315, 1975.
17. **Holly, F. J.**, Novel methods of studying polymer surfaces by employing contact angle goniometry, in *Physicochemical Aspects of Polymer Surfaces*, Vol. 1, Mittal, K. L., Ed., Plenum Press, New York, 1983, 141.
18. **Ratner, B. D., Weathershy, P. K., Hoffman, A. S., Kelly, M. A., and Scharpen, L. H.**, Radiation-grafted hydrogels for biomaterial applications as studied by the ESCA technique, *J. Appl. Polym. Sci.*, 22, 643, 1978.
19. **McLaren, A. D. and Rowen, J. W.**, Sorption of water vapor by proteins and polymers — a review, *J. Polym. Sci.*, 7, 289, 1951.
20. **White, H. J. and Eyring, H.**, The adsorption of water by swelling high polymeric materials, *Text. Res. J.*, 17, 523, 1947.
21. **Andrade, J. D., Gregonis, D. E., and Smith, L. M.**, Polymer-water interface dynamics, in *Physicochemical Aspects of Polymer Surfaces*, Vol. 2, Mittal, K. L., Ed., Plenum Press, New York, 1983, 911.
22. **Schwartz, A. M.**, Contact angle hysteresis: a molecular interpretation, *J. Colloid Interface Sci.*, 75, 404, 1980.
23. **Capo, C., Bongrand, P., Benoliel, A. M., and Depieds, R.**, Non-specific recognition in phagocytosis: ingestion of aldehyde-treated erythrocytes by rat peritoneal macrophages, *Immunology*, 36, 501, 1979.
24. **Albertsson, P. A.**, Partition of cell particles and macromolecules in polymer two-phase systems, *Adv. Protein Chem.*, 24, 309, 1979.
25. **Schürch, S., Gerson, D. F., and McIver, D. J. L.**, Determination of cell: medium interfacial tension from contact angles in aqueous polymer systems, *Biochim. Biophys. Acta*, 640, 557, 1981.
26. **Dobry, A. and Boyer-Kawenoki, F.**, Phase separation in polymer solution, *J. Polym. Sci.*, 2, 90, 1947.
27. **Walter, H.**, Partition of cells in two-polymer aqueous phases — a surface affinity method for cell separation, in *Methods of Cell Separation*, Vol. 1, Catsimpoolas, N., Ed., Plenum Press, New York, 1977, 307.
28. **Miner, K. M., Walter, H., and Nicolson, G. L.**, Subfractionation of malignant variants of metastatic murine lymphosarcoma cells by countercurrent distribution in two-polymer aqueous phases, *Biochemistry*, 20, 6244, 1981.
29. **Omenyi, S. N. and Neumann, A. W.**, Thermodynamic aspects of particle engulfment by solidifying melts, *J. Appl. Phys.*, 47, 3956, 1976.
30. **Omenyi, S. N., Snyder, R. S., van Oss, C. J., Absolom, D. R., and Neumann, A. W.**, Effect of zero van der Waals and zero electrostatic forces on droplet sedimentation, *J. Colloid Interface Sci.*, 81, 402, 1981.
31. **Rowlinson, J. S. and Widom, B.**, *Molecular Theory of Capillarity*, Clarendon Press, Oxford, 1984.
32. **Hansen, J. P. and McDonald, I. R.**, *Theory of Simple Liquids*, Academic Press, New York, 1976.
33. **Verlet, L.**, Computer "experiments" on classical fluids. II. Equilibrium correlation functions, *Phys. Rev.*, 165, 201, 1968.
34. **Miyazaki, J. and Barker, J. A.**, A new Monte-Carlo method for calculating surface tension, *J. Chem. Phys.*, 64, 3364, 1976.
35. **Rao, M. and Levesque, D.**, Surface structure of a liquid film, *J. Chem. Phys.*, 65, 3233, 1976.
36. **Eisenberg, D. and Kauzmann, W.**, *The Structure and Properties of Water*, Clarendon Press, Oxford, 1969.
37. **Horne, R. A., Ed.**, *Water and Aqueous Solutions. Structure, Thermodynamics and Transport Processes*, John Wiley & Sons, New York, 1972.
38. **Stillinger, F. H. and Rahman, A.**, Improved simulation of liquid water by molecular dynamics, *J. Chem. Phys.*, 60, 1546, 1974.

39. **Jorgensen, W. L., Chandrasekhar, J., Madura, J. D., Impey, R. W., and Klein, M. L.,** Comparison of simple potential functions for simulating liquid water, *J. Chem. Phys.,* 79, 926, 1983.

40. **Matsuoka, O., Clementi, E., and Yoshimine, M.,** CI study of the water dimer potential surface, *J. Chem. Phys.,* 79, 1351, 1976.

41. **Schuster, P.,** The fine structure of the hydrogen bond, in *Intermolecular Interactions: From Diatomics to Biopolymers,* Pullman, B., Ed., John Wiley & Sons, New York, 1978, 363.

42. **Rahman, A. and Stillinger, F. H.,** Molecular dynamics study of liquid water, *J. Chem. Phys.,* 55, 3336, 1971.

43. **Lee, C. Y., McCammon, J. A., and Rossky, P. J.,** The structure of water at an extended hydrophobic surface, *J. Chem. Phys.,* 80, 4448, 1984.

44. **Croxton, C. A.,** Molecular orientation and interfacial properties of liquid water, *Physica,* 106A, 239, 1981.

45. **Girifalco, L. A. and Good, R. J.,** A theory for the estimation of surface and interfacial energies. I. Derivation and application to interfacial tension, *J. Phys. Chem.,* 61, 904, 1957.

46. **Good, R. J.,** Surface free energy of solid and liquids: thermodynamics, molecular forces and structures, *J. Colloid Interface Sci.,* 59, 398, 1977.

47. **Kinloch, A. J.,** The science of adhesion. I. Surface and interfacial aspects, *J. Mater. Sci.,* 15, 2141, 1980.

48. **Fowkes, F. M.,** Determination of interfacial tensions, contact angles and dispersion forces in surfaces by assuming additivity of intermolecular interactions in surfaces, *J. Phys. Chem.,* 66, 382, 1962.

49. **Andrade, J. D., Ma, S. M., King, R. N., and Gregonis, D. E.,** Contact angles at the solid-water interface, *J. Colloid Interface Sci.,* 72, 488, 1979.

50. **Wu, S.,** Polar and non-polar interactions in adhesion, *J. Adhes.,* 5, 39, 1973.

51. **Marcus, Y.,** *Introduction to Liquid State Chemistry,* John Wiley & Sons, New York, 1977, 105.

52. **Dann, J. R.,** Forces involved in the adhesive process. I. Critical surface tensions of polymeric solids as determined with polar liquids, *J. Colloid Interface Sci.,* 32, 302, 1970.

53. **Neumann, A., Good, R. J., Hope, C. J., and Sejpal, M.,** An equation-of-state approach to determine surface tensions of low-energy solids from contact angles, *J. Colloid Interface Sci.,* 49, 291, 1974.

54. **Gerson, D. F.,** Interfacial free energies of cells and polymers in aqueous media, in *Physicochemical Aspects of Polymer Surfaces,* Mittal, K. L., Ed., Plenum Press, New York, 1983, 229.

55. **Matsunaga, T. and Ikada, Y.,** Dispersive component of surface free energy of hydrophilic polymers, *J. Colloid Interface Sci.,* 84, 8, 1981.

56. **King, R. N., Andrade, J. D., Ma, S. M., Gregonis, D. E., and Brostrom, L. R.,** Interfacial tensions at acrylic hydrogen-water interfaces, *J. Colloid Interface Sci.,* 103, 62, 1985.

57. **Wrobel, A. M.,** Surface free energy of plasma-deposited thin polymer films, in *Physicochemical Aspects of Polymer Surfaces,* Mittal, K. L., Ed., Plenum Press, New York, 1983, 197.

58. **van Oss, C. J. and Gillman, C. F.,** Phagocytosis as a surface phenomenon. I. Contact angles and phagocytosis of non-opsonized bacteria, *J. Reticuloendothelial Soc.,* 12, 283, 1972.

59. **Dahlgren, C. and Sunqvist, T.,** Phagocytosis and hydrophobicity: a method of calculating contact angles based on the diameter of sessible drops, *J. Immunol. Methods,* 40, 171, 1981.

60. **Mege, J. L., Capo, C., Benoliel, A. M., Foa, C., and Bongrand, P.,** Non-specific cell surface properties: contact angle of water on dried cell monolayers, *Immunol. Commun.,* 13, 211, 1984.

61. **Gerson, D. F.,** Cell surface energy, contact angles and phase partition. I. Lymphocyte cell lines in biphasic aqueous mixtures, *Biochim. Biophys. Acta,* 602, 269, 1980.

62. **Gerson, D. F. and Akit, J.,** Cell surface energy, contact angles and phase partition. II. Bacterial cells in biphasic aqueous mixtures, *Biochim. Biophys. Acta,* 602, 281, 1980.

63. **McIver, D. J. L. and Schurch, S.,** Interfacial free energies of intact and reconstituted erythrocyte surfaces. Implications for biological adhesion, *Biochim. Biophys. Acta,* 691, 52, 1982.

64. **Gerson, D. F., Capo, C., Benoliel, A. M., and Bongrand, P.,** Adhesion, phagocytosis and cell surface energy. The binding of fixed human erythrocytes to rat macrophages and polymethylpentene, *Biochim. Biophys. Acta,* 692, 147, 1982.

65. **Napper, D. H.,** Steric stabilization, *J. Colloid Interface Sci.,* 58, 390, 1977.

66. **Schrader, M. E.,** On adhesion of biological substances to low energy solid surfaces, *J. Colloid Interface Sci.,* 88, 296, 1981.

67. **Baier, R. E. and Weiss, L.,** Demonstration of the involvement of adsorbed proteins in cell adhesion and cell growth on solid surfaces, *Adv. Chem. Ser.,* 145, 300, 1975.

68. **Soderquist, M. E. and Walton, A. G.,** Structural changes in proteins adsorbed on polymer surfaces, *J. Colloid Interface Sci.,* 75, 386, 1980.

69. **Jonsson, U., Malinquist, M., and Ronnberg, I.,** Adsorption of immunoglobulin G, protein A and fibronectin in the submonolayer region evaluated by a combined study of ellipsometry and radiotracer techniques, *J. Colloid Interface Sci.,* 103, 360, 1985.

70. **Klebe, R. J., Bently, K. L., and Schoen, R. C.,** Adhesive substrates for fibronectin, *J. Cell. Physiol.,* 109, 481, 1981.

71. **Dea, I. C. M., McKinnon, A. A., and Rees, D. A.,** Tertiary and quaternary structure in aqueous polysaccharide systems which model cell wall cohesion: reversible changes in conformation and association of agarose, carrageenan and galactomannans, *J. Mol. Biol.,* 68, 153, 1972.

72. **Garrouste, F., Capo, C., Benoliel, A. M., Bongrand, P., and Depieds, R.,** Non-specific binding by macrophages: different modulation of adhesive properties of rat peritoneal cells after plating on a glass or a plastic surface, *J. Reticuloendothelial Soc.,* 31, 415, 1982.

73. **Capo, C., Benoliel, A. M., Bongrand, P., and Depieds, R.,** Existence de plusiers systèmes de reconnaissance non specifique chez le macrophages, *C.R. Acad. Sci. Ser D,* 289, 729, 1979.

74. **Bongrand, P. and Golstein, P.,** Reproducible dissociation of cellular aggregates with a wide range of calibrated shear forces: application to cytolytic lymphocyte-target cell conjugates, *J. Immunol. Methods,* 58, 209, 1983.

75. **Neumann, A. W., Abosolom, D. R., van Oss, C. J., and Zingg, W.,** Surface thermodynamics of leukocyte and platelet adhesion to polymer surfaces, *Cell Biophys.,* 1, 79, 1979.

76. **Gerson, D. F. and Zajic, J. E.,** The biophysics of cellular adhesion, in *Immobilized Microbial Cells,* Venkatsubramanian, K., Ed., ACS Symp. Ser. American Chemical Society, Washington, D.C., 1979, 29.

77. **Curtis, A. S. G. and Forrester, J. V.,** The competitive effects of serum proteins on cell adhesion, *J. Cell Sci.,* 71, 17, 1984.

78. **Grinnell, F. and Feld, M. K.,** Fibronectin adsorption on hydrophilic and hydrophobic surfaces detected by antibody binding and analyzed during cell adhesion in serum-containing medium, *J. Biol. Chem.,* 257, 4888, 1982.

79. **Margolis, L. B., Tikhonov, A. N., and Vasilieva, E. Y.,** Platelet adhesion to fluid and solid phospholipid membranes, *Cell,* 19, 189, 1980.

80. **Maroudas, N. G.,** Chemical and mechanical requirements for fibroblast adhesion, *Nature (London),* 244, 353, 1973.

81. **Maroudas, N. G.,** Polymer aggregation and cell adhesion, in *Cell Shape and Surface Architecture,* Revel, J. P., Henning, V., and Fox, C. F., Eds., Alan R. Liss, New York, 1977, 511.

82. **Rich, A. and Harris, A. K.,** Anomalous preferences of cultured macrophages for hydrophobic and roughened substrata, *J. Cell Sci.,* 50, 1, 1981.

83. **Fenn, W. O.,** The theoretical response of living cells to contact with solid bodies, *J. Gen. Physiol.,* 4, 373, 1922.

84. **Mudd, S., Lucké, B., McCutcheon, M., and Strumia, M.,** On the mechanism of opsonin and bacteriotropin action. I. Correlation between changes in bacterial surface properties and in phagocytosis caused by sera of animals under immunization, *J. Exp. Med.,* 49, 779, 1929.

85. **Torza, S. and Mason, S. G.,** Coalescence of two immiscible liquid drops, *Science,* 163, 813, 1968.

86. **van Oss, C. J. and Gillmann, C. F.,** Phagocytosis as a surface phenomenon. II. Contact angles and phagocytosis of encapsulated bacteria before and after opsonization by specific antiserum and complement, *J. Reticuloendothelial Soc.,* 12, 497, 1972.

87. **Stendahl, O., Tagesson, C., and Edebo, M.,** Partition of *Salmonella typhimurium* in a two-polymer aqueous phase system in relation to liability to phagocytosis, *Infect. Immun.,* 8, 36, 1973.

88. **Edebo, L., Kihlstrom, E., Magnusson, K. E., and Stendahl, O.,** The hydrophobic effect and charge effects in the adhesion of enterobacteria to animal cell surfaces and the influence of different immunoglobulin classes, in *Cell Adhesion and Motility,* Curtis, A. S. G. and Pitts, J. D., Eds., Cambridge University Press, London, 1980, 65.

89. **Rabinovitch, M.,** Attachment of modified erythrocytes to phagocytic cells in absence of serum, *Proc. Soc. Exp. Biol. Med. Sci.,* 124, 386, 1967.

90. **Beukers, H., Deierkauf, F. A., Blom, C. P., Deierkauf, M., and Riemersma, J. C.,** Effects of albumin on the phagocytosis of polystyrene spherules by rabbit polymorphonuclear leucocytes, *J. Cell. Physiol.,* 97, 29, 1978.

91. **Absolom, D. R., van Oss, C. J., Zingg, W., and Neumann, A. W.,** Phagocytosis as a surface phenomenon: opsonization by aspecific adsorption of IgG as a function of bacterial hydrophobicity, *J. Reticuloendothelial Soc.,* 31, 59, 1982.

92. **Huber, H., Polley, M. J., Linscott, W. D., Fudenberg, H. H., and Muller Eberhard, H. J.,** Human monocytes: distinct receptor sites for the third component of complement and for immunoglobulin G, *Science,* 162, 1281, 1968.

93. **Lay, W. and Nussenzweig, V.,** Receptors for complement on leukocytes, *J. Exp. Med.,* 128, 991, 1968.

94. **Weir, D. M.,** Surface carbohydrates and lectins in cellular recognition, *Immunol. Today,* 1, 45, 1980.

95. **Sharon, N.,** Surface carbohydrates and lectins in cellular recognition, *Immunol. Today,* 5, 143, 1984.

96. **Unkeless, J. C.,** Characterization of a monoclonal antibody against mouse macrophage and lymphocyte Fc receptors, *J. Exp. Med.,* 150, 580, 1979.

97. **Fearon, D. T.,** Identification of the membrane glycoprotein that is the C_{3b} receptor of the human erythrocyte, polymorphonuclear leukocyte, B lymphocyte and monocyte, *J. Exp. Med.,* 152, 20, 1980.

98. **Wileman, T. E., Lennartz, M. R., and Stahl, P. D.,** Identification of the macrophage mannose receptor as a 175 kDa membrane protein, *Proc. Natl. Acad. Sci. U.S.A.,* 83, 2501, 1986.

99. **Ezekowitz, R. A. B., Sim, R. B., MacPherson, G. G., and Gordon, S.,** Interaction of human monocytes, macrophages and polymorphonuclear leukocytes with zymosan *in vitro, J. Clin. Invest.,* 76, 2368, 1985.

100. **Benoliel, A. M., Capo, C., Bongrand, P., Ryter, A., and Depieds, R.,** Nonspecific binding by macrophages: existence of different adhesive mechanisms and modulation by metabolic inhibitors, *Immunology,* 41, 547, 1980.

101. **Chi, L., Vogel, J. E., and Shelokov, A.,** Selective phagocytosis of nucleated erythrocytes by cytotoxic amoebae in cell culture, *Science,* 130, 1762, 1959.

Chapter 4

MECHANICS OF CELL DEFORMATION AND CELL-SURFACE ADHESION

Evan Evans

TABLE OF CONTENTS

I. INTRODUCTION

Surface recognition and adhesion are prominent processes in biology. In general, these events involve specific molecular binding and cross-bridging reactions; however, nonspecific processes (without identifiable molecular reaction sites) also promote cell aggregation and adhesion. Formation of either type of adhesive contact induces stresses in the cells that eventually limit the extent of contact. Subsequent separation of adherent cells by physical force also creates stresses that are transmitted through the cell cortex to the contact zone. The common view in biology is that factors which influence cell adhesion simply alter the chemical attraction between opposing membrane surfaces and that this attraction is the sole determinant of adhesion. However, other physical factors play equally important roles in the mechanics of cell adhesion and separation. Most obvious are external forces which act to disrupt cell-cell contacts, e.g., shear stresses in a convecting suspension. Implicitly significant, although less obvious, is the mechanical rigidity of cells, because cell stiffness directly opposes the adhesion process. Clearly stated, "rigid" bodies can form strong adhesive contacts only if they fit together perfectly, whereas "flaccid" bodies can easily deform to create large contact areas with minimal work.[1] Thus, it is important to have a conceptual understanding of the deformability properties of biological cells in order to properly analyze and predict the behavior of cell-adhesion processes. Likewise, even though the details of the forces involved in surface-surface adhesion remain obscure because of the small range over which they act, it is equally important to consider how the mechanics of the adhesion process depend on general properties of the attractive forces, e.g., surface distribution, effective range, and mobility of the contact sites.

Biological cells have many common structural features: a bounding plasma membrane, subsurface cortical gel, internal granular or other organellar bodies, reticular elements, nuclei, etc. Even with this structural complexity, many cells remain nearly spherical in suspension. When stressed by external forces, the cells deform continuously, albeit lethargically, but subsequently recover their original spherical shape after the forces are removed. This behavior indicates that the cell can be modeled as a complex liquid suspension, encapsulated by a cortical shell made up of a composite of plasmalemma and subsurface gel.[2,3] The interior of such a cell limits only the rate of deformation in response to applied forces; the cortical shell establishes the static or limiting balance of stresses. Further, for motile cells, an active locomotory apparatus (a polymerizable, actin filament solution) is concentrated in the cortical layer adjacent to the cell surface.[4,5] Thus, the mechanical abstraction of cells as complex liquid interiors surrounded by cortical shells can be a starting point for examination of cell deformation properties. In fact, the mechanical model for membrane-bound liquid capsules is a good approximation to the deformation response of blood phagocytes and is an exact representation of mammalian red blood cells and synthetic (phospholipid bilayer) vesicles. These capsules provide excellent laboratory "substrates" for testing membrane-membrane adhesion. However, when cells have structural connections that span the cytoplasm, simple models will not be appropriate; in suspension, these cells will usually not be spherical in shape. In general, the intrinsic mechanical properties of a cell must be determined by direct mechanical experiment and cannot be deduced from shape alone.

The nature of adhesion, specific or nonspecific, also has an important impact on the mechanics of contact formation and separation. As previously noted, specific adhesion processes are those associated with reactions between discrete receptor sites located on the membrane surfaces. Here, adhesion is established by molecular cross-bridges that create a sort of "molecular-scale roughness". The effective character of the roughness can be smoothed out by molecular motions and thus is modulated by receptor mobility. Localization of molecular contacts leads to a marked difference between the minimal stresses induced in cells by contact formation and the large stresses produced by the external forces necessary

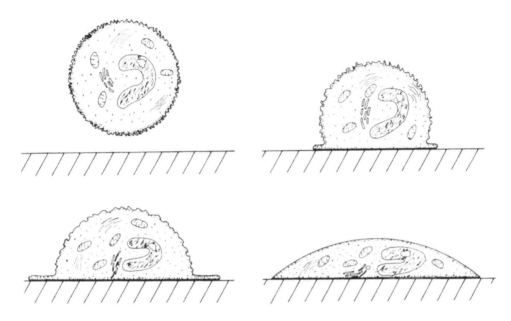

FIGURE 1. Schematic of cell adhesion to a solid substrate. Key features: cell body deforms as a sequence of spherical-like segments; plasma membrane-subsurface gel spreads outward from the cell base to form an annular skirt; ruffled plasma membrane is eventually smoothed out, which limits the extent of contact.

to separate an adherent region. By comparison, nonspecific adhesion processes are characterized by uniform spreading of one surface on another until limited by the stresses in the substrates. In general, these processes exhibit reversible energetics where the free energy per unit area of contact formation is equal to the work required to separate the contact. Thus, from the viewpoint of mechanics, these disparate situations can be represented by two distinct models: the first represents an adhesion process promoted by fixed, discrete sources of interaction between the surfaces; the second represents adhesion as caused by uniformly distributed or continuous interactions between the surfaces.[6,7]

II. MECHANICS OF CELL DEFORMATION

Consistent with the reasoning outlined in the introduction, many of the essential features of cell deformation can be considered in the context of an abstract representation of cells as complex liquid cores surrounded by composite plasma membrane-cortical gel layers. Even though the interior of the cell is a liquid ''slurry'' of suspended organelles and nucleus, which as a whole is nonuniform and may be non-Newtonian in character, the contents only resist the *rate* of deformation and the body will exhibit a uniform isotropic stress when stationary (i.e., a uniform hydrostatic pressure). The static shape and stress distribution for a cell of this type are determined by the plasma membrane-cortical gel subject to a uniform internal pressure.[3] Consider the spreading (adhesion) of such a cell at a solid surface. A sequence of shapes often observed is schematically illustrated in Figure 1. The cell is initially spherical with a ruffled plasmalemma envelope. As adhesion to the surface progresses, the plasmalemma smooths out to form large contact regions. In conjunction with the subsurface cortical layer, the plasmalemma flows outward to form a lamellar skirt around the cell body. Throughout adhesion, the cell body deforms as a series of approximately spherical segments (indicating a nearly uniform stress state or hydrostatic pressure in the interior). Some important implications can be deduced from this figure: (1) the form of the cell not in contact with the solid substrate is determined by opposition of the stresses in the cortical shell to the internal pressure, (2) the total plasmalemma area (initially in the form of a ruffled surface) and the total cell volume establish a constraint or ultimate limit to the cell-surface contact area which is reached when the plasmalemma is completely smoothed out, and (3) the subsurface or cortical layer must be cohesive in order to form the lamellar skirt around the cell body (otherwise the two plasmalemma surfaces would separate to give a spherical segment shape like the limiting shape illustrated in Figure 1). Figure 2 shows an example of this behavior in a sequence of video micrographs of a blood granulocyte as it spreads on an endothelial cell surface (here, additional complexity arises because the spreading process involves active, contractile stresses in the cortex). Clearly, the dynamics of deformation of a cell with this type of structure will be determined by the properties of the plasmalemma-cortical gel composite and the effective viscosity of the interior suspension.

Analysis of cell deformation involves three independent and distinct developments which are the bases for the mechanics of continuous media: (1) quantitation of deformation and rate of deformation for changes in geometric shape (independent of substance and forces), (2) balance of forces (independent of deformation and substance), and (3) material properties of the substance. Knowledge of any two of the three can be used to predict the third aspect, e.g., observation of deformation and rate of deformation in response to controlled forces can be analyzed to give the material properties (e.g., elastic and viscous coefficients). We will outline each of these components in the following paragraphs (see Evans and Skalak[8] for more detailed aspects).

A. Deformation and Stress Distribution in the Liquid Core

The simplest approach to the mechanics of the liquid-like core is to model the response of the cell contents as a uniform Newtonian liquid with a constant viscosity. The equations of motion reduce to the "creeping" flow form — low Reynolds' number limit — of the Navier-Stokes equations for a fluid.[22,23] Inertial forces (due to acceleration of the fluid) are completely negligible because of the small size of cells and the slow flow rates. Thus, the equations of mechanical equilibrium balance viscous forces against pressure forces. For example, consider axisymmetric flow with spherical boundaries; the equations of mechanical equilibrium are expressed in spherical polar coordinates (R, θ, ϕ) by:

$$0 = \frac{1}{R^2} \cdot \frac{\partial(R^2 \cdot \sigma_{RR})}{\partial R} + \frac{1}{R \cdot \sin\theta} \cdot \frac{\partial(\sin\theta \cdot \sigma_{\theta R})}{\partial \theta} - \frac{\sigma_{\theta\theta}}{R} - \frac{\sigma_{\phi\phi}}{R}$$

$$0 = \frac{1}{R^2} \cdot \frac{\partial(R^2 \cdot \sigma_{R\theta})}{\partial R} + \frac{1}{R \cdot \sin\theta} \cdot \frac{\partial(\sin\theta \cdot \sigma_{\theta\theta})}{\partial \theta} - \frac{\sigma_{\theta R}}{R} - \frac{\cot\theta \cdot \sigma_{\phi\phi}}{R} \tag{1}$$

where σ_{RR}, $\sigma_{\theta\theta}$, $\sigma_{R\theta}$ ($\sigma_{\theta R}$), $\sigma_{\phi\phi}$ are components of the stress field in the core (the first subscript defines the direction along which the force acts — either radial R, meridional θ, or azimuthal ϕ — and the second subscript denotes the surface on which the stress is defined). For a Newtonian liquid, the constitutive behavior is represented by proportionality between stresses and rates of deformation:

$$\sigma_{RR} = -P + 2\bar{\eta} \cdot \frac{\partial v_R}{\partial R}; \quad \sigma_{R\theta} = \bar{\eta}\left(\frac{1}{R} \cdot \frac{\partial v_R}{\partial \theta} + \frac{\partial v_\theta}{\partial R} - \frac{v_\theta}{R}\right)$$

$$\sigma_{\theta\theta} = -P + 2\bar{\eta}\left(\frac{1}{R} \cdot \frac{\partial v_\theta}{\partial \theta} + \frac{v_R}{R}\right); \quad \sigma_{\phi\phi} = -P + \frac{2\bar{\eta} \cdot v_R}{R} \tag{2}$$

where the hydrostatic pressure P is introduced because of fluid incompressibility:

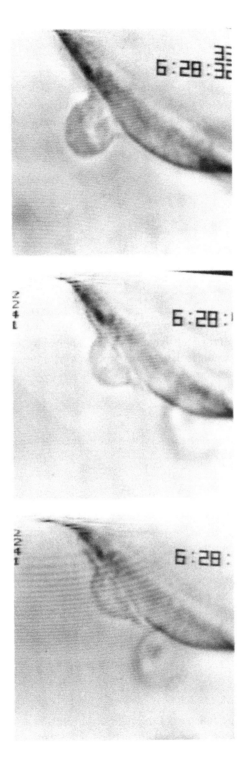

FIGURE 2. Video micrographs of a blood granulocyte as it adheres to a vascular endothelial cell cultured on a microcarrier bead. The undeformed spherical diameter of the granulocyte is about 8×10^{-4} cm. Note the similarity to the shapes shown in Figure 1.

$$O = \frac{1}{R^2} \cdot \frac{\partial(R^2 \cdot v_R)}{\partial R} + \frac{1}{R \cdot \sin\theta} \cdot \frac{\partial(\sin\theta \cdot v_\theta)}{\partial \theta} \tag{3}$$

The coefficient $\bar{\eta}$ is the effective viscosity of the cell contents; v_R and v_θ are the velocity components along radial and meridional directions, respectively. The solution for axisymmetric flow inside a spherical boundary is expressed in terms of infinite series expansions in angular harmonic functions (Y_n, Y'_n):[23]

$$v_R = -\sum_{n=2}^{\infty} (a_n \cdot R^{n-2} + c_n \cdot R^n) \cdot Y_{n-1}(\cos\theta)$$

$$v_\theta = \sum_{n=2}^{\infty} [n \cdot a_n \cdot R^{n-2} + (n+2) \cdot c_n \cdot R^n] \cdot Y'_n(\cos\theta)/\sin\theta$$

$$P = -\bar{\eta} \cdot \sum_{n=2}^{\infty} \left[\frac{2(2n+1)}{n-1} \cdot c_n \cdot R^{n-1} \right] \cdot Y_{n-1}(\cos\theta) + c_o \tag{4}$$

With the appropriate boundary conditions at the shell interface (for stresses or velocities), the coefficients a_n and c_n in these equations are fully specified.

The objective is to determine the dynamic stresses applied by the liquid core to the cortical shell. In this spherical example, the stresses are given by a dynamic pressure P' normal to the cortex and a shear stress σ_m tangent to the shell:

$$-P' = -P + 2\bar{\eta} \cdot \frac{\partial v_R}{\partial R}, \quad (R = R_o)$$

$$\sigma_m = \bar{\eta}\left(\frac{1}{R} \cdot \frac{\partial v_R}{\partial \theta} + \frac{\partial v_\theta}{\partial R} - \frac{v_\theta}{R} \right), \quad (R = R_o)$$

These fluid dynamic stresses are opposed by material forces in the shell. This approach has been used to model the flow response of the interior contents of blood phagocytes in micropipette aspiration tests.[10,14] Analysis of the creeping flow of a spherical liquid drop into a tube has shown that the stress applied normally to the sphere surface is essentially uniform over the spherical segment exterior to the tube. Hence, for pipette aspiration, the pressure on the cortex can be approximated by a constant and an abrupt pressure drop inside the core, introduced at the pipette entrance in proportion to rate of entry.[3] (The pressure drop at the pipette entrance is proportional to the entry-flow rate scaled by the pipette radius $[\dot{L}/R_p]$):

$$\Delta P = \bar{\eta} \cdot \left(\frac{\dot{L}}{R_p} \right) \cdot f(R_p/R_o)$$

multiplied by an algebraic function of the ratio of pipette size to cell diameter (R_p/R_o).

In general, simultaneous solution of the equations of mechanical equilibrium for the cortical shell with those for the liquid core is required. The coupling between these sets of equations is provided by the dynamic stresses (P', σ_m) that act on the shell. The mechanics of shell deformation and force equilibrium will be discussed next.

B. Intrinsic Deformation and Rate of Deformation of a Thin Shell

Changes in overall shape or conformation of a shell can be viewed conceptually as the superposition of local deformations of "imaginary" differential elements of the shell. If the

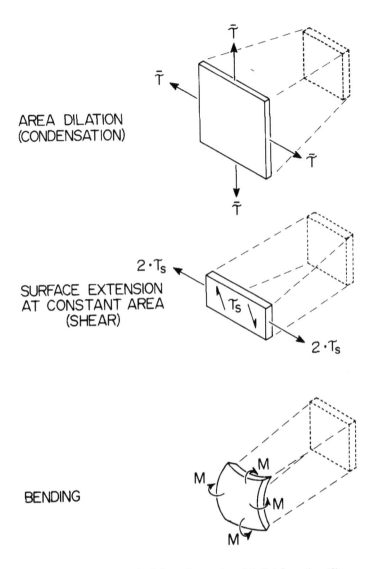

FIGURE 3.   Schematic of the independent modes of shell deformation: (1) area
dilation, (2) shear or extension at constant area, and (3) bending. Also shown are
the force and moment resultants associated with each mode of deformation.

shell is thin in comparison to the radii of curvature that describe the contour of the shell,
then the local deformation of each small element can be represented by three independent
modes of deformation: (1) area dilation or condensation without changing shape or curvature
of the element, (2) extension of the element without change in surface area or curvature,
and (3) changes in element curvature without change in element area or aspect ratio (the
modes of deformation are illustrated in Figure 3). These geometric changes are quantitated
by the fractional change in element area α, the in-plane extension ratio λ (at constant surface
density), and changes in element principal curvatures $(1/R_1)$ and $(1/R_2)$. Similarly, the
intrinsic rates of deformation are defined by the rate of area dilation or condensation, the
rate of extension at constant element area, and the rate of change of curvatures. Since
deformation is defined locally, the small element of the shell surface can be considered
"flat" with dimensions described by orthogonal surface coordinates that form a plane tangent
to the shell surface. Thus, rectangular deformations of the element are characterized by the

ratios (λ_1, λ_2) of instantaneous dimensions of the element to the initial dimensions in some reference state. Based on these (extension) ratios, the fractional change in area is given by:

$$\alpha = (\lambda_1 \cdot \lambda_2) - 1 \qquad (5)$$

and the measure of in-plane extension at constant area is given by a single extension ratio:

$$\bar{\lambda} = \lambda_1/\lambda_2 \qquad (6)$$

Simple rectangular deformation of an element embodies surface shear deformation which is maximal along lines in the surface at $\pm 45°$ to the extension axis; the maximum shear is represented by the Eulerian shear strain e_s:

$$e_s = |\bar{\lambda}^2 - \bar{\lambda}^{-2}|/4(1 + \alpha) \qquad (7)$$

Hence, extension at the constant area creates pure shear deformations. Changes in element curvature give rise to differential deformations between strata of the element (e.g., for an increase in convexity, outer layers expand and inner layers condense relative to the midplane). Measures of bending deformation are obtained by integration of the differential deformations between strata over the shell thickness; these bending strains are proportional to the element thickness times the changes in principal curvatures:

$$|\delta\alpha| \sim (1 + \alpha) \cdot h \cdot [\Delta(1/R_1) + \Delta(1/R_2)]$$

$$|\delta\bar{\lambda}| \sim \bar{\lambda} \cdot h \cdot [\Delta(1/R_1) - \Delta(1/R_2)] \qquad (8)$$

Equations 5 to 8 describe the instantaneous state of deformation at a local position on the shell. Similarly, intrinsic rates of deformation can be derived that quantitate the local rates of change of element geometry. The fractional rate of expansion or condensation of the element area is given by:

$$V_\alpha = \frac{\partial \ln(1 + \alpha)}{\partial t} \qquad (9)$$

The maximum rate of in-plane extension (or surface shear rate) is given by:

$$V_s = \frac{\partial(\ln \bar{\lambda})}{\partial t} \qquad (10)$$

Finally, the rates of bending deformations are given by:

$$|\delta V_s| \sim h \cdot \frac{\partial[(1/R_1) - (1/R_2)]}{\partial t}$$

$$|\delta V_\alpha| \sim h \cdot \frac{\partial[(1/R_1) + (1/R_2)]}{\partial t} \qquad (11)$$

In the previous discussion, deformation was evaluated in a domain small enough so that the material element could be considered flat and locally uniform. This approach leads to intensive quantification of shape changes, i.e., the variables are point definitions which do not depend on the size or shape of the shell as a whole. However, deformation and rate of

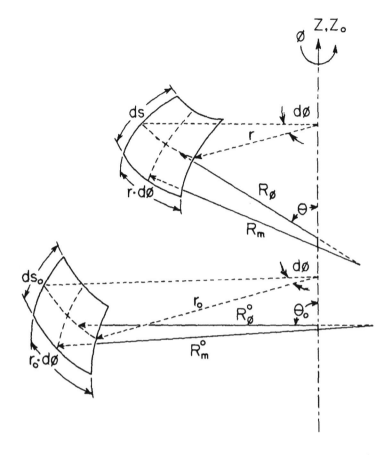

FIGURE 4. Diagram of initial and instantaneous (deformed) elements of an axi-symmetric shell. Deformation is measured by the ratios of element dimensions (deformed to undeformed); $\lambda_m = ds/ds_o, \lambda_\phi = r/r_o$.

deformation variables may vary over the shell contour. The continuous variation is derived from differential geometry. For example, with axisymmetric deformations, the shell geometry can be described by curvilinear coordinates (s, θ, ϕ) and (s_o, θ_o, ϕ_o) for the instantaneous (deformed) and initial (reference) contours, respectively, as shown in Figure 4. The distance along a meridian (contour line) is defined as s, the coordinate θ is defined as the angle between the outward normal to the surface and the axis of symmetry, and the coordinate ϕ is the azimuthal angle. By transformation, these intrinsic coordinates are related to the spatial coordinates (r, z, ϕ) for the instantaneous geometry and (r_o, z_o, ϕ) for the reference contour. Extension ratios, which define local changes in element geometry, are given by the ratio of differential lengths (deformed:undeformed) along a meridian and around a latitude circle for the surface elements as follows:*

$$\lambda_m = \frac{ds}{ds_0} = \lambda_1$$

$$\lambda_\phi = r/r_0 = \lambda_2 \tag{12}$$

These extension ratios specify the positional variation of surface deformation for the axi-

---

* The subscripts (m, ϕ) replace the index values (1, 2) since the former represent the principal directions in the case of axial symmetry.

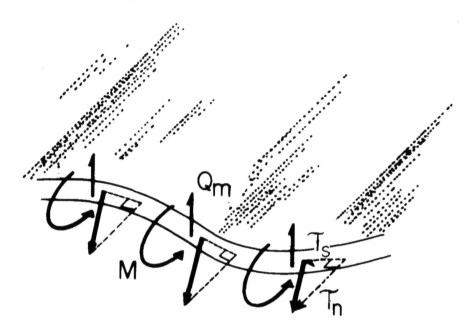

FIGURE 5.   Schematic of the cumulated stresses in the shell: in-plane force resultants (τ_n tension, τ_s shear); transverse shear resultant Q_m; bending moment resultant M. These resultants act along contours in a mathematical (conceptual) surface that replaces the shell.

symmetric shell. Rates of deformation are derived from local time derivatives of extension ratios as given in Equations 9 to 11. In spatial coordinates, the rates of deformation are also given in terms of velocity components that characterize motion of points on the shell contour:

$$V_\alpha = \frac{\partial v_s}{\partial s} + \frac{v_s}{r} \cdot \frac{\partial r}{\partial s} + v_n \cdot (1/R_m + 1/R_\phi)$$

$$2 \cdot V_s = \frac{\partial v_s}{\partial s} - \frac{v_s}{r} \cdot \frac{\partial r}{\partial s} + v_n \cdot (1/R_m - 1/R_\phi) \tag{13}$$

where the velocity components are v_s and v_n, tangent to the contour and normal to the contour, respectively.

C. Shell Mechanical Equilibrium

The general approach to the mechanics of thin shells is to cumulate the actions of forces that are distributed in the shell by integration over the shell thickness. This yields equivalent force and moment resultants which act at a mathematical surface that replaces the shell.[8,9] Figure 5 schematically illustrates these effective reactions of the shell that oppose the dynamic stresses (P', σ_m) applied normal and tangent to the shell surface. Integration of the "lateral" stresses over the shell thickness yields forces per unit length along edges of the element that act in the plane of the shell, normal and tangent to the edge (i.e., tension and shear force, respectively). Deviation of lateral stresses from the mean values of stress gives rise to force couples that cumulate into moment resultants which act around the edge of the element. Finally, the shear-stress component parallel to the outward normal cumulates into the transverse shear resultant, which acts to "cut" the shell as shown in Figure 5.

Examination of the in-plane force resultants shows that these forces can be partitioned into two independent parts: a mean or isotropic force resultant $\bar{\tau}$, and a deviatoric or maximum shear force resultant τ_s (that acts maximally along lines at $\pm 45°$ to the axes for purely tensile or compressive stresses, as shown in Figure 3). Similarly, moment resultants can be decomposed into isotropic and deviatoric parts as well.

$$\bar{\tau} \equiv (\tau_m + \tau_\phi)/2; \quad \tau_s \equiv (\tau_m - \tau_\phi)/2$$

$$\overline{M} \equiv (M_m + M_\phi)/2; \quad \breve{M} \equiv (M_m - M_\phi)/2$$

In general, the sum of all forces on the shell is equal to the inertial force due to acceleration. Because of the small scale of biological structures and low accelerations, inertial forces are totally negligible. Hence, mechanical equilibrium is established when the sum of forces on the body is equal to zero. Equations of mechanical equilibrium are not equivalent to thermodynamic statements of equilibrium. For example, in mechanical equilibrium, forces may be dissipative (nonconservative) as well as elastic (conservative). Thermodynamic equilibrium is based on reversibility of the physical and chemical processes that occur at the atomic and molecular level during deformation and, consequently, only represents elastic forces.

For an axisymmetric shell (illustrated in Figure 4) subject to a net pressure differential P′ and a net tangential traction or shear stress σ_m along the meridian of the shell contour, two differential equations must be satisfied for the local balance of forces:

$$O = r \cdot \sigma_m + r \cdot \frac{\partial(\bar{\tau} + \tau_s)}{\partial s} + 2 \cdot \tau_s \cdot \frac{\partial r}{\partial s} + \frac{(r \cdot Q_m)}{R_m}$$

$$P' = \bar{\tau} \cdot (1/R_m + 1/R_\phi) + \tau_s \cdot (1/R_m - 1/R_\phi) - \frac{1}{r} \cdot \frac{\partial(r \cdot Q_m)}{\partial s} \qquad (14)$$

Likewise, the balance of moments for the shell element is given by:

$$r \cdot Q_m = r \cdot \frac{\partial \overline{M}}{\partial s} + \frac{1}{r} \cdot \frac{\partial(r^2 \breve{M})}{\partial s} \qquad (15)$$

Here, the equations of equilibrium have been written in terms of surface-isotropic and maximum shear force resultants plus moments to demonstrate the explicit dependence of mechanical equilibrium on these independent actions. The stresses applied to the cortex are obtained by simultaneous solution of the equations for "creeping" flow of the liquid core (as discussed previously).

D. General Constitutive Equations for Thin Shell Materials

The principles of thermodynamics can be used to derive phenomenological recipes for the storage and dissipation of energy when work is done on a material by displacement of external forces.[8] The differential work associated with deformation of an element of the shell can be expressed per unit surface area of the shell as a local density function:

$$d\bar{W} = \bar{\tau} \cdot d\alpha + 2 \cdot \tau_s \cdot (1 + \alpha) \cdot d(\ln\bar{\lambda})$$

$$+ \overline{M} \cdot (1 + \alpha) \cdot d[(1/R_m) + (1/R_\phi)] + \breve{M} \cdot (1 + \alpha) \cdot d[(1/R_m) - (1/R_\phi)] \quad (16)$$

Here, it is seen that the work is made up of contributions from dilation and extension in the plane of the shell plus a contribution from bending or curvature changes, all given by products of intensive forces multiplied by conjugate deformations. For reversible thermodynamic processes, work can be attributed to conservative forces which are derivatives of a potential function that depends only on the independent modes of deformation. For a shell whose properties do not depend on directions tangent to the shell (i.e., isotropic in the plane of the surface), an elastic potential can be postulated by:

$$(d\bar{F})_T \simeq K \cdot d(\alpha^2/2) + \mu \cdot d[(\bar{\lambda}^2 + \bar{\lambda}^{-2})/2] + \bar{B} \cdot d[(1/R_m + 1/R_\phi)^2/2]$$

$$+ \tilde{B} \cdot d[(1/R_m - 1/R_\phi)^2/2] \tag{17}$$

where the coefficients in this equation are elastic properties: area compressibility modulus K, the surface extension (shear) modulus μ, and bending or curvature elastic modulus B. From the first law of thermodynamics for a reversible process, work is equal to the change in the elastic potential which leads to proportionality between material forces and deformation. Thus, mean or isotropic force resultants are proportional to area dilation or condensation:

$$\bar{\tau} = K \cdot \alpha \tag{18}$$

Shear forces in the plane of the shell are proportional to extensional deformation:

$$\tau_s = \mu \cdot (\bar{\lambda}^2 - \bar{\lambda}^{-2})/2 = 2 \cdot \mu \cdot e_s \tag{19}$$

Bending moments are proportional to changes in curvature:

$$\overline{M} = \bar{B} \cdot \Delta(1/R_m + 1/R_\phi)$$

$$\tilde{M} = \tilde{B} \cdot \Delta(1/R_m - 1/R_\phi) \tag{20}$$

In contrast with elastic behavior, irreversible processes create nonconservative forces that depend on rate of deformation as well as the instantaneous deformation state. For example, a solid shell structure will be characterized by elastic levels of stress at static equilibrium, but the time-dependent response of the shell to applied forces will involve greater stresses because of viscous dissipation within the material. Further, shell materials may exhibit liquid-like or plastic behavior which is characterized by continuous flow in response to prolonged exposure to stress when the magnitude of the stresses are sufficiently large. For such non-conservative stresses, mechanical power is dissipated in the material as heat and the density of mechanical power is given by the local time derivative of the work density:

$$\frac{\partial \tilde{W}}{\partial t} = \bar{\tau} \cdot \frac{\partial \alpha}{\partial t} + 2 \cdot \tau_s \cdot (1 + \alpha) \cdot \frac{\partial(\ln\bar{\lambda})}{\partial t} + \overline{M} \cdot (1 + \alpha)$$

$$\cdot \frac{\partial[(1/R_m) + (1/R_\phi)]}{\partial t} + \tilde{M} \cdot (1 + \alpha) \cdot \frac{\partial[(1/R_m) - (1/R_\phi)]}{\partial t} \tag{21}$$

For ideal liquids, dissipation of mechanical power is a quadratic function of the rates of deformation. Hence, force and moment resultants are simply inelastic and proportional to rates of deformation:

$$\bar{\tau} = \kappa \cdot \frac{\partial\ln(1 + \alpha)}{\partial t}; \quad \tau_s = 2\eta \cdot \frac{\partial(\ln\bar{\lambda})}{\partial t}$$

$$\overline{M} = \bar{\nu} \cdot \frac{\partial[(1/R_m) + (1/R_\phi)]}{\partial t}; \quad \tilde{M} = \tilde{\nu} \cdot \frac{\partial[(1/R_m) - 1/R_\phi)]}{\partial t} \tag{22}$$

where proportionality is established by coefficients of viscosity for the appropriate type of deformation. In general, material behavior can only be approximated by solid or liquid-like constitutive relations, and usually is characterized by more complicated rheological equations that must be defined empirically by experiment.

A reasonably general representation for the mechanical response of materials is given by constitutive relations for three ideal regimes: solid, semisolid, and plastic or liquid. Solid-like materials are represented by viscoelastic equations that model parallel superposition of elastic and viscous processes as originally envisioned by Kelvin:

$$\tau_s = \mu \cdot \left[(\tilde{\lambda}^2 - \tilde{\lambda}^{-2})/2 + 2 \cdot t_e \cdot \frac{\partial(\ln\tilde{\lambda})}{\partial t} \right] \tag{23}$$

given here for the in-plane shear resultant. The key feature for this type of behavior is that the dynamic response is characterized by a time constant t_e, which is the ratio of the coefficient of viscosity to the elastic modulus (η_e/μ). The transition from solid to liquid-like behavior depends on the magnitude and duration of forces applied to the material. This intermediate regime is characterized by creep and relaxation phenomena, i.e., the material exhibits slow, continuous deformation when the force is held constant, whereas the material responds elastically to rapid changes in applied force. This behavior can be modeled by the serial addition of elastic (conservative) and inelastic (plastic) processes as originally developed by Maxwell. Here, the in-plane extensional constitutive relation for creep and relaxation is represented by:

$$\frac{\partial(\ln\tilde{\lambda})}{\partial t} = \frac{1}{\mu} \left[\frac{1}{2\sqrt{(\tau_s^2/\mu^2 + 1)}} \cdot \frac{\partial\tau_s}{\partial t} + \frac{\tau_s}{2 \cdot t_c} \right] \tag{24}$$

where the characteristic time constant t_c is given by the ratio of a different coefficient of viscosity to the elastic modulus (η_c/μ). As such, the characteristic material time constants for viscoelastic solid and semisolid material regimes represent different molecular relaxation processes. Finally, the simplest form of material transition from solid to liquid behavior is that of an ideal plastic modeled by elastic solid behavior below a yield threshold and liquid behavior above the yield:

$$\frac{\partial(\ln\tilde{\lambda})}{\partial t} = (\tau_s - \hat{\tau}_s)/(2 \cdot t_p), \qquad \tau_s > \hat{\tau}_s \tag{25}$$

as introduced by Bingham in 1922 and generalized 10 years later by Hohenemser and Prager. Two material properties are necessary to characterize the behavior of a plastic body: a yield-force resultant and a viscosity ($t_p \equiv \eta_p/\hat{\tau}_s$). For a perfect liquid, the yield threshold is identically zero. For both the semisolid and plastic types of material behavior, deformations of the material are irrecoverable in that the shell is left permanently deformed after the forces are removed. The constitutive relations given above outline general regimes of material behavior, but are not unique. However, these relations characterize the essential features of mechanical response of materials and transitions in behavior; the time constants (t_e, t_c, t_p) are especially useful as guidelines for predicting the appropriate response.

Equations for material behavior, i.e., the relation between intensive force (and moment) resultants and intensive deformation and rate of deformation variables, are used with the equations of mechanical equilibrium to predict the evolution of shell geometry in response to applied forces. The necessary ingredients are the material properties (elastic and viscous coefficients) for the actual shell materials and the effective viscosity of the liquid core. In the following paragraphs, a brief overview of the known properties of synthetic (phospho-lipid) bilayer and cell membranes are given, as well as a discussion of the behavior anticipated for plasma membrane-cortical gel composite materials. In the context of these properties and constitutive relations, the detailed mechanics of cell-surface adhesion are examined.

E. Material Properties of Lipid Bilayer and Red Blood Cell Membranes

From the viewpoint of mechanics, membranes are simply very thin shells, so the effects of moment resultants (due to bending ridigity) on the mechanical equilibrium can be neglected in comparison to the contributions from force resultants in the plane of the shell. In biology, the term membrane usually represents a molecularly thin structure that separates one region of a cell or subcellular body from another and is often synonomous with the concept of a plasmalemma or lipid bilayer structure. Because of molecular thinness, bilayer membrane structures must in general be supported by stiff adjacent materials. The unifying feature of all bilayer membranes is the preferential assembly of the lipid amphiphiles into two-dimensional, condensed liquid bilayers which form tight, cohesive chemical insulators. Subsurface or superficial layers provide "scaffolding" and support for the liquid bilayer; these differ from cell to cell but are essential for maintenance of cell shape and strength. For simple capsules like red blood cells, the lipid bilayer is supported by an adjacent network of filamentous and globular proteins which is strongly associated with the bilayer. As such, the red-cell envelope is a thin, trilamellar membrane composite which is smooth in contour and exhibits recoverable elastic deformations.[8] On the other hand, cells like blood phagocytes appear to possess a gelatinous layer adjacent to the ruffled outer plasmalemma which is forced onto the spherical form. The difference between smooth and ruffled configurations is more readily understandable after examination of the specific properties characteristic of phospholipid bilayer membranes.

Lipid bilayer membranes can exist either as thin solid or liquid materials where the polar head groups are anchored to the water interfaces and the hydrocarbon polymer chains form a thin double layer between the polar interfaces.[10,11] In biological cells, these hydrocarbon chains are in the fluid state so the cell surfaces behave as two-dimensional liquids (i.e., $\mu \equiv 0$). Because of the relatively large energies required to expose the hydrocarbon interior to water, the bilayer surfaces have very small area compressibility (i.e., $K \sim 10^2$ dyn/cm). Thus, the molecules remain closely packed even when subjected to tensions that lead to rupture. A spherical bilayer capsule with fixed internal volume is essentially rigid because any deformation of the sphere requires an increase in area, whereas deflation of the sphere leaves the capsule completely flaccid (easily deformable) stabilized in shape by the extremely small bending rigidity (i.e., $B \sim 10^{-12}$ dyn-cm) of the bilayer. Bilayer bending energies become comparable to surface dilation energies only when the membrane radii of curvature are less than 10^{-6} cm.[1] For example, osmotic deflation of a vesicle (shown in Figure 6) allows the vesicle to be aspirated by a micropipette with very low suction pressure ($>10^{-6}$ atmosphere) to an extent determined by vesicle area and reduced volume. When this limit is reached, large suction pressures are required to produce small displacements of the vesicle projection into the pipette, which leads to vesicle rupture.[10,11] On the other hand, when compressed, the membrane readily folds and wrinkles unless subsurface structure exists to provide a stiff support for the bilayer. Hence, the plasmalemma of cells offers no resistance to deformation until the surface area is required to dilate, upon which the bilayer membrane becomes rigid to expansion. The bilayer is simply a fixed-area envelope that limits the expansion and distribution of the interior contents. Since the interior contents of cells has a high osmotic activity, pressures created by cell deformation are insufficient to change cell volume, which adds a second important constraint to cell deformation. Departure from the constant plasmalemma surface area and constant cell volume requirements can occur only if there is a transport of materials out of the cell interior or incorporation of amphiphilic

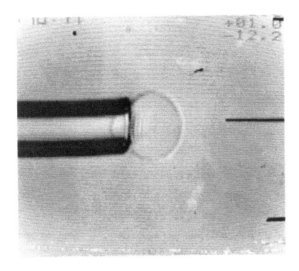

FIGURE 6. Video micrograph of a single-walled (phospholipid bilayer) vesicle aspirated with low suction pressure into a pipette to the extent permitted by the fixed area envelope and internal volume. Further pressure increase up to the point of rupture only produces very small displacement of the aspirated projection. The vesicle diameter is about 2×10^{-3} cm.[10]

molecules into the plasmalemma. It is important to note that large amounts of surface may be present in the wrinkled and ruffled conformation of the plasmalemma. For example, micropipette aspiration of blood phagocytes has shown that more than 110% excess surface area (over that of a sphere of equivalent volume) is in the form of surface ruffles.[2] When the ruffles are pulled smooth, the cell envelope becomes tight and rigid; further expansion leads to cell lysis.

The red blood cell is unique in that the surface contour of the plasmalemma is kept smooth by the subsurface protein meshwork. The red-cell membrane composite normally behaves as an elastic solid material which again has a very small area compressibility ($K \sim 10^2$ to 10^3 dyn/cm) but with a measurable nonzero extensional modulus ($\mu \sim 10^{-2}$ dyn/cm) and a small bending rigidity similar to lipid bilayers.[8,12-14] The extensional rigidity accounts for the static resistance to deformation (shown in Figure 7) as well as the elastic recovery of the red blood cell from large deformations (shown in Figure 8). It is important to note that the red-cell membrane extensional (shear) modulus is four orders of magnitude lower than the area compressibility modulus. Bending rigidity of the red cell is much smaller yet in its effect, but is sufficient to maintain the red-cell membrane contour in a smooth form until compressional stresses cause the surface to "buckle" or fold.[14] Extensional and bending rigidities are significant and important parameters in the analysis of red-cell adhesion to surfaces. Red-cell membranes also exhibit a wide range of dynamic and plastic transition properties which are important measures of material structure and viability.[8] However, it is usually sufficient to represent the red cell as a viscoelastic body with a solid membrane structure, characterized by a response time t_e equal to 0.1 sec and the elastic coefficients just described; the internal hemoglobin solution offers negligible dynamic resistance to deformation at normal cell concentrations ($\sim$32 g/dℓ).

F. Properties of a Plasmalemma-Cortical Gel Composite Shell with Liquid Core

For a plasmalemma-cortical gel composite, the plasmalemma is simply a ruffled envelope that bounds the gel on one side but offers no resistance to deformation until the ruffles are pulled smooth where further expansion is not possible without rupture. Hence, within the

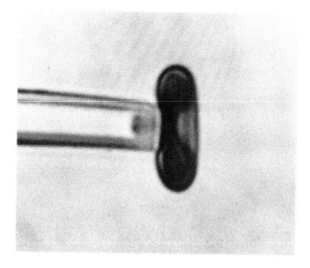

FIGURE 7. Video micrograph of a human red blood cell held in a pipette at fixed length by the suction pressure. The static length in the pipette increases linearly with the increase in suction pressure. The major dimension of the red cell is about 8×10^{-4} cm.[13]

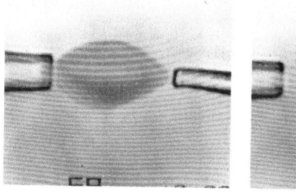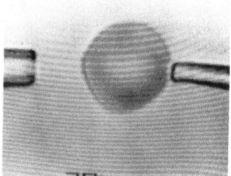

FIGURE 8. Video micrographs of elastic recovery of a red blood cell from end-to-end extension by pipettes. The characteristic time for the recovery process is on the order of 0.1 sec.[8]

limits of this constraint, the deformation of the cortical composite is determined by the properties of the subsurface gel layer. In general, there is a minimum of six parameters (three elastic properties K, μ, B, and three viscous properties κ, η, ν) for the gel which may be of comparable magnitude, unlike the properties of simple bilayer or red-cell membranes where there are orders of magnitude differences between parameters. The simplest case is to assume that the gel is isotropic (in the three local dimensions) like an amorphous, bulk material. Then, the following relations can be used to reduce the number of parameters to single elastic and viscous properties:

$$K \simeq 3 \cdot \mu; \quad B \simeq K \cdot h^2/12$$

$$\kappa \simeq 3 \cdot \eta; \quad \nu \simeq \kappa \cdot h^2/12 \qquad (26)$$

where h is the thickness of the gel. For most cells, there is also a net passive contractility in the cortex which is represented by a constant mean stress $\bar{\tau}_0$. An example of this behavior is shown in Figure 9 which presents a sequence of video micrographs of a blood phagocyte subjected to micropipette aspiration and then released.[15] Here, the continuous flow of the cell into the pipette was proportional to suction pressure in excess of a small threshold. When the pressure was lowered to the threshold value, the flow ceased and the aspirated portion of the cell inside the pipette remained stationary. The shape of the cell exterior to the pipette remained close to spherical throughout aspiration, and when the cell was released from the pipette, it slowly but eventually recovered its original spherical form. Even with the simplified approximation for gel properties, solution of the equations of mechanical equilibrium coupled with the constitutive relations requires complicated numerical computations.[3] Figure 10 shows predictions of cell shapes for micropipette aspiration of a spherical cell where the cortex was assumed to be an ideal gel but with semisolid (Maxwell) character:

$$\bar{\tau} = \bar{\tau}_0 + \kappa_1 \cdot V_\alpha + K_2 \cdot \int_0^t V_\alpha \cdot e^{-K_2 \cdot (t-t')/\kappa_2} \cdot dt', \quad \bar{\tau} = \bar{\tau}_0 + (\kappa_1 + \kappa_2) \cdot V_\alpha$$

$$\tau_s = 2 \cdot \eta_1 \cdot V_s + 2\mu_2 \cdot \int_0^t V_s \cdot e^{-\mu_2 \cdot (t-t')/\eta_2} \cdot dt', \quad \tau_s = 2(\eta_1 + \eta_2) \cdot V_s$$

The cell interior was modeled as a highly viscous liquid. As such, the continuous flow into the pipette closely models the behavior of blood granulocytes.[2] Correlation of the model of a viscous core surrounded by a plasma membrane-cortical gel composite with pipette aspiration measurements yields values on the order of 3×10^{-2} dyn/cm for the passive contractile stress and 10^3 to 10^4 dyn-sec/cm^2 (poise) for the effective viscosity of the cell contents. Although these properties depend strongly on temperature and activity, phagocytic cells are obviously highly viscous and possess low levels of cortical stress when inactive. The dynamic rigidity (resistance to rapid deformation) for these cells is many orders of magnitude greater than for membrane capsules like red blood cells.

III. MECHANICS OF CELL ADHESION

A. Deformation of Cells or Membrane Capsules Caused by Adhesion to Solid Substrates

Since the adhesive forces that cause a cell or capsule to spread on a surface are very short range compared with dimensions of the body, and since the bending rigidities of the outer shell are usually small, deformation of the cell or capsule is most prominent near the contact zone. Here, surface contours are highly curved along the direction normal to the edge of the contact, i.e., the bend is usually sharp at the edge of the contact zone. Consequently, mechanical equilibrium and deformation close to the edge of the contact zone are dominated by bending stiffness of the shell, whereas away from the contact, shell force resultants oppose the excess pressure of the interior contents with little contribution from bending rigidity. Hence, cell or capsule adhesion can be analyzed approximately by consideration of two separate problems: (1) deformation of the cell or capsule to form contact with the surface without resistance to bending, and (2) local deformation of the cell close to the contact zone modeled as pure bending. The local bending deformation forms a boundary layer or narrow zone peripheral to the contact; this zone is required to be continuous with the overall deformation of the cell or capsule. Hence, for axisymmetric shapes, the angle θ, which defines the orientation of the contour, and the shell force resultant τ_m, which acts along the contour, must be continuous with values for these variables derived from analysis of the narrow "bending zone". The first step is to analyze the deformation of the cell or capsule without resistance to bending. For lipid bilayer vesicles, the macroscopic contact

FIGURE 9. Video micrographs of the continuous flow of a
blood granulocyte into a pipette when the suction pressure exceeds
the small threshold established by the contractile stress in the cell
cortex. Also shown is the subsequent recovery to its original
spherical shape.[15]

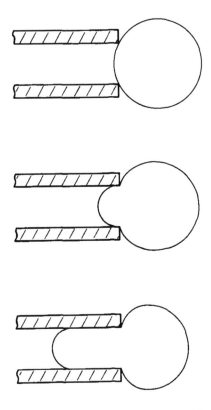

FIGURE 10. Sequence of shapes determined by numerical computation for the model of a cell as a viscous liquid encapsulated by cortical gel.

angle is fixed by the geometric constraints of constant surface area and volume of the vesicle; therefore, the force resultant or membrane tension is arbitrary and takes on whatever value necessary to balance adhesive stresses at the fixed macroscopic contact angle (Figure 11A).[1] Red blood cells, on the other hand, develop membrane tensions which depend on deformation. Thus, when a red-cell membrane is forced to conform to a spherical surface (Figure 11B), the tension close to the contact increases in proportion to the fractional coverage of the sphere ($x \equiv z_c/2R_s$):

$$\tau_m \simeq \tau_m^o + \frac{\mu}{2} \cdot \frac{x}{(1 - x)} \tag{27}$$

because of the elastic extension (shear) of the red-cell membrane.[1] This relation is valid provided the interior contents of the cell are not pressurized by the deformation. If the limit established by cell volume and surface area is reached, the membrane stress is augmented by an arbitrary membrane tension $\bar{\tau}_m^o$ as for the lipid bilayer vesicle in order to balance stresses.

As diagrammed in Figure 1 (and shown in Figure 2), more complex cells behave differently. For cells with highly viscous liquid interiors, adhesion will cause the cell body to deform very slowly because of the small cortical stresses. Thus, the plasmalemma and cortex may flow away from the cell core to create an annular ''skirt'' at the base of the cell. Here, the cell body can be approximated by an evolution of segments determined from the shell model with a liquid interior where the shell is forced to form flat contact with the substrate at various angles. The apparent contact angle of the shell (discontinuous at the intersection

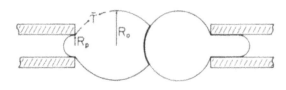

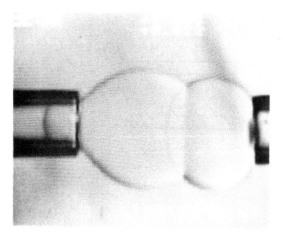

A

FIGURE 11.   (A) Schematic of vesicle geometry and bilayer tension when stable adhesive contact is formed with a rigid spherical substrate. (Below) A video micrograph of two phospholipid bilayer vesicles that have formed an adhesive contact by van der Waals' attraction in aqueous salt solution.[17] (B) Schematic of red-cell contour and membrane tension when adhesive contact is formed by encapsulation of a rigid spherical particle. The contour was derived from minimization of the elastic free energy for red-cell deformation.[8] (Below) A video micrograph of a red cell that was induced to encapsulate a spherical particle by high molecular weight dextran polymers.[19]

with the annular skirt) will depend on rate of deformation of the cell body. If the rate is small, then the cell shapes will be nearly spherical segments maintained by a passive mean shell tension; the pressure inside the cell will decrease. The outward flow of the membrane cortex from the cell is driven by adhesive interactions, but is opposed by drag at the substrate surface and viscous dissipation in the thin annular skirt surrounding the cell. If the membrane-substrate attachment is rigid, then resistance to outward flow of the membrane-gel layer above the surface will be dominated by dissipation in the gel between plasmalemma surfaces and presumably will be driven by membrane tension as the plasma membrane adheres to the substrate. Figure 12 illustrates the radial flow of the annular region at the base of the cell. Tension in the superficial membrane at the lead edge of the skirt will exceed the value of the tension at the base of the cell; a crude estimate of the tension difference is given by:

$$\Delta\tau_m \simeq \bar{\eta} \cdot v_r \cdot \Delta r/h \qquad (28)$$

where v_r is the radial velocity of the upper surface and Δr is the width of the annular region. The coefficient $\bar{\eta}/h$ represents a drag or frictional coefficient for the upper layer sliding over the attached lower layer. The important feature is that the tension at the lead edge must

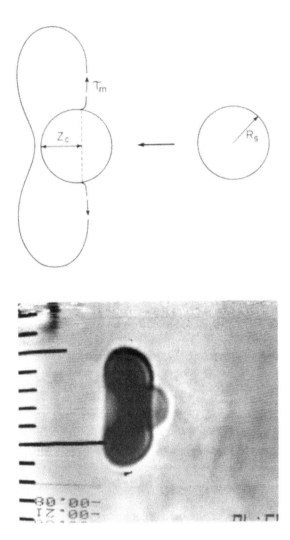

FIGURE 11B.

exceed the tension at the base of the cell; the difference is caused by viscous dissipation in the gel layer. Thus, adhesion may cause little increase in the cortical stress in the region that surrounds the cell body until the final spherical segment limit is reached. As mentioned in the introduction, there must be cohesive interactions between lamelli of the skirt in order to resist separation by the normal component of the shell stress at the intersection between the skirt and base of the cell. The cohesive energy per unit area of the interface between lamellas of the skirt must obey the following inequality:

$$\bar{\epsilon} > \tau_m \cdot (1 + \cos\theta) \tag{29}$$

This example shows that observation of cell shape during adhesion and estimates of cortical tension when the cell is nonadherent may be insufficient for derivation of adhesion properties. More detailed analysis is required. Geometric shapes can be misleading in other ways as well. For example, spreading and adhesion may be dominated by the mechanics of deformation (smoothing) of the submicroscopic surface ruffles. The stiffness of the ruffles is primarily derived from membrane bending rigidity and the high local curvature (analogous to stiffening of thin materials by corrugation); however, there can also be filamentous

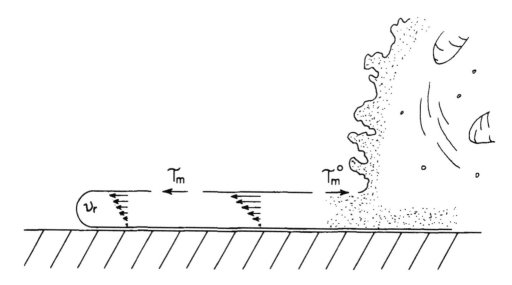

FIGURE 12.   Schematic of the outward flow of an annular skirt (formed by a layer of cortical gel between plasma membrane surfaces) from the base of the cell caused by adhesion of the plasma membrane to the solid substrate.

structure inside the ruffle that resists deformation. Ultrastructural data are essential considerations here. Another example (where cell shape can lead to a misleading deduction of mechanical structure) comes from the observation of cells that have completely spread out on a substrate. In this case, the surface contour approaches that illustrated in the last frame of Figure 1. However, because of an irregular pattern of adhesion at the cell periphery, the cell will not become a smooth, spherical segment. Instead, the surface will buckle to form radial pleats or folds that give the impression of "stress fibers" which act to anchor the cell body to the substrate. Identification of stress fibers from geometric form can be erroneous; additional evidence is required before such conclusions can be accepted.

The first step in the analysis of adhesion of cells or capsules to surfaces has been outlined. The key results are the values of the macroscopic contact angle and cortical stress at the edge of the contact zone, as well as the excess pressure inside the cell which will act over the contact region. (Note: the excess pressure inside the cell is usually negligible over the contact region in comparison to the strong, short-range forces involved in surface-surface adhesion.) The next step is to analyze the mechanics of bending the shell to form the sharply curved contour between the cell body and adhesion zone. Definition of the range and distribution of forces that cause adhesion are required in this analysis.

B. Bending Deformation of a Cell Cortex or Capsule Membrane at the Edge of an Adhesion Zone

In the highly curved region adjacent to the adhesion zone, displacements of the shell from the substrate (to which it adheres) approach the scale of the short-range forces that cause adhesion. Except for a few types (e.g., electrostatic double-layer repulsion, van der Waals' attraction, and hydration repulsion),[16] the functional form and physical origin of the forces involved in adhesion are not well established. Based on knowledge of simple interactions and intuition, attraction is expected to increase as the surfaces are brought together from large separation to a maximum value established by shorter-range repulsion. Then, attraction will diminish with closer approach and eventually be overwhelmed by rapidly increasing repulsive forces. This is illustrated schematically in Figure 13. When the surfaces are separated by distances greater than the position of maximum attraction, a small increase in force will cause separation to infinity. On the other hand, when the surfaces are closer than

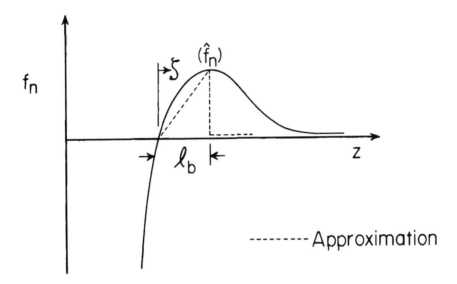

FIGURE 13. Conceptual illustration of adhesion force vs. separation for specific intermolecular bonds.

this distance, a small increase in force will only slightly expand the separation gap but will not disrupt the contact. Hence, the position of maximum attraction can be used to differentiate regions into two categories: free (nonadherent) and adherent. A conservative approximation to this concept is also shown in Figure 13, where the "adherent" range is approximated by linear proportionality between force and displacement, and the "free" range is established by a step discontinuity from maximum attraction to zero force. The significant parameters become immediately apparent: the maximum attractive force $\hat{f}_n$ and the range of the attraction l_b. Although crude, this approximation makes it possible to analyze the mechanics of adhesion at the edge of the contact zone, simply and with sufficient richness, to relate major features of the submicroscopic short-range interactions to macroscopic measurements.

Another important property of adhesive interactions is the distribution of short-range forces over the surface. Forces due to specific molecular cross bridges between surface receptor sites will be discrete on a time scale less than the off rate for the binding reaction. Localization vs. dispersion of these forces will depend on mobility of the surface receptors. Clearly, limiting representations for surface-surface forces are given by (1) a continuous (infinitely dense) distribution, and (2) a discrete (sparse) distribution of fixed sites of action. The former characterizes a kinetically free (full equilibrium) interaction, whereas the latter characterizes a kinetically trapped (nonequilibrium) interaction. Figure 14 illustrates the conceptual views embodied in the previous discussion, i.e., short-range attraction leads to definition of free/adherent regions of the shell contour, and sources of adhesion may be either continuous or localized on the surface. Also shown in Figure 14 are the force and moment resultants supported by the shell in response to adhesion.

1. Analysis of the Free Zone Proximal to the Adhesive Contact

The cortical shell curvature proximal to the contact will usually be quite large for an arc normal to the edge of the contact zone, whereas the curvature of an arc tangent to (or concentric with) the contact zone will be very small. As illustrated in Figure 14, analysis of bending depends only on the curvilinear coordinates (s,θ) of the shell in the meridional plane normal to the edge of the contact and the intensive forces supported by the shell: a principal tension τ_m and a transverse shear Q_m.[6] The transverse shear is only appreciable in the region adjacent to and just inside the adherent zone. In the macroscopic region away

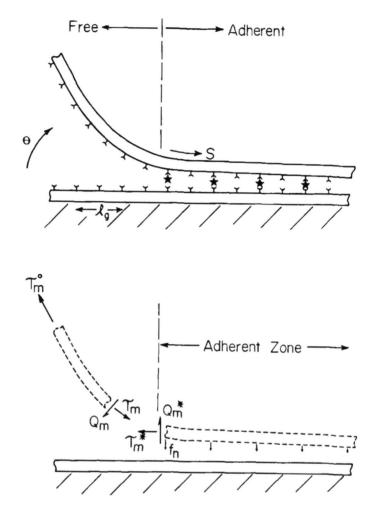

FIGURE 14.   Illustration of the adherent and free (nonadherent) zones proximal to the edge of the contact zone. The attractive forces between surfaces (shown schematically) represent imaginary cross bridges. In addition to the attractive forces f_n, the shell supports a principal tension τ_m that acts tangent to the plane of the shell and a transverse shear Q_m that acts normal to the shell surface.

from the adherent zone, it is assumed that the shell is a plane structure under uniform tension τ_m^o; here, the shell forms a macroscopic (observable) contact angle of θ_o. These variables are derived from the analysis of deformation of the cell body as discussed previously.

The local mechanical equilibrium of the shell in the free (nonadherent) region is given by Equations 14 and 15 with a change in sense (sign) of the coordinate s, i.e., the balance of forces tangent to the surface:

$$\frac{d\tau_m}{ds} - Q_m \cdot k_m = 0 \qquad (30)$$

and the balance of forces normal to the surface:

$$\tau_m \cdot k_m + \frac{dQ_m}{ds} = 0 \qquad (31)$$

where k_m is the local curvature of the meridional arc. In order to solve these two equations for the unknowns (τ_m, Q_m, k_m), the elastic, constitutive behavior for the shell must be introduced. In this region of high curvature, the bending or curvature elasticity determines the shell shape. The elastic constitutive relation for bending yields the relation that the transverse shear is proportional to the gradient of the curvature:

$$Q_m = -B \cdot \frac{dk_m}{ds} \tag{32}$$

A solution to these equations is easily obtained as a function of curvilinear angle θ and the macroscopic tension applied to the shell at the intersection with the cell body:

$$\tau_m = \tau_m^o \cdot \cos(\theta_0 - \theta)$$

$$Q_m = \tau_m^o \cdot \sin(\theta_0 - \theta)$$

$$(k_m)^2 = \left(\frac{2 \cdot \tau_m^o}{B}\right) \cdot [1 - \cos(\theta_0 - \theta)] + (k_m^o)^2 \tag{33}$$

It is apparent that the transverse shear and shell curvature increase to maximums which depend on the microscopic contact angle θ^* at the edge of the contact zone. The tension, transverse shear, curvature, and microscopic contact angle must be continuous with values for these variables in the adherent zone.

2. Analysis of the Contact Zone for Continuously Distributed Adhesion Forces

In the adhesion zone, it is assumed that the forces of adhesion act normal to the shell surface. For adhesive interactions which are infinitely dense, the force per unit area of the shell surface is a normal stress σ_n that must be included in the equations of mechanical equilibrium. Hence, the local balance of normal forces becomes:

$$\tau_m \cdot k_m + \frac{dQ_m}{ds} = -\sigma_n \tag{34}$$

The tangential-force balance remains the same (Equation 30). At this point, a constitutive relation must be introduced for the normal stress in terms of displacement of the shell from equilibrium (planar) contact with the substrate.

In the limit that the adhesion forces can be considered as a continuous distribution, the normal stress is modeled by the product of the surface density of cross bridges $\bar{n}$ times the force-displacement relation (Figure 13). Displacement of the shell from planar contact is represented by the variable ζ; the attractive stress is approximated by a linear relation that depends on the ultimate strength of the attachment $\hat{\sigma}_n$ and the range of attraction l_b. For this model, the work per unit area for formation or separation of an adhesive contact is given by:

$$\gamma = \hat{\sigma}_n \cdot l_b/2 = (\bar{\eta} \cdot \hat{f}_n \cdot l_b)/2 \tag{35}$$

where γ is the adhesion energy. In terms of the adhesion energy, the normal stress relation can be written as:

$$\sigma_n = (2\gamma/l_b^2) \cdot \zeta = (\bar{\eta} \cdot \hat{f}_n/l_b) \cdot \zeta \tag{36}$$

For contour angles (measured relative to planar contact) that are less than about 30°, the angle and curvature are well approximated by the first and second spatial derivatives of the displacement from the equilibrium plane. With these derivatives plus the constitutive relation for bending, the local balance of normal forces (Equation 34) becomes:

$$-\tau_m \cdot \frac{d^2\zeta}{ds^2} + B \cdot \frac{d^4\zeta}{ds^4} = -\left(\frac{2\gamma}{l_b^2}\right) \cdot \zeta \tag{37}$$

which in general must be solved simultaneously with the balance of tangential forces. The tension term makes Equation 37 nonlinear but is a higher-order (smaller) term in comparison to the bending-stress term. Consequently, the variation in tension can be neglected and the tension can be assumed constant just inside the contact zone as a first-order approximation; this will allow Equation 37 to be solved analytically. The solution predicts the shell displacement from planar contact in the adherent zone:

$$\zeta = e^{-a_1 \cdot s} \cdot [C_1 \cdot \sin(a_1 \cdot s) + C_2 \cdot \cos(a_2 \cdot s)] \tag{38}$$

where the spatial variations are governed by the parameters a_1 and a_2, and:

$$a_1 \equiv (b + \theta_\alpha^2)^{0.5}/l_b; \quad a_2 \equiv (-b + \theta_\alpha^2)^{0.5}/l_b$$

$$\theta_\alpha \equiv (\gamma \cdot l_b^2/2 \cdot B)^{0.25}; \quad b \equiv (\tau_m \cdot l_b^2/4 \cdot B)$$

The displacement is maximum at the edge of the contact zone and decreases exponentially with distance into the contact. There is only a small distance into the adhesion zone or boundary layer where the forces of attraction are appreciable. If adhesive molecules are involved, bonds will be stretched in this region near the edge of the contact zone where the width is given by:

$$\delta \sim l_b/\theta_\alpha = (2 \cdot B \cdot l_b^2/\gamma)^{0.25} \tag{39}$$

It is apparent that the width δ will be large when the shell is "stiff" (i.e., large, bending modulus) and will be small when the adhesion energy is strong (i.e., strong attractive forces between the surfaces).

From the contour given by Equation 38, the microscopic contact angle, curvature, and transverse shear at the edge of the contact zone are specified by the first, second, and third spatial derivatives, respectively:

$$\theta^* = a_1 \cdot C_1 - a_2 \cdot C_2; \quad k_m^* = (a_2^2 - a_1^2) \cdot C_2 + 2a_1 \cdot a_2 \cdot C_1$$

$$Q_m^*/B = [(a_2^2 - a_1^2) \cdot a_1 + 2a_1 \cdot a_2^2] \cdot C_2 + [(a_1^2 - a_2^2) \cdot a_2 + 2a_1^2 \cdot a_2] \cdot C_1 \tag{40}$$

The displacement at the edge of the contact zone is equal to the position of maximum attraction l_b; thus, there are only two unknown variables, θ^* and C_1. These are determined by the requirements of continuity with the solution previously derived for the free (nonadherent) zone.

The continuity at the edge of the contact zone of the solutions derived for the free (nonadherent) and adherent regions can be satisfied only for specific values of the macroscopic tension applied to the shell τ_m° and the microscopic contact angle θ^*. Specific values for these variables are obtained as a function of the dimensionless parameter θ_α which represents the ratio of adhesion to bending energies. In the absence of bending stiffness and for attractive forces with zero range, the classical Young equation predicts an ideal relation for macroscopic tension in terms of the macroscopic contact angle θ_0:

$$\tau_m^0/\gamma = (1 - \cos\theta_0) \tag{41}$$

The macroscopic tension determined by the continuity requirements at the edge of the contact zone show that the Young equation is also valid for finite range adhesion forces and nonzero microscopic contact angles. The results for the microscopic contact angle as a function of the ratio of adhesion to bending energies show that the microscopic contact angle is appreciable when the strength of adhesion is large or the range of attraction is large; however, when the shell is "stiff", the microscopic contact angle approaches zero. For shallow macroscopic contact angles (e.g., $\leq 30°$), the microscopic contact angle approaches the macroscopic value as the ratio of adhesion to bending energies becomes large. Also, when the microscopic contact angle becomes large, the width of the boundary layer δ is small; thus, there will be few "bonds" stretched at the edge of the contact zone if adhesive molecules are responsible for the contact.[6]

3. Analysis of the Contact Zone for Discrete, Sparsely Distributed Adhesion Forces

In the previous section, the mechanics of shell-substrate adhesion were examined for the case where the adhesive forces between the surfaces were distributed continuously over the surfaces, but with a finite range of attraction. For the continuum model of attraction, the tension necessary to oppose spreading of the contact is equal to the minimum level of tension required to separate adherent surfaces. Also, the macroscopic tension applied to the shell in the free region away from the contact is given by the Young equation that relates the free-energy reduction per unit area for formation of planar contact to the shell tension and macroscopic contact angle. The continuum model has been shown to be a valid representation for adhesion and separation of synthetic (phospholipid) bilayer membrane vesicles which were allowed to aggregate either via van der Waals' attraction[16] or in high molecular weight glucose polymer (dextran) solutions.[20] It also appears to be an appropriate model for red-cell rouleaux formation.[18,19]

In contrast with this classical model, experiments often show that there is little or no tendency for adhesive contact to spread, even though the tension required to subsequently separate the contact is very large.[20] The deviation of the level of tension induced in the shell by contact formation from that required to separate the contact appears to be due to the sparse distribution of strong molecular cross bridges. Thus, the purpose of this section is to consider the mechanics of adhesion and separation for discrete adhesion forces.[7] The approach is to determine the shell contour that minimizes the total free energy (elastic energy of deformation plus adhesion energies) in the contact zone. Two disparate values of the macroscopic tension are found: (1) the level of tension that stresses the first cross bridge at the edge of the contact zone to near the breaking point, and (2) the reduced level of tension that permits the shell to approach the substrate sufficiently for the next cross bridge to form adjacent to the contact zone. Deviation between these two tensions can be very large and is shown to depend strongly on the surface density of cross bridges.

As in the previous development, the analysis here considers the cross-bridging (adhesion) forces as finite range interactions and the shell as an elastic material. In the adherent zone, the total free-energy functional includes bending elastic plus cross-bridge energies. The functional is minimized in conjunction with the work required to displace forces at the boundaries; the variation is taken with respect to the parameters that characterize the shell contour in the contact zone. The same linear approximation to the intermolecular force $\hat{f}_n$ is employed here as used previously, although more general models could easily be treated. The contribution to the total free energy from extension of the cross bridges is given by the approximation:

$$F_{CB} = \frac{(\hat{f}_n \cdot l_b)}{2} \cdot \sum_{i=0,1}^{N} (\zeta_i)^2 \tag{42}$$

where N is the number of cross bridges involved in the adhesive contact and the product F_b = $\hat{f}_n \cdot l_b/2$ is the free-energy change (work) which results from formation ($-$) or breakage ($+$) of a single cross bridge. The initial value for the sum of cross-bridge energies (i = 0 or 1) is determined by one of two situations: i = 0 when a cross bridge is maximally stretched at the edge of the contact zone, i.e., separation of the adherent contact, or i = 1 when a cross bridge is about to be formed at the edge of the contact (given by the lattice position i = 0), i.e., minimal spreading of the contact.

The free-energy functional for elastic deformation of the shell is given by:

$$F_D = \left(\frac{B \cdot l_g}{2}\right) \cdot \int_0^{(N \cdot l_g)} (k_m - k_m^o)^2 \cdot ds + l_g \cdot \int_0^{(N \cdot l_g)} (\tau_m \cdot e) ds \tag{43}$$

where the first term is the bending or curvature elastic energy, and the second term is the work for in-plane extension of the shell. The principal curvature of the contour is k_m and e is the measure of strain along the contour (i.e., the fractional extension of shell elements). The free energy equations for shell deformation and cross-bridge extension (Equations 42 and 43) represent the total energy of a strip of shell which has a width of l_g. The parameter l_g is the average distance between cross-bridge sites. In the region of high curvature which characterizes the shell contour close to and within the adherent zone, bending energy determines the shape of the shell so the in-plane (extensional) elasticity can be neglected. As such, the shell tension acts as a Lagrange multiplier in Equation 43 with the auxiliary requirement that the local strain approaches zero. The result is that the density of cross-bridge sites remains uniform and constant.

Mechanical equilibrium is defined by the expression that the variation in total free energy is equal to the variation in work required to displace forces at the boundaries. The virtual work required to displace forces at the boundaries reduces to the virtual displacement of the macroscopic tension τ_m^o supported by the shell at the intersection with the cell body away from the contact zone:

$$\delta W = \tau_m^o \cdot \delta s_0 \tag{44}$$

where δs is the virtual displacement of the membrane in the direction tangent to the surface. Equation 44 presumes that the shell contour in the free (nonadherent) zone is prescribed and that the shell surface is inextensible.

The contour is defined by spatial displacements (z, r) of the shell from planar contact (z normal to the plane of the adhesive contact, and r parallel to the plane of the contact). Solutions to the variational expression for mechanical equilibrium are restricted to the class of contours that are locally inextensible:

$$\frac{dr}{ds} \cdot \left[1 + \frac{1}{2}\left(\frac{dr}{ds}\right)\right] = -\frac{1}{2}\left(\frac{dz}{ds}\right)^2 \tag{45}$$

Hence, lateral displacements are determined by the normal displacements, and a single function z(s) completely specifies the contour. For this class of contours, the variational statement of equilibrium reduces to:

$$0 = \frac{(\hat{f}_n \cdot l_b)}{B} \cdot \sum_{i=0,1}^{N} (z_i \cdot \delta z_i) + l_g \cdot \int_0^{(N \cdot l_g)} (k_m - k_m^o) \cdot \delta k_m \cdot ds - \tau_m^o \cdot \delta s_0 \tag{46}$$

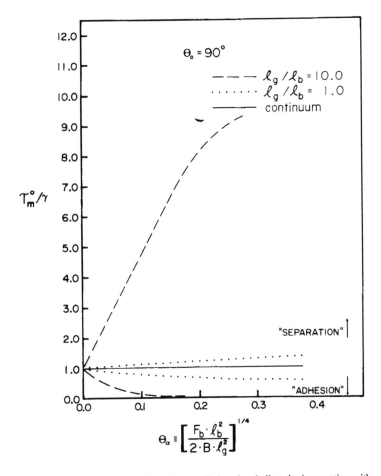

FIGURE 15. The macroscopic tension applied to the shell at the intersection with the cell away from the contact zone, normalized by the adhesion energy per unit area, is plotted vs. the dimensionless parameter that represents the ratio of adhesion to bending (deformation) energies. These results are for the specific case where the macroscopic contact angle is 90° and for two ratios l_g/l_b of cross-bridge spacing to bond length (1:1 and 10:1). For each cross-bridge density (l_g/l_b), two levels of tension are predicted: (1) an upper value which represents the minimum level of tension required to separate the adhesive contact, and (2) a lower value which characterizes the maximum level of tension that will just allow the contact to spread. Even for moderately dense cross bridges ($l_g/l_b = 1$), the tension for separation exceeds the level of tension which will allow the contact to spread. The solid line is the level of tension predicted from the classical Young equation where there is no deviation between the minimum tension required to separate adhesive contact and the tension that will allow the contact to spread.[6] (From Evans, E. A., *Biophys. J.*, 48, 185, 1985. With permission.)

The method of solution utilizes a Newton-Raphson iterative computation of Equation 46 where the displacement function z(s) is a piece-wise continuous function with continuous first and second derivatives.[7]

Continuity requirements at the intersection with the free zone yield two specific values of the macroscopic tension for each macroscopic contact angle: an upper value which represents the minimum level of tension required to separate the adhesive contact, and a lower value which characterizes the maximum level of tension that will just allow the contact to spread. Figure 15 shows the two levels of tension determined for a macroscopic contact angle of 90° plus two ratios l_g/l_b of cross-bridge spacing to bond length (1:1 and 10:1). The tension values depend on the dimensionless parameter θ_α:

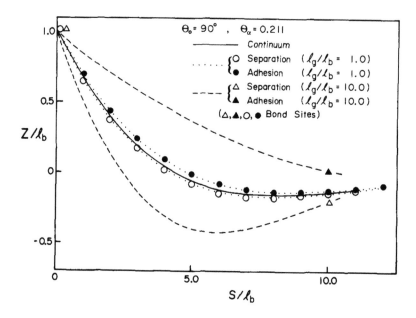

FIGURE 16.   *Shell contours are shown for two ratios* l_g/l_b *of cross-bridge spacing to bond length (1:1 and 10:1), a specific value of the parameter for adhesion to bending energy ratio (0.211), and a macroscopic contact angle of 90°. Also shown is the contour (solid line) derived previously from the continuum model where attractive forces were assumed to be infinitely dense.[6] For relatively dense populations of cross bridges, the contours deviate only slightly from that predicted by the continuum model. However, for low densities (i.e.,* l_g/l_b *= 10), there is significant deviation of the contours from the continuum solution. Note: no bonds are formed at the first receptor site (i.e., no solid circle and no star) for the situation where the contact is about to spread to the next site. The displacement axis is greatly enlarged in comparison to the axis for the curvilinear distance along the membrane. (From Evans, E. A., Biophys. J., 48, 185, 1985. With permission.)*

$$\theta_a \equiv [F_b \cdot l_b^2/(2 \cdot B \cdot l_g^2)]^{0.25} \tag{47}$$

which represents the ratio of adhesion (based on the energy F_b of a single cross bridge) to bending (deformation) energies. As expected, the two levels of tension approach the ideal value from the Young equation when the density of cross-bridge sites becomes large.

There is a specific shell contour and microscopic contact angle associated with each equilibrium level of tension. Figure 16 shows the contours at the entrance to the adherent zone determined for two ratios l_g/l_b of cross-bridge spacing to bond length (1:1 and 10:1) and a specific value of the ratio of adhesion to bending energy. Also shown is the contour derived previously from the continuum model where cross bridges were assumed to be infinitely dense. Figure 17 shows the progressive increase in microscopic contact angle at the edge of the contact zone as the ratio of adhesion to bending energies is increased. It is apparent that the contours (one for separation and the other for spreading of the contact) and microscopic contact angles deviate only slightly from the continuum solution when the cross bridges are relatively dense (i.e., l_g/l_b = 1) whereas the contours and angles deviate appreciably from the continuum solution when the cross bridge density is low (i.e., l_g/l_b = 10). It is also apparent that when the cross-bridge density is low, only the first cross bridge is significantly stressed. Also, when the sites are far apart, the tension must be reduced to nearly zero in order to permit the next cross bridge to form; thus, there is little or no tendency for the contact to spread unless the surfaces are forced together. Since nearly all of the stress is taken by the first cross-bridge site at the edge of the contact zone when the density is low, the tension required to separate the contact greatly exceeds the value anticipated from

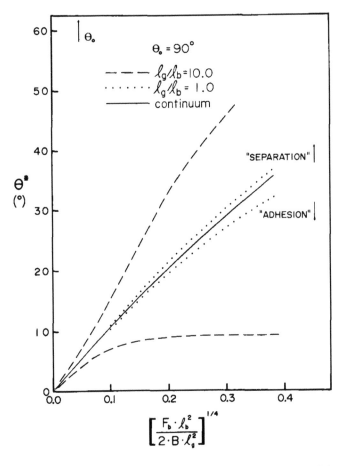

FIGURE 17. Values for the microscopic contact angle at the edge of the contact zone are plotted as a function of the parameter which represents the ratio of adhesion to bending energies for a specific value of the macroscopic contact angle of 90°. The solid line is the microscopic contact angle derived from the continuum solution where attractive forces are assumed to be infinitely dense.[6] There are two broken curves plotted for each ratio l_g/l_b of cross-bridge spacing to bond length (1:1 and 10:1). The upper broken curves represent the situation where contact is separated and the lower curves are appropriate to the situation where contact is about to spread. (From Evans, E. A., *Biophys. J.*, 48, 185, 1985. With permission.)

the Young equation. Further, when the cross-bridge density is low, the level of tension required to separate the contact approaches the value predicted by the Young equation multiplied by the ratio of the cross-bridge spacing to bond length.

IV. COMMENT

In conclusion, it is apparent that mechanical stresses produced in cells by contact formation or separation depend on the distribution of the short-range interactions between the surfaces as well as the strength and range of the attraction. Because of the paucity of receptors in biological membrane surfaces, mechanical properties of the substrates will always interfere with initial recognition and adhesion processes. Once contact is established, subsequent separation may be very difficult and even require forces sufficient to fragment the cell before separation can be effected. Obviously, cells can increase the contact area by active spreading, by passive relaxation of mechanical stiffness, or by accumulation of membrane receptors proximal to the contact zone. Application of mechanical models for adhesion and separation of cell-cell contacts to living growing cells must be undertaken with caution since the stresses that oppose contact depend on time and cellular chemistry. However, these model calculations are extremely useful for the analysis of controlled adhesion experiments which involve simple membrane capsules. Hence, information about molecular forces and energies can be obtained directly. Examples of such experiments will be given in a later chapter.

ACKNOWLEDGMENTS

This research was supported in parts by the U.S. National Institutes of Health, grant HL26965, and the Medical Research Council of Canada, grant MT7477.

REFERENCES

1. **Evans, E. A. and Parsegian, V. A.,** Energetics of membrane deformation and adhesion in cell and vesicle aggregation, *Ann. N.Y. Acad. Sci.,* 13, 416, 1983.
2. **Evans, E., Kukan, B., and Yeung, A.,** Mechanics of a plasma membrane-cortical gel composite: model for passive deformability of blood phagocytes, *Biophys. J.,* submitted.
3. **Evans, E. and Yeung, A.,** Limiting models for passive flow of liquid-like spherical cells into micropipets: (1) Uniform viscous droplet with constant interfacial tension; (2) plasma membrane-cortical gel composite surrounding an inviscid fluid interior, *Biophys. J.,* submitted.
4. **Southwick, F. S. and Stossel, T. P.,** Contractile protein in leukocyte function, *Semin. Hematol.,* 20, 305, 1983.
5. **Valerius, N. H., Stendahl, O., Hardwig, J. H., and Stossel, T. P.,** Distribution of actin-binding protein and myosin in polymorphonuclear leukocytes during locomotion and phagocytosis, *Cell,* 24, 195, 1981.
6. **Evans, E. A.,** Detailed mechanics of membrane-membrane adhesion and separation. I. Continuum of molecular cross-bridges, *Biophys. J.,* 48, 175, 1985.
7. **Evans, E. A.,** Detailed mechanics of membrane-membrane adhesion and separation. II. Discrete, kinetically trapped molecular cross-bridges, *Biophys. J.,* 48, 185, 1985.
8. **Evans, E. A. and Skalak, R.,** *Mechanics and Thermodynamics of Biomembranes,* CRC Press, Boca Raton, Fla., 1980.
9. **Flugge, W.,** *Stresses in Shells,* Springer-Verlag, New York, 1966.
10. **Evans, E. and Needham, D.,** Giant vesicle bilayers composed of mixtures of lipids, cholesterol and polypeptides: thermomechanical and (mutual) adherence properties, *Faraday Discuss. Chem. Soc.,* 81, 267, 1986.
11. **Evans, E. and Needham, D.,** Physical properties of lipid bilayer membranes: cohesion, elasticity, and (colloidal) interactions, *J. Phys. Chem.,* 91, 4219, 1987.
12. **Evans, E. and Waugh, R.,** Osmotic correction to elastic area compressibility modulus of red cell membrane, *Biophys. J.,* 20, 307, 1977.
13. **Waugh, R. and Evans, E. A.,** Thermoelasticity of red blood cell membrane, *Biophys. J.,* 26, 115, 1979.
14. **Evans, E. A.,** Bending elastic modulus of red blood cell membrane derived from buckling instability in micropipet aspiration test, *Biophys. J.,* 27, 43, 1983.
15. **Evans, E. and Kukan, B.,** Large deformation recovery after deformation, and activation of granulocytes, *Blood,* 64, 1028, 1983.
16. **Verweg, E. J. W. and Overbeek, J. Th. G.,** *Theory of the Stability of Lymphobic Colloids,* Elsevier/North-Holland, Amsterdam, 1948; **Parsegian, V. A., Fuller, N., and Rand, R. P.,** Measured work of deformation and repulsion of lecithin bilayers, *Proc. Natl. Acad. Sci. U.S.A.,* 76, 2750, 1979.
17. **Evans, E. A. and Metcalfe, M.,** Free energy potential for aggregation of large, neutral lipid bilayer vesicles by van der Waals' attraction, *Biophys. J.,* 46, 423, 1984.
18. **Evans, E. A. and Metcalfe, M.,** Free energy potential for aggregation of mixed PC:PS lipid vesicles in glucose polymer (dextran) solutions, *Biophys. J.,* 45, 715, 1984.
19. **Buxbaum, K., Evans, E. A., and Brooks, D. E.,** Quantitation of surface affinities of red blood cells in dextran solutions and plasma, *Biochemistry,* 21, 3235, 1982.
20. **Skalak, R., Zarda, P. R., Jan, K.-M., and Chien, S.,** Mechanics of rouleau formation, *Biophys. J.,* 35, 771, 1981.

21. **Evans, E. A. and Leung, A.,** Adhesivity and rigidity of red blood cell membrane in relation to WGA binding, *J. Cell Biol.,* 98, 1301, 1984.
22. **Landau, L. D. and Lifshitz, E. M.,** *Fluid Mechanics,* Pergamon Press, Oxford, 1959.
23. **Happel, J. and Brenner, H.,** *Low Reynolds Number Hydrodynamics,* Noordhoff International, Leyden, Netherlands, 1973.

Section II
Selected Experimental Methods and Results

Chapter 5

USE OF HYDRODYNAMIC FLOWS TO STUDY CELL ADHESION

Pierre Bongrand, Christian Capo, Jean-Louis Mege, and Anne-Marie Benoliel

TABLE OF CONTENTS

I. INTRODUCTION

In the first section of this book, intermolecular forces and cell-surface properties likely to play a role in adhesion were described. However, more specific data are required before we try to build quantitative models for cell adhesion and subject them to experimental testing.

The present section is devoted to the description of several complementary experimental approaches allowing a quantitative study of cell adhesion. In this chapter and the following one by Segal, general methods of quantitating cell binding are presented. Also, selected experimental results are described in order to provide an intuitive feeling for the attachment process. Further data included in the chapters by Evans and Foa should facilitate a mechanistic interpretation of the aforementioned results.

First, we shall analyze a thought experiment in order to examine the definition and significance of cell adhesion. The following steps are considered:

1.   Two cells are pushed against each other with force F_1 during time t_1; F_1 is usually a result of well-known components (e.g., a centrifugal force) and more complex interactions (e.g., hydrodynamic repulsive forces resulting from the fluid egress from the intercellular space, or "thermal" forces generated by random collisions between the cell and surrounding molecules).
2.   The newly formed cell doublet is left at rest for time t_2.
3.   The cell doublet is subjected to a disruptive force F_3 for time t_3. A suitable assay is then performed to determine whether cells remained in contact or not after experiencing the disruptive force. Discriminating between bound and separated cells is usually an easy task.

If the above experiment is repeated, it is possible to calculate the fraction of encounters resulting in cell association after step 3. A logical name for this fraction may be the *collision efficiency*[1] or *capture efficiency*[2] (which may be related to the *stability ratio*[3], discussed further on). Several points must be emphasized:

First, the occurrence of time-dependent intercellular repulsive forces makes important the order of magnitude of the collision duration. Indeed, cell-surface molecules may undergo conformational changes to decrease a strong initial repulsion (as a suitable order of magnitude, the formation of a helix out of a coil state requires about 10^{-5} to 10^{-7} sec for a polypeptide[4]). Also, adhesion may require that "repulsive" molecules diffuse out of the interaction area. The characteristic time for this process is about r_o^2/D, where r_o is the linear size of the interaction area and D is the diffusion constant of considered molecules. Since r_o is about 0.01 to 0.1 μm (see Chapter 8 electron microscopic data) and a reasonable value for D is between 10^{-12} and 10^{-10} cm²/sec,[5] the redistribution time for membrane molecules may range between 10^{-2} and 100 sec. A similar time may be required for ligand molecules to

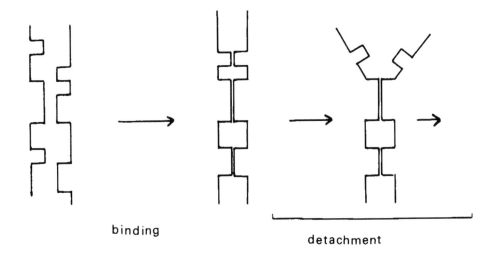

binding

detachment

FIGURE 1. Cell adhesion and detachment are time-dependent processes.

concentrate in the interaction area, as was first discussed by Bell.[6] Finally, the establishment of an extensive intercellular contact area may require bulk membrane deformations. Modeling human leukocytes as viscoelastic bodies, Schmid-Schönbein and colleagues[7] obtained values of 100 to 500 dyn/cm² for elastic constants and 50 to 100 dyn.sec/cm² for viscous constants. Hence, passive cell deformations are expected to be achieved much more easily when a strain is exerted for more than about 0.2 to 0.5 sec (see also Chapter 4). This time is also a suitable order of magnitude for active cell deformations such as are involved in phagocytosis.[8]

Second, cell deformations and molecular redistribution may also play a role in cell-cell separation. Indeed, rapid detachment requires that many bonds be broken at the same time, whereas a slower process may be achieved with a smaller separation force if bonds are broken sequentially[9] (see Figure 1 and Chapter 7).

Another point is about the significance of the bond strength. As pointed out by Weiss,[10] some mixing of cell-surface molecules may occur during adhesion, and separation may not restore the initial configuration. Also, cell-cell separation may involve some tearing of plasma membranes instead of breaking only intercellular bonds. Further information on this possibility is detailed in Section IV.

II. BASIC RESULTS ON CELL MOVEMENTS AND COLLISIONS IN FLUID FLOW

Before reviewing current models of cell aggregation, we shall summarize basic data on diffusion and Brownian motion, since this process is always superimposed on cell movements generated by fluid flows. However, Brownian motion is not thought to play an important role when large particles (say, several microns in diameter) are considered.

A. Brownian Motion in Resting Medium
1. Diffusion Constant
Let us consider a suspension of identical cells. Define $c(x,y,z,t)$ as the average cell concentration around a point M of Cartesian coordinates (x,y,z) at time t. If c is nonuniform, random thermal movements result in the generation of a net cell flow $\vec{J}$ that is expected to be parallel to the concentration gradient, namely:

$$\vec{J} = -D \overrightarrow{\text{grad}} c \qquad (1)$$

This defines the diffusion constant D. Further, Einstein[11] provided a simple and informative relationship between D and the hydrodynamic properties of the cell flow through the surrounding medium. When a cell moves with low velocity $\vec{v}$ through a viscous medium, a resistance force $\vec{R}$ proportional to $\vec{v}$ is generated:

$$\vec{R} = -f\,\vec{v} \tag{2}$$

The frictional constant is f. According to Stokes' law,[12] in an incompressible fluid of viscosity μ, the frictional constant of a sphere of radius a is

$$f = 6\,\pi\,\mu\,a \tag{3}$$

Now, Einstein's argument may be sketched as follows: if diffusive particles are immersed in a potential field V, the equilibrium concentration follows Boltzmann's law:

$$c(x, y, z) = c_o \exp(-V/kT)$$

where T is the absolute temperature, k is Boltzmann's constant, and c_o is a constant depending on the total particle concentration. Since at equilibrium the net particle flow is zero, we may equate to zero the sum of a diffusive flow ($-D\overrightarrow{\mathrm{grad}}c = (cD/kT)\,\overrightarrow{\mathrm{grad}}V$) and a flow generated by the force field $-\overrightarrow{\mathrm{grad}}V$ (which is the product of the cell concentration c and mean velocity $-\overrightarrow{\mathrm{grad}}V/f$). We obtain:

$$D = kT/f \tag{4}$$

In aqueous medium, the diffusion constant of a sphere of 5 μm radius is found to be 4.1 $\times$ 10^{-14} cm^2/sec.

Another result may give a better understanding of the significance of D.[13] Assuming that particles are concentrated at some point 0 at time zero, we write that the number of particles remains constant, which yields:

$$\partial c/\partial t = -\mathrm{div}\,\vec{J} = D\Delta c$$

Since the system possesses spherical symmetry, Δc is expected to depend only on time t and the radial coordinate r. After simple algebraic manipulation, the diffusion equation reads:

$$\partial c/\partial t = (1/r)D\ \partial^2/\partial r^2\ (rc) \tag{5}$$

Multiplying both sides of Equation 5 by r^2, we obtain, after a double integration by parts of the right side and division of both sides by the total number of particles:

$$\partial/\partial t\ (<r^2>) = 6D \tag{6}$$

Where $<>$ is for "mean value". Equation 6 shows that the mean squared distance between the positions of a particle at time zero and time t is 6Dt. Using the above results, we find that the displacement of a cell of 5 μm radius is about 0.2 μm after 1 sec. An important consequence of this process is that diffusion limits the time allowed for bond formation, even in the absence of imposed fluid flow. However, more information is required before

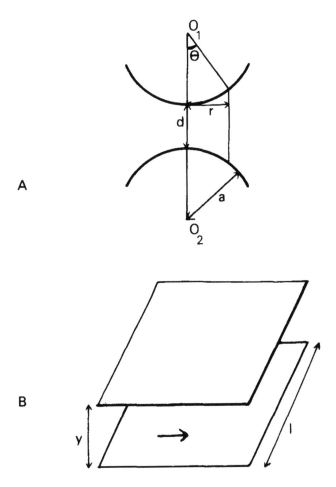

FIGURE 2. Hydrodynamic cell interactions. (A) The distance between the surfaces of two spheres of radius a is d. Varying d generates an axisymmetric flow. (B) Flow of a viscous fluid between two parallel plates of width l and distance y much smaller than l.

we address this point quantitatively. Indeed, an important phenomenon is the hydrodynamic interaction between approaching cells.

2. Hydrodynamic Forces between Smooth Spheres

When two spheres approach each other in a viscous medium, they exert a mutual repulsion. The problem of determining the hydrodynamic interaction between two spheres in resting medium or under shear flow was studied in considerable detail (see, e.g., Brenner and O'Neill[14] for a general analysis of the problem, and Bossis and Brady[15] for recent references). However, the complexity of calculations and results may obscure the extent of errors involved in modeling cells as smooth spheres, thus neglecting surface roughness and the presence of membrane-bound macromolecules. Hence, we shall describe an approximate treatment of hydrodynamic intercellular forces in order to convey an intuitive view of basic phenomena. More complex results will be detailed further on (see Appendix).

Let us consider two spheres of radius a separated by a gap of width d much smaller than a. Decreasing d requires the generation of a fluid flow out of the intercellular zone. If Q(r) is the net flow rate out of the cylinder of radius r and axis $0_1 0_2$ (see Figure 2A) we may write, when the fluid is incompressible:

$$Q(r) = -\pi r^2 (dd/dt) \tag{7}$$

Now, it is well known from classical hydrodynamics that the pressure gradient associated with the flow rate Q of a viscous fluid between two parallel plates of width 1 separated by distance y is (Figure 2B):

$$\Delta P/\Delta x = 12\mu Q/ly^3 \tag{8}$$

where μ is the fluid viscosity. In a first approximation, we calculate the hydrodynamic force F by integrating P over the interaction area. After integrating by parts and noticing that P is zero far from the spheres, we obtain:

$$F = \int P(r) \cdot 2\pi r dr = -\int \pi r^2 [dP(r)/dr] dr$$

Approximating $(1-\cos\theta)$ as $r^2/2a^2$, we obtain:

$$F = 3\pi\mu a^2/2d \cdot dd/dt \tag{9}$$

A similar formula can be obtained by using Equation 11 and Table II of Reference 16 and taking the limit for vanishingly small distance between spheres. See Section II.C for more details.

It is fairly instructive to use Equations 3 and 9 to study the interaction between cells modeled as perfectly smooth spheres of 5 μm radius. Clearly, hydrodynamic forces play a dominant role as compared to Stokes' drag when the distance between cells is comparable to the range of intermolecular forces (the ratio a to d is 100 when d is 50 nm). However, Equation 9 may become grossly inaccurate when d is very low because of two major effects that we shall now consider.

3. Hydrodynamic Forces between Cells at Distances Compatible with Molecular Interactions
a. Importance of Cell Surface Asperities

It is well known that nucleated cells are endowed with highly folded membranes. The excess surface area is about 50 to 100% as compared with the area of a sphere of similar volume, and this may act as a reserve allowing constant-volume cell deformation and spreading.[17,18] A typical value of the radius of curvature of the tip of a cell-surface asperity (or microvillus) is about 0.1 μm,[18] in accordance with the view of a membrane fold surrounding a bundle of microfilaments.[19] Intercellular contact may thus occur through the tips of microvilli (see Chapter 8 on electron microscopy), reducing the hydrodynamic force by a factor on the order of 2500 (this figure was obtained by replacing the cell radius with the microvillus curvature radius in Equation 9).

b. Importance of the Cell Surface Glycocalyx

When one considers the cell-cell interaction area at a scale comparable to the range of intermolecular forces (see Chapter 1), it appears that as soon as a contact occurs between membrane molecules the gap separating phospholipid bilayers can no longer be treated as empty space, since membrane-associated molecules are likely to interfere with the fluid flow during the cell-to-cell approach.

The problem of fluid flow through porous macromolecular systems is fairly complicated (see the review by Wiegel[20]). Modeling cell-surface molecules as homogeneously distributed spheres of radius r_o and density n, the relationship between the fluid velocity $\vec{v}$ and pressure gradient $\vec{\text{grad}}P$ may be approximated as:[21]

$$-\overrightarrow{\text{grad}}P = 6 \ \mu r_o n \overrightarrow{v} \tag{10}$$

where μ is the fluid viscosity. Two important assumptions used to obtain this result were that Reynolds' number be low enough (see Appendix) and the fraction of space occupied by spheres be small as compared to unity.[20]

Replacing the flow rate Q with lyv in Equation 8, the relationship between the fluid velocity and pressure gradient in absence of interfering membrane molecules would be

$$-\overrightarrow{\text{grad}}P = 12\mu \overrightarrow{v}/d^2 \tag{11}$$

(where d is the intermembrane distance). In order to compare Equations 10 and 11, we use reasonable guesses of 3 nm for r_o and 100 μm^{-3} for n (see Chapter 1), which yields:

$$r_o n = 0.3 \ \mu m^{-2}$$

This is much smaller than $1/d^2$ at a separation range of a few tens of nanometers. Hence, the effect of cell-surface-bound glycoproteins on hydrodynamic interactions may be low. However, more accurate estimates would be needed to assess the generality of this conclusion.

The above results may now be taken together to get a quantitative view of the properties of Brownian cell-cell encounters.

4. General Features of Brownian Collisions

Two parameters of a collision are of particular relevance to adhesion. These are

1. The collision duration, i.e., the length of the time during which the intercellular distance is compatible with bond formation (this is time t_2 of step 2 described in the introduction of this chapter.
2. The energy involved in separating cells after a collision (this sets the minimum number of bonds required to maintain prolonged contact).

We shall examine these points sequentially.

The range of intercellular distances compatible with bond formation is obviously dependent on the nature of adhesive molecules. If these are rigid and binding sites are located at a fixed distance from the phospholipid bilayers, bonding may require that the width of the gap separating membranes be fixed with accuracy of order 1 Å. However, if adhesion involves fairly flexible macromolecules such as immunoglobulins,[21] fibronectin,[22] or carbohydrate chains and lectin-like receptors,[23] binding may occur when the intercellular distance spans an interval several tens of angstroms wide. Therefore, it seems reasonable to choose $\delta l = 10$ Å as a suitable order of magnitude for the width of the binding region. The time t required for a Brownian particle to span this region is about $\delta l^2/D$, where D is the diffusion constant. In the absence of hydrodynamic intercellular forces, the above estimate of D for a sphere of 5 μm radius yields a value of 2.5×10^{-5} sec for t. Now, introducing hydrodynamic intercellular forces (Equation 9) with an intercellular distance d of about 100 Å results in a 500-fold increase of t. However, if cells interact through microvilli of about 0.1 μm radius, t is expected to remain near 10^{-5} to 10^{-4} sec.

A point of caution is needed: we used "macroscopic" diffusion equations to describe the motion of the average position of cell centers. It is important to ensure that the range 1 of microscopic fluctuations is much smaller than δl. Although parameter 1 is not as clearly defined as the "mean free path" of molecules in a dilute gas, a crude modeling of Brownian motion as a random walk made of elementary paths of length 1 allows a rough estimate of l with basic results of statistical mechanics. For a cell of radius 5 μm and density close to 1 g/cm³, l is a few angstrom units.[24]

Another point is the energy involved in Brownian motion. It is well known that the mean kinetic energy of a rigid sphere is about 3kT/2 or 6×10^{-21} J. This is comparable to the free energy of a single noncovalent bond (Chapter 1).

An interesting parameter is the force required to prevent two cells from moving apart by more than δl (i.e., 10 Å) as a consequence of Brownian motion. Following Boltzmann's law, F is about:

$$F \approx kT/\delta l = 4.1 \times 10^{-12} \text{ N}$$

This is tenfold higher than the gravitational force exerted on a typical cell of radius a = 5 μm and density ρ = 1.07 g/cm³ in aqueous medium, namely:

$$F = 4/3 \, \pi \, a^3 \, (\rho - \rho_w) \cdot g = 3.5 \times 10^{-13} \text{ N}$$

(g is 9.81 m/sec² and ρ_w is the density of water). Hence, a centrifugation as low as 10 g may substantially improve cell adhesion.

We shall now briefly review some models of Brownian coagulation.

B. Models for Brownian Coagulation

This is called perikinetic coagulation, in contrast with shear-induced or "orthokinetic" coagulation. Much work was based on Von Smoluchowski's[26] model. Let us consider a suspension of particles of different species i, j . . . with radius a_i, a_j . . . , diffusion constant D_i, D_j . . . , and concentration c_i, c_j . . . . The rate of collision between particles of types i and j is evaluated by solving the diffusion equation (Equation 5) in a mobile reference frame centered on a type i particle. Neglecting hydrodynamic interaction, the diffusion coefficient of type j particles in this frame is $D_i + D_j$ (this may be shown by using the interpretation of the diffusion coefficient described in Section A.1 and noticing that the root mean square length of the sum of two vectors of length l_i and l_j and random orientation is $(l_i^2 + l_j^2)^{1/2}$. This yields:

$$\partial/\partial t \, [r \cdot c_j(r, t)] = (D_i + D_j)\partial^2[r \cdot c_j(r, t)]/\partial r^2$$

This equation is solved using as boundary conditions:

$$\begin{cases} \lim_{r \to \infty} c_j(r, t) = c_j \\ c_j(r, t) = 0 \quad \text{when} \quad r \leq r_i + r_j \end{cases}$$

The influx of type j particles into the sphere of equation $r = r_{i + r_j}$ is then determined by means of Equation 11, which yields:

$$J = 4\pi(D_i + D_j)(r_i + r_j)c_j[1 + (r_i + r_j)/(\pi(D_i + D_j)t)^{1/2}] \tag{12}$$

Usually, $(r_i + r_j)/[\pi(D_i + D_j)t]^{1/2}$ is neglected. When particles are spheres of 5 μm radius, this requires that t be much higher than 400 sec. The rate of collision between type i and type j particles is then simplified as:

$$b_{ij} = (2kT/3\mu)(r_i + r_j)(1/r_i + 1/r_j)c_i \cdot c_j \tag{13}$$

Now, the coagulation of an assembly of identical particles may be represented by Equation 13, denoting as c_i the concentration of i uplets (or clusters of i particles). However, determination of the functions $c_i(r,t)$ requires ad hoc approximations about the diffusive behavior of clusters. A simple approximation is obtained by using Equation 13 and writing:

$$(r_i + r_j)(1/r_i + 1/r_j) \approx 4 \tag{14}$$

This is expected to hold when r_i is not too different from r_j. This is the case on the beginning of coagulation experiments, when most collisions involve single cells. Equations 13 and 14 readily yield:

$$dc/dt = -4kTc^2/3\mu \tag{15}$$

$$1/c(t) - 1/c(0) = 4kTt/3\mu \tag{16}$$

where $c(t)$ is the total particle concentration at time t (a factor of $1/2$ is required to avoid counting the same particle twice).

The above treatment implied that every collision resulted in adhesion. Introducing the collision efficiency E (i.e., the fraction of collisions resulting in adhesion), Equation 15 reads:

$$dc/dt = -4kTEc^2/3\mu \tag{17}$$

This equation was checked experimentally by Swift and Friedlander,[27] who studied the coagulation of aqueous suspensions of oil droplets of approximately 0.5 μm diameter. The linear relationship between time and $1/c(t)$ was found to hold, and the collision efficiency was of order unity.

Several refinements were proposed for Equation 15. Von Fuchs[3] solved the diffusion equation in the presence of an interaction potential V(r) between colliding particles, which yielded an analytical expression for the coefficient E of Equation 17. This was called the stability ratio:

$$E = 2a \int_{2a}^{\infty} \exp[V(r)/kT]dr/r^2 \tag{18}$$

Also, Spielman[28] took account of both hydrodynamic forces and interaction potential V(r), which yielded:

$$E = 2a \int_{2a}^{\infty} [D'(r)/2D]\exp[V(r)/kT]dr/r^2$$

where $D'(r)$ is the distance-dependent diffusion constant of a particle in a mobile reference frame centered on another particle.

More recently, the validity of a "macroscopic" treatment of Brownian coagulation was analyzed by different authors.[29,30] However, we shall not give further details on this topic since Brownian coagulation is expected to play a minor role in most cellular systems (platelets may constitute an important exception). Indeed, using Equation 15 and considering cell suspensions concentrated at $10^7/m\ell$ (which is a fairly high value in most cases), we find:

$$1/c \, dc/dt \approx 6.10^{-5}$$

Hence, a minimal time of order 10^4 sec is required to obtain significant diffusion-driven coagulation. During such a period of time, substantial cell sedimentation would have occurred, thus invalidating the above treatment.

Hence, the main interest of the above data is to convey an intuitive feeling for cell behavior in liquid media.

C. Cell Behavior in Linear Shear Flows

Two problems are of interest with respect to cell adhesion. Determining the interaction between a sphere and a plane surface may help in understanding cell-substrate adhesion. The mutual interaction between two spherical particles is a model relevant to intercellular adhesion.

1. Interaction between a Sphere and a Flat Surface

In view of the linearity of basic flow equations (see Appendix), we shall consider separately the movements oriented parallel and perpendicular to a given surface.

a. A Model for Cell-to-Substrate Approach

A first point is to know whether contact between a plane and a cell subjected to 1-g sedimentation may be substantially delayed by hydrodynamic forces. Applying the simple theory described in Section II.A.2 to the interaction between a sphere and a flat surface, it is readily found that the width d(t) of the gap between the surface of a sphere of radius a and density ρ and the plane is

$$d(t) = d(0) \cdot \exp[-2(\rho - \rho_0)agt/9\mu] \qquad (19)$$

where ρ_0 is the medium density and μ the medium viscosity. Since the above equation is valid when d is substantially smaller than the cell radius, it is readily found that the time required for d to decrease from 5 μm to 10 Å is on the order of 4 sec (taking $\rho - \rho_0 = 0.1$ g/cm^3, a = 5 μm and g = 9.81 msec^{-2}). Hence, in most cases hydrodynamic interaction should play no role in experimental studies on cell-substrate adhesion.

b. Cell-Substrate Interaction in Shear Flows

The slow viscous motion of a sphere parallel to a plane wall was thoroughly investigated by Goldman et al.[31] We refer to their paper for further details. Two situations are of interest:

i. Induction of Adhesion

The following results may help understanding leukocyte-endothelium interaction in blood vessels. Further, they provide a basis for experimental studies on the rapidity of bond formation. The basic problem is described in Figure 3. We define a fixed Cartesian reference frame Oxyz with unit vectors $\vec{i}, \vec{j}, \vec{k}$. The horizontal plane y = 0 is exposed to a flowing medium located in the half space y > 0. The undisturbed flow velocity is $\vec{u} = G.y.\vec{k}$, where the shear rate G is a constant. A sphere of radius a and center 0_1 is separated from the plane by a small gap of width d and subjected to no external force. The sphere motion is entirely defined by the center velocity $\vec{U}$ (parallel to $\vec{k}$) and rotation $\vec{\omega}$. It may be noticed that when d is zero the sphere will not slip on the wall if ω is equal to U/a. Now, Goldman and colleagues found that the sphere motion was critically dependent on d/a. Indeed, ω and U were found to be zero in absence of a gap between the cell and the wall (d = 0). However, when d was as small as $10^{-8} \times$ a (i.e., 0.05 Å if a was 5 μm!), the sphere velocity was not zero. In view of the discussion of Section II.A.3, d may be expected to range between a few hundred angstrom units (i.e., the glycocalyx thickness) and about 0.5 μm (i.e., the

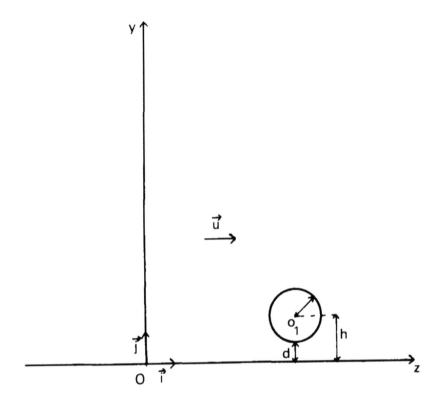

FIGURE 3. Cell-substrate interaction under simple shear flow.

length of surface microvilli); d/a would thus range between 10^{-3} and 10^{-1}. The corresponding boundary values for U/aG, ω/G, and aω/U are, respectively, (0.40, 0.75), (0.21, 0.39), and (0.54, 0.45).

The duration of the contact between some region of the cell surface and the substrate may thus be evaluated by introducing the aforementioned estimate of 10 Å for δl (i.e., the flexibility of binding sites, Section II.A.4). Assuming that the cell behaves as a rigid body on the considered time scale, the lateral displacement of point C (Figure 3) during time t is about t. (U − ωa), whereas the vertical displacement is a(1 − cosωt) aω^2t^2/2. The contact duration is therefore derived from the following inequalities:

$$t \approx \delta l/(U - \omega a); \quad t \approx (2\delta l/a\omega^2)^{1/2}$$

$$t \approx 0.07/G \tag{20}$$

Hence, the interaction time is about 0.1/G for an "average cell".

A final point of caution that was emphasized by Goldman and colleagues is that it is very difficult to check their theoretical results experimentally, since it is not easy to monitor the width of the gap separating the sphere from the plane wall. However, it may be pointed out that basic equations were subjected to considerable experimental check, and it seems acceptable to rely on the order of magnitude of our estimates.

ii. Cell-Substrate Separation by Shearing Forces

Many experimental results demonstrated that the ease of bond formation and strength of adhesion often behaved as independent parameters.[32,33] Also, the equilibrium between endothelium-bound ("marginated") and free leukocytes in blood vessels may be heavily dependent on local shearing conditions.

A point of caution may be useful. Whereas living cells are spherical when they are maintained in suspension, adhesion to a plane substrate often induces dramatic flattening of the cell.[34,35] Hence, substrate-bound cells may not always be modeled as spheres. Two limiting cases will be considered:

Flat disk model — In case of complete spreading, a cell may be modeled as a flat disk of radius a. When the wall shear rate is G, the force exerted on the cell is

$$F = \mu G \pi a^2 \tag{21}$$

whereas the torque is zero.

Sphere model — The force and torque exerted on a sphere bound to a plane wall were calculated by Goldmann and colleagues[31] for various values of the gap width d. No discontinuity occurred at zero gap width. The limiting force F and torque T of hydrodynamic forces were

$$F = 32.05 \ \mu a^2 G$$

$$T = 11.86 \ \mu a^3 G \tag{22}$$

It may be noticed that the use of Stokes' Equation 3 with complete neglect of the wall effect would have resulted in less than 50% error.

A final point is that cell spreading may result in marked increase of the apparent radius a. Hence, the resultant force F is only weakly dependent on the cell shape. However, the torque T may play a dominant role in cell detachment by multiplying the force exerted on cell-substrate bonds by a factor of order a/σ, where a is the cell radius (using the sphere model) and σ is the size of the contact area.

2. Cell-Cell Interaction in Shear Flows

In order to convey an intuitive feeling of the behavior of two spherical cells subjected to a linear shear flow, we shall first review the approximate Von Smoluchowski[26] theory that met with some success in accounting for experimental data.[27]

As shown in Figure 4, we consider a Cartesian reference frame Oxyz with unit vectors $\vec{i}$, $\vec{j}$, $\vec{k}$. The undisturbed fluid is subjected to a linear shear flow with velocity $\vec{u} = Gy\vec{k}$. We consider a sphere of center O_1 and radius a located on axis Oz (O_1 is fixed since intercellular hydrodynamic forces are neglected) and a second sphere of center O_2 and similar radius a passing through the plane Oxy. We use polar coordinates (r, θ) to describe the location of O_2 in the plane Oxy. Now, neglecting intercellular forces, O_2 is assumed to move parallel to Oz with velocity $G.r.\sin\theta$. A collision will occur if r is smaller than 2a, resulting in the formation of a doublet that will rotate with a period of order 1/G (modeling the doublet as an ellipsoid of axis ratio 2, and using a theoretical study by Jeffery,[36] one obtains a value of $5\pi/6G = 2.62/G$ for the rotation period). The collision frequency per unit volume f is obtained after a simple integration (the factor $^1/_2$ is to avoid counting the same collision twice):

$$f = (1/2)n^2 \int_0^{2a} \int_0^{2\pi} Gr\sin\theta \ rd\theta dr = 16n^2 a^3 G/3 \tag{23}$$

where n is the cell number per unit volume. This simple theory gives no information on the separative force exerted on cell doublets during the rotation phase. Order of magnitude estimates may be obtained by using Stokes' law (Equation 3), assuming that the spheres are subjected to uniform flows with a velocity difference of 2aG, which yields:

$$F \approx 12\pi\mu a^2 G \tag{24}$$

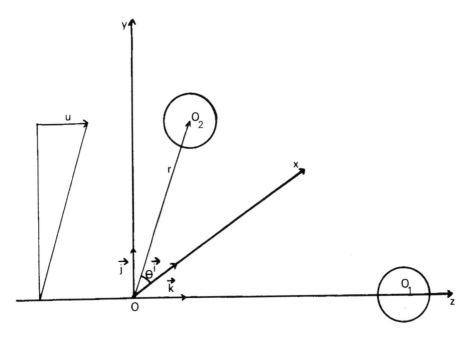

FIGURE 4. Orthokinetic coagulation. According to the Von Smoluchowski theory, the trajectory of the sphere of center O_2 is a straight line parallel to Oz until collision occurs. The figures shows O_2 passing through the plane Oxy.

(see, e.g., Reference 37). This estimate is more a dimensional analysis than an actual determination. We shall now describe more accurate theories and data.

Curtis and Hocking[1] were the first authors to use a rigorous theory of the low Reynolds' number behavior of two spheres in a simple shear flow to study the outcome of the interaction between two colloid particles exerting electrostatic repulsion and electrodynamic attraction, according to the Derjaguin-Landau-Verwey-Overbeek (DLVO) theory of colloid stability.[38,39] They rightly pointed out that hydrodynamic forces prevented actual contact in the absence of attractive forces. Further, they determined numerically the cell trajectories that were conducive to adhesion for various values of the DLVO interaction parameters. This allowed numerical derivation of the Hamaker constant from binding measurements. Unfortunately, these numerical results are difficult to apply to different situations. Also, there is a problem with their definition of the collision efficiency.[2]

More recently, Brenner and O'Neill[14] developed a general formalism allowing simpler expression and use of the results of hydrodynamic calculations on the behavior of two spheres in a linear shear flow (see Appendix). The parameters they defined became of wide use, and numerical results of direct relevance to the problem of two equal spheres were clearly exposed by Arp and Mason.[16] The interaction of unequal spheres was described by Adler,[40] and the behavior of rigid or flexible dumbbells was studied by Adler and colleagues[41] in view of its practical importance (see Section IV). These theories were used to elaborate on collision theories[42] and were subjected to experimental testing.[43,44] We shall now sketch the main results.

a. Collision Efficiency

It would have been tempting to define the collision efficiency E as the fraction of intercellular collisions resulting in the formation of stable doublets. However, as emphasized by Curtis and Hocking[1] and Arp and Mason,[42] it is difficult to define collisions rigorously since no contact is expected to occur in the absence of intercellular forces.

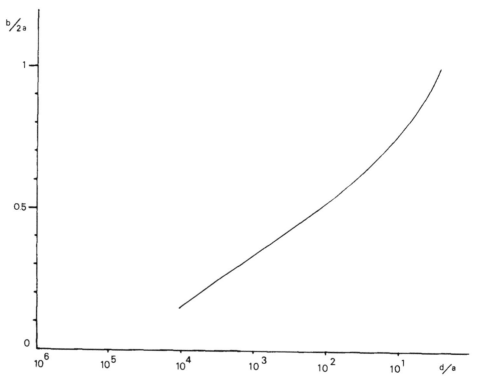

FIGURE 5.   Effect of hydrodynamic forces on cell-cell approach. This curve was derived by simple transformation of results from Table 5 of Reference 16. Abscissa: d/a is the ratio between the distance d separating the cell surfaces and the cell radius a. Ordinate: b/2a is half the ratio between the impact parameter b (i.e., the distance between O_2 and axis Oz in Figure 4) and a. $(b/2a)^2$ is the ratio between the frequency of cell-cell approaches to a distance smaller than d and the "theoretical" number of collisions predicted by the Von Smoluchowski theory.

The simplest procedure may be to choose as a reference the number of collisions predicted by the simple Von Smoluchowski theory (i.e., Equation 23). The initial rate of decrease of the concentration of independent kinetic units (isolated cells or aggregates) is therefore:

$$dn/dt \; (t = 0) = -(16/3) \cdot n(0)^2 a^3 GE \qquad (25)$$

which defines the collision efficiency E as a result of hydrodynamic and other forces. Further, assuming that the collision efficiency is the same for all cell aggregates, and modeling these aggregates as spheres with a volume equal to the total volume of the cells they are made of, the following formula is readily obtained for n(t) (see, e.g., Reference 24):

$$n(t) = n(0) \; exp[-16n(0)a^3GEt/3]$$

Now, E may be written as a product of two parameters E_1 and E_2, where E_1 is the ratio between the frequency of cell-cell encounters resulting in approach compatible with bond formation and the theoretical collision rate predicted by the Von Smoluchowski theory, and E_2 is the fraction of these encounters "compatible with bond formation" that actually yield a stable doublet. However, the definition of E_1 and E_2 is somewhat arbitrary. Referring to the foregoing discussion on cell asperities and cell-surface glycocalyx, a reasonable range for the maximum width d_m of intercellular gap compatible with adhesion is between 200 Å and 0.5 μm. Using dimensionless quantities, d_m/a must range between 4×10^{-3} and 0.1. As shown in Figure 5, the corresponding range for E_1 is between 0.2 and 0.6. Hence, hydrodynamic forces are not expected to reduce dramatically the efficiency of bond formation.

b. Rotation of Cell Doublets

As was demonstrated theoretically and checked experimentally by Adler and colleagues,[41,43] the period of rotation of a doublet of equal spheres in a linear shear flow is markedly dependent on the rigidity of intercellular bonds and the distance between the sphere centers, with a period of 15.62/G for a rigid dumbbell. The problem is that even a relative increase as small as 10^{-3} of the distance between the sphere centers may induce marked variation of the period of rotation. Hence, this theory may not be used with irregular shaped cells.

c. Forces Experienced by Bound Cells

Of major interest is the evaluation of the force and torque experienced by cells during rotation of the doublet. This may be done with the results reported by Arp and Mason,[16] by calculating the limit of the force required to hold two spheres together at zero distance between their surfaces (see Appendix). The results are as follows.

When two cells encounter each other, they are driven together by hydrodynamic forces. Then, after a rotation, they experience a separating force F and a torque T (that is perpendicular to the line joining the sphere centers). The maximum values of F and T are

$$F_m = 19.2 \ \mu a^2 G$$

$$T_m = 3.4 \ \mu a^3 G \tag{26}$$

where a is the sphere radius, G is the shear rate, and μ is the medium viscosity.

In conclusion, the behavior of cell doublets in simple shear flows is well understood. However, more work is required to obtain a quantitative view of the effect of cell surface irregularities on their behavior in these simple flows.

We shall now review some practical means of subjecting cell suspensions to controlled shear in order to study cell adhesion and separation. As shown in the last section, it is necessary to generate a very wide range of shear rates in order to achieve an exhaustive study of the clustering and dissociation of a variety of cell types. Indeed, adhesion is usually obtained with shear rates of 1 to 100 $\sec^{-1}$ whereas separation requires shear rates of 10^4 to 10^5 $\sec^{-1}$.

III. GENERATION OF CALIBRATED SHEAR FLOWS

A. Study of Cell-Substrate Adhesion

As will be shown, cell-substrate adhesion is more easily amenable to accurate experimental study than cell-cell adhesion. We shall describe methods for inducing adhesion and breaking cell-substrate bonds.

1. Induction of Adhesion

The basic idea is to let cell suspensions flow through a suitable duct with a wall shear rate of a few $\sec^{-1}$. Adhesion may be monitored by microscopic observation,[45] cinematography,[46,47] or by counting bound cells at the end of the experiment.[33] Most easily available ducts are tubes with a circular section. However, flat chambers may allow easier microscopic examination (Figure 6).

The relationship between the flow rate Q, the fluid velocity u and the pressure difference ΔP between the inlet and the outlet may be found in every standard textbook on fluid mechanics (e.g., Reference 12). When the tube section is a disc of radius r_o, we find:

$$Q = \Delta P \ \pi r_o^4 / (8 \mu L)$$

$$u(r) = 2Q\ (r_o^2 - r^2)/(\pi r_o^2)$$

$$G_w = 4Q/(\pi r_o^2) \tag{27}$$

where L is the duct length, μ is the medium viscosity (about 10^{-3} Poiseuille or $kg.m^{-1}sec^{-1}$ in water at room temperature), u(r) is the flow velocity at distance r from the duct axis, and G_w is the wall shear rate (i.e., du/dr at $r = r_o$). It may be noticed that the flow velocity is zero at $r = r_o$, in accordance with the usual "nonslip" boundary condition. The shear rate is zero on the duct axis and it is maximum near the wall.

When the inlet is rectangular with height h much smaller than the width W, we obtain:

$$Q = \Delta P \cdot Wh^4/(12\mu L)$$

$$u(x) = 6Q(h^2/4 - x^2)/(Wh^3)$$

$$G_w = 6Q/(Wh^2) \tag{28}$$

where u(x) is the flow velocity at distance x from the symmetry plane parallel to Oy.

Equations 27 and 28 allow complete characterization of the flow and the data described in Section II.C.1, giving a quantitative description of cell-substrate interaction. Here are some practical points:

1. Microscopic examination of circular-sectioned ducts may be difficult due to optical distortion (although this may be feasible with low magnification[33]). It may thus be useful to include these ducts in water-filled chambers (Figure 6A).
2. The following setup may be useful when radioactive (e.g., chromium-labeled) cells are used to quantify adhesion: an extremity of capillary tubes is dipped into the cell suspension and the other end is connected to a pump through a disposable plastic tube as shown in Figure 6B. It is thus possible to determine the exact amount of suspension passing through adhesion chambers by mere radioactive counting.
3. The flow may be generated by a motor-driven syringe or a peristaltic pump. In the latter case (and even in the former case if the flow rate is very low), the flow velocity may not be constant. Constant low flow rates may be achieved by means of an inexpensive device made of a syringe body, a needle (acting as flow regulator following Equation 27), and a capillary tube dipping into a liquid reservoir to avoid unpredictable decrease of the flow by surface forces (Figure 6C).
4. Finally, it may be pointed out that adhesion can be studied on different substrates by preexposing glass surfaces to different molecules (e.g., fibronectin, laminin, or polylysine). Adhesion was also studied on endothelial cell monolayers.[48]

2. Dissociating Cell-Substrate Bonds

When one tries to separate cells from a substrate with shearing flows, it appears that very high shear rates are required (G must be on the order of 10,000 sec^{-1} or more). In this case, the flow Equations 27 or 28 may no longer be valid due to the appearance of turbulence. An empirical procedure to evaluate this possibility is to calculate the so-called Reynolds' number:

$$R = ud\rho/\mu$$

where u is the fluid velocity, d is a characteristic distance of the system, ρ is the fluid density, and μ is the fluid viscosity. Note that R is dimensionless. A somewhat equivalent expression for the Reynolds' number is $G\rho d^2/\mu$, where G is the shear rate. The limiting

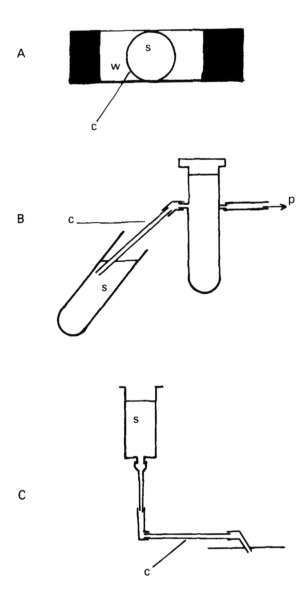

FIGURE 6. Practical problems in studying cell-substrate adhesions. c: capillary tube, s: cell suspension, w: water, p: pump. (A) Embedding a circular-sectioned tube in a water chamber allows better microscopic observation; (B) collecting cells after their passage through the adhesion chamber allows accurate measurement of the actual flow rate; (C) simple generation of slow continuous flows.

value of R for laminar shear flows is between 10^3 and 10^4 for a circular pipe, depending on the regularity of the inlet.[12] That a flow is laminar may be assessed by checking that the flow rate is proportional to the driving pressure. Taking for d the radius of the pipe (about 10^{-3}m), the corresponding limit for G is between 1000 and 10,000 sec^{-1}. Since this value is too low to allow efficient detachment of bound cells, special procedures were used to delay the appearance of turbulence.

a. Using Low d: The Plane Plate Viscometer

Weiss[49,50] made fibroblasts adhere to plane surfaces, then generated a shearing flow by rotating a disc parallel to the substrate at a very small distance (say, 0.05 cm or less). The shear rate at distance r from the disc axis is given by:

$$G = \omega r/d \qquad (29)$$

where ω is the angular velocity (expressed in rd/s). This apparatus was found to yield laminar (i.e., turbulence-free) flow when Reynolds' number was lower than 10^4.

b. Increasing the Medium Viscosity

Mège and colleagues[33] achieved high shear rates in a capillary tube by replacing conventional media with concentrated dextran solutions (e.g., 10% by weight dextran of 500,000 mol wt), resulting in about 40-fold increase of the viscosity. Hence, higher shear rates could be achieved, and the force exerted on cells for a given shear rate was also increased.

A practical point about pressure generation: a very simple and efficient procedure was to use a 5 to 30 mℓ glass syringe set vertically with a holder and depositing a known weight on the piston (100- or 250-mℓ Erlenmeyer flasks partially filled with water proved quite satisfactory).

B. Generation of Hydrodynamic Flows to Study Intercellular Adhesion
1. Induction of Adhesion
a. Couette Viscometer

The experimental study of cell aggregation under controlled hydrodynamic flow as pioneered by Curtis,[52] who made use of a Couette viscometer. Other authors used the same technique.[53,54] The basic apparatus is made of two coaxial cylinders of radius R and R + h (with h much smaller than R). Cell suspensions are located between cylinders, one of which is rotating with angular velocity ω, thus generating a laminar shear flow of uniform shear rate:

$$G = \omega R/h$$

At various times after the onset of the experiment, aliquots are examined for determination of the number of aggregates. Equation 25 may then be used to determine the collision efficiency E.

b. The Traveling Microtube

The interest of Couette flow is that all cells are subjected to the same shear rate. However, this does not allow any study of individual cells. This may result in some difficulty in interpreting experimental data, since it is well known that even cell lines usually display extensive structural and functional heterogeneity. Also, one has to rely on theoretical flow equations. These difficulties were overcome with the very elegant "traveling microtube" technique described and used by Goldsmith and Mason.[55] The principle is as follows: a thin glass capillary tube (50 to 200 μm diameter) is embedded in a liquid sheath of similar refraction index as the cell suspension medium (Figure 6A) and maintained vertically on a motor-driven stage. The tube is examined with a microscope mounted horizontally while a slow laminar flow is generated in the tube. Further, the stage is moved vertically in order that examined cells remain in the same microscope field. This technique allowed very accurate observation of individual collisions, and was used to study the adhesion of inert and biological particles.

2. Dissociation of Intercellular Bonds

The aforementioned devices could not yield sufficient shear rates to dissociate cell aggregates. This was achieved with a very simple method consisting of driving a suspension of cell clusters through a syringe needle with graded pressure.[32,51,56] Since the duct radius was as low as a few tens of millimeters, shear rates higher than 100,000 sec^{-1} could be achieved with laminar flows.[51] Several problems are raised by these experiments:

1. A possible problem might be the simultaneous formation and dissociation of cell clusters in the needle, yielding hardly interpretable data. However, it was checked that no significant adhesion was found when aggregate-free cell suspensions were sheared, in accordance with the view that only low shear collisions (say, $G \approx 100$ sec^{-1}) may be conducive to adhesion in most experimental systems.
2. As was pointed out, the shear rate at any point is dependent on its distance to the duct axis. Hence, different aggregates may be subjected to different shear forces. This difficulty may be overcome by shearing several times a given cell suspension in order to increase the probability for a given cell cluster to be subjected to high shear (near the needle wall). Since this procedure did not dramatically increase dissociation,[51] it may be hypothesized that uncontrolled movements perpendicular to the needle axis ensure that each cell cluster will pass through the high shear region with a range of orientations.
3. It was found that the mechanical resistance of cell clusters might span a very wide range, thus requiring generation of very different flows. This could be achieved with very simple techniques that were detailed in a recent report.[56] The main points are briefly described here.

A practical problem is often to measure the inner radius of a syringe needle. This may be easily performed by measuring the time required for a vertical syringe body containing 10 to 20 mℓ of water to empty through the needle, the tip of which must be dipped in a large reservoir (i.e., a reservoir with a constant water level) to avoid capillary effects. The measured time t is

$$t = [8\mu LV/(\rho g r_o^4 (h_1 - h_2))] \cdot \ln(h_1/h_2) \tag{30}$$

where μ is the medium viscosity, L and r_o are the needle length and inner radius, V is the volume of water passing through the needle during time t, h_1 and h_2 are the height differences between the water level in the syringe and the reservoir before and after the flow, ρ is the fluid density, and g is 9.81 m.sec^{-2}.

Whereas the deposition of a sufficient weight on the piston of a glass syringe allows fairly accurate generation of a pressure higher than about 10,000 N/m^2, it is difficult to obtain smooth movements of the piston with too-low weights. A simple means of overcoming this problem is to adapt a syringe needle to the tip of an automatic pipette (after cutting this tip to a suitable length and diameter). The pressure generated by the pipette is not constant. However, maximal values may be determined by measuring the initial flow rate during aspiration of a sufficiently viscous solution and using Poiseuille's formula (Equation 27). The pressures we obtained with a Gilson® P1000 pipette set at 200 $\mu\ell$ and a P200 pipette set at 200 $\mu\ell$ were 3900 and 19,000 N/m^2, respectively. Still lower pressures can be obtained by interposing a reservoir between the pipette tip and the needle, but this was not found useful in the experimental systems we studied.

IV. SELECTED EXPERIMENTAL RESULTS

The purpose of the present section is to review selected experiments yielding quantitative data on adhesion in order to convey some intuitive feeling for binding efficiency and binding strength.

A. Binding Efficiency

1. Cell-Substrate Interaction

Turitto and colleagues[57,58] studied the adhesion of platelets to subendothelium exposed to flowing blood. For this purpose, everted arterial walls were fixed along the axis of cylindrical flow chambers. Endothelial cells were removed to allow platelet adhesion (since platelets do not bind to healthy endothelium). The authors solved the diffusion equations in this system and they measured the extent of platelet deposition at various shear rates G. They concluded that platelet deposition was essentially a diffusion-limited process when G was lower than about 1000 sec^{-1}. However, when G was higher than 1000 to 10,000 sec^{-1}, platelet binding was lower than expected, which suggested a decrease of the binding efficiency. Hence, platelet adhesion was found to require less than 0.1 msec (see Equation 20).

In a series of controlled experiments, Doroszewski[46,47] studied the adhesion of L1210 murine leukemia cells to the glass surface of flow chambers in culture medium supplemented with 10 to 20% fetal calf serum. The use of microcinematography allowed several interesting conclusions:

1.   Nonadherent cells that were in contact with the wall flowed with a velocity U of about 5.G_w (where G_w is the wall shear rate). Since the radius a of L1210 cells is close to 5 μm, the ratio U/aG_w was thus about 1, which is indicative of relatively high cell-substrate distance according to the sphere model reviewed in Section II.C.1.
2.   The ratio between the cell angular velocity and the shear rate ω/G_w was on the order of 0.5 when G_w was 5 to 10 sec^{-1}, which was consistent with a rolling motion.
3.   Cell adhesions were quite rare events, since the fraction of cells binding to the substrate during their passage through a microscope field was of order 1 to 2%, and this decreased with increasing shear rate.
4.   Adhesion was a very rapid process. The cell velocity fell from 25 μm/sec to zero within less than 0.25 sec.
5.   After a period of time varying between 1 sec or less and 10 min or more, cells separated from the substrate either instantaneously or first slowly, then more rapidly when they had moved by a few cell diameters. The last process was consistent with the formation of thin tethers out of the cell membrane, as was described by Hochmuth and Mohandas,[59] who reported on the adhesion of red blood cells to artificial substrates.

Clearly, adhesion in this model required a minimal time of order 0.01 sec (Equation 20), and cell-substrate detachment involved forces on the order of 4×10^{-12} N (see Equation 22). It must be pointed out that the studied model was a very poor adhesive system.

Mège and colleagues[33] studied the adhesion of P388D1 murine macrophage-like cells to glass capillary tubes in serum-deprived culture medium. It may be pointed out that P388D1 cells are considered as glass adherent, although they bind to artificial substrates much less efficiently than normal macrophages.

Since the usual definition of binding efficiency does not hold in this system, the authors modeled cell-substrate encounter as a rolling motion with probability p.dz for a cell to adhere on any elementary path of length dz. The parameter p was determined for various values of the wall shear rate G_w ranging between 7 and 220 sec^{-1}; p displayed a dramatic decrease when G_w became higher than about 50 sec^{-1}. Hence, the minimal contact time required for binding initiation was of order 1 msec (Equation 20).

Houdjik and colleagues[60] studied the adhesion of human platelets to collagen-coated glass surfaces under flow conditions. Substantial adhesion was obtained at a wall shear rate of 800 sec^{-1}; further, when cells were suspended in factor VIII-defective plasma, adhesion was similar to that of controls when G_w was 490 sec^{-1}, but it was abnormally low when G_w was further increased. Hence, factor VIII was effective in increasing platelet adhesion at high shear rate. Fibronectin was also found to enhance adhesion. Hence, in accordance with aforementioned studies, binding required a time of contact on the order of 0.1 msec.

2. Cell-Cell Interaction

Several authors studied cell-cell adhesion with a Couette viscometer. Curtis[52] measured the collision efficiency E on chicken embryo cells subjected to various shear rates. When suspensions of 7-day embryo retinal cells in 199 medium were studied, E decreased from 0.2 to 0.12 when G was increased from 2.3 to 17 sec^{-1}.

Evans and Proctor[54] measured collision efficiencies on lymphoid cells (thymocytes or lymphocytes) subjected to various shear rates. Using a theoretical model and assuming that a subpopulation of 15% of cells displayed particularly high adhesiveness, they gave estimates of 0.01, 0.0026, and 0.0002 for E when the shear rate was 10, 100, and 1000 sec^{-1}, respectively. It must be pointed out that lymphoid cells are poorly adhesive as compared to the embryo cells used by Curtis. The physiological significance of the latter adhesion experiments is also difficult to assess.

Goldsmith and colleagues[61] used the traveling microtube technique to study the collisions between blood platelets, sphered erythrocytes, or rabbit granulocytes. Since these authors based their interpretations on the rigorous theory of the interaction between two spheres under shear flow, of particular interest are their studies on erythrocytes made spherical by exposure to hypotonic media, then fixed with glutaraldehyde.[61,62] The suspending medium was a saline solution that might be supplemented with glycerol.

By measuring the rotation period of cell doublets, the authors estimated the minimum distance of approach between cells at about 400 Å, in contrast with predictions from the DLVO theory. The authors ascribed this discrepancy to the presence of surface asperities they were able to detect with scanning electron microscopy. They concluded that the DLVO theory could not be applied to biological surfaces. Further, they studied the interaction between antibody-coated red cells.[62] Although adhesion occurred, the value of the rotation period of erythrocyte doublets was suggestive of the possibility of relative rotation of bound cells. Also, the maximum shear rate compatible with adhesion was about 400 sec^{-1}. The authors interpreted these results by suggesting that the separative force experienced by cell doublets (about 7.10^{-11} N) was higher than antigen-antibody bonds. It is not known whether prolonged incubation of erythrocyte doublets would have resulted in higher mechanical resistance. Also, the critical shear rate compatible with adhesion was lowered when the antibody concentration was decreased.

Collisions between granulocytes were also studied.[61] Experiments were difficult to interpret since cells displayed marked morphological heterogeneities. When doublets of fairly spherical cells were examined, their trajectories were suggestive of rigid dumbbells even when cells did not seem to be in close contact. The authors suggested that adhesion might be mediated by cell projections.

ADP-induced platelet aggregation was also studied with the same apparatus.[63,64] In a representative experiment, the collision efficiency decreased from 0.27 (at a mean shear rate of 7.9 sec^{-1}) to 0.04 (at a shear rate of 53.5 sec^{-1}).

Recently, Duszyk and colleagues[65] studied Ricina communis lectin-mediated agglutination of rat thymocytes subjected to shear flow. When the lectin concentration was higher than 1 μg/mℓ, each collision was conducive to adhesion up to a shear rate on the order of 500 to 1000 sec^{-1}. However, the collision efficiency decreased for increasing shear rate when the lectin concentration was 0.1 μg/mℓ. Interestingly, the authors gave a quantitative interpretation of their findings. They estimated the rate of bond formation with Bell's[6] theory and they assumed that adhesion occurred when the number of bonds formed during the rotation of a cell doublet was sufficient to oppose disruptive hydrodynamic forces.[66] Although the choice of interaction areas and intercellular distances compatible with adhesion may raise some problems, their theoretical estimates were consistent with experimental findings.

The general conclusion that may be drawn from all these experiments is that binding efficiency is highly dependent on the duration of intercellular contact. The limiting shear rates vary between about 5 and 1000 sec^{-1}, corresponding to a molecular contact duration ranging between 10 and 0.1 msec (Equation 20). Since the force required to separate cells that remained bound for a few minutes correspond to much higher shear rates (see below), it seems that this limit of 5 to 1000 sec^{-1} is a consequence of kinetic limitations to bond formation.

A final point is that it is difficult to compare data on cell-cell and cell-substrate adhesion, since in the latter case contact may be rather durable, and there is some arbitrariness in the definition of collision efficiency under flow conditions (Section IV.A.1).

B. Mechanical Strength of Cell Adhesion
1. Cell-Substrate Separation

Weiss separated fibroblasts from glass substrates by shearing flows generated by a rotating disc. His main conclusions are as follows:

1. The wall shear rate required to detach a substantial amount of cells after a fairly prolonged period of shearing (about 200 sec) is on the order of 500 sec^{-1}.[49,50] The applied force was thus of order 4.10^{-11} N (Equation 21).
2. Binding strength displayed high heterogeneity. In a representative set of experiments, the fraction of detached cells was increased from 0.21 ± 0.03 to 0.47 ± 0.03 when the shear rate was increased from 370 to 3650 sec^{-1}.[50]
3. Cell separation was increased when the shearing time was increased.
4. Binding strength was dependent on the duration of cell culture[49] when this was varied between several hours and several days.
5. Cell-substrate separation probably involved membrane rupture. Indeed, a similar fraction of fibroblasts (0.25 to 0.30) were separated from a variety of artificial substrates (perspex, teflon, polyethylene, nylon, terylene) by a given shear, suggesting that the binding strength was representative of the membrane mechanical resistance rather than of cell-substrate bonds.[10] Also, using a mixed agglutination technique, Weiss and Coombs[67] were able to demonstrate the presence of cell-related antigenic material on sheared substrates.

Mège and colleagues[33] studied the separation of P388D1 macrophage-like cells from glass surfaces. The use of viscous dextran solutions allowed generation of high shear rates and shearing forces. The main conclusions were the following:

1. A wall shear stress of about 100 N/m^2 was required to detach 50% of bound cells (this was about 20-fold higher than the values reported by Weiss on fibroblasts). When the shear duration was about 20 sec, the applied force was on the order of 8.10^{-9} N.
2. The fraction of detached cells was markedly dependent on the shearing time. At a wall shear stress of 100 N/m^2, 25% of cells were detached within a few seconds, whereas 60% of them were separated from the substrate after 100 sec.
3. Finally, the binding strength was found to increase during the first 15 min following the onset of adhesion.

Schmid-Schönbein and colleagues[68] estimated the shear force imposed on endothelium-bound leukocytes in venules of rabbit omentum. The estimated force varied between 4×10^{-10} and 2×10^{-8} N, which must give an order of magnitude of the binding strength since adherent leukocytes are in equilibrium with circulating cells.

Clearly, all the above results demonstrate the wide variability of binding strengths.

Now, before reviewing some results on cell-cell adhesion, we shall briefly describe some experimental studies on cell-substrate adhesion done with alternative methods.

As early as 1944, Coman[69] used a glass microneedle held by a micromanipulator to measure the force required to detach epithelial cells from the surrounding tissues. He observed microscopically the needle bending immediately before cell separation and compared it to the bending caused by microweights fixed to the needle tip. His purpose was to compare the cohesion of normal and tumoral tissues to determine whether metastasis might be ascribed to decreased adhesiveness of malignant cells. The force required to detach a cell from surrounding tissues was on the order of 10^{-5} N. However, it is not known whether the pericellular matrix played a role in maintaining cohesion. Similar figures were reported by Zeidman[70] in a later study.

More recently, McKeever[71] used a similar method to study the adhesion of rabbit alveolar macrophages to glass coverslips. He reported binding strengths on the order of 10^{-7} N, i.e., 100-fold lower than those obtained with epithelial cells.

Other authors subjected substrate-bound cells to centrifugal forces. Easty and colleagues[72] plated murine carcinoma cells on glass surfaces and exposed these cells to tangential centrifugal fields ranging between $100 \times g$ and $1000 \times g$. They reported a wide heterogeneity of binding strengths since the fraction of cells removed by $100 \times g$ centrifugation ranged between 0 and 1. Adhesion was decreased in the presence of serum, and some (not all) cell lines lost their adhesiveness when they were heat-killed.

Assuming a cell density of 1.07 kg/m³ and a volume of 523 μm³, the overall force (centrifugal force minus Archimedes force) corresponding to an acceleration of $100 \times g$ was 3.6×10^{-11} N.

More recently, McClay and colleagues[73] prepared monolayers of chicken-embryo retinal cells or sheep erythrocytes by depositing them in polylysine-coated microwells. Then they deposited radioactive cells on these monolayers, and after incubation periods of varying length, they inverted the plates for centrifugation and measurement of the fraction of radioactivity remaining bound after this treatment.

They concluded that the strength of the bonds between retinal cells was on the order of 10^{-10} N, whereas erythrocytes bound with concanavalin A (0.5 μg/mℓ) resisted only 10^{-11} N.

Interestingly, the adhesion between retinal cells was markedly strengthened on incubation at 37°C, and this strengthening was partially abolished by cold or metabolic inhibitors.

Charo and colleagues[48] used a similar technique to measure the adhesion of human polymorphonuclear leukocytes to endothelial monolayers. Binding strengths were also on the order of 4×10^{-11} to 4×10^{-10} N, since cell detachment began at a centrifugal acceleration of about $200 \times g$ and was not completed at $1200 \times g$. Also, the adhesion strength was maximum after about 15 min, and it was increased when cells were exposed to the chemotactic oligopeptide N-formyl-methionyl-leucyl-phenylalanine.

Other authors studied the binding of tumor cells by activated macrophages.[75,76] This model is of interest since activated macrophages were reported to bind and "specifically" destroy tumoral cells, which may play a role in host-tumor relationship. Interestingly, whereas activated and nonactivated macrophages could bind malignant or nonmalignant cells with low strength (about 4×10^{-11} N), only aggregates of activated macrophages and neoplastic cells resisted tenfold-higher forces. Further, when kinetic studies were performed on the latter system, binding strength exhibited 15-fold increase (from 1.6×10^{-11} to 2.4×10^{-10} N) during a 60 to 90 min incubation under conditions allowing metabolic activity.

Clearly, the centrifugation method is an attractive alternative to shearing techniques. The only problem is that the duration of exposure to disruptive forces cannot be lowered below at least 10 sec (or even 1 min with a conventional centrifuge) with the latter method.

2. Mechanical Resistance of Cell Aggregates

Cell aggregates are more difficult to break under controlled conditions than cell-substrate bonds. Although centrifugation techniques and micromanipulation might be used, in principle, for this purpose, it seems that experiments were essentially done with flow techniques.

Rat peritoneal macrophages were made to bind[38] sheep erythrocytes that had been treated with (1) glutaraldehyde, (2) immunoglobulin G, (3) complement, and (4) concanavalin A. These erythrocytes were bound by different classes of macrophage membrane receptors, and all of these except concanavalin A-binding structures triggered phagocytosis at 37°C. Binding experiments were thus done at 4°C to prevent the endocytic process, and conjugates were exposed to a shear rate on the order of 200,000 sec^{-1} in a syringe needle. The fraction of detached erythrocytes was 0.22, 0.68, 0.40, and 0.78, respectively, with the four erythrocyte batches tested. The binding strength was thus on the order of 2×10^{-8} N (assuming a radius of 2.5 μm for sheep erythrocytes). Interestingly, the most weakly bound erythrocytes (i.e., concanavalin A-coated red cells) were also the least efficient in triggering phagocytosis.

In another study,[9] rat thymocytes were agglutinated with different concentrations of concanavalin A, then exposed to shear of increasing intensity. The binding strength was markedly dependent on the lectin concentration. Indeed, when thymocytes were incubated with 0.125 μg/mℓ lectin (thus binding about 10^4 molecules per cell), 64% of adhesions were broken by a mean shear rate of 80,000 sec^{-1}. When the lectin concentration was raised to 2 μg/mℓ (each cell bound about 10^5 molecules) only 28% of adhesions were disrupted by the same treatment. In view of the measured value of the binding strength and a previous estimate of the maximal strength of a single bond, the authors concluded that at least 1000 to 3000 bonds were involved in thymocyte adhesion. Since they performed electron microscopic studies to estimate intercellular contact areas, and also measured the surface density of lectin molecules, they concluded that ligand molecules must gather into the interaction area. Direct demonstration of this phenomenon was recently achieved by direct quantification of fluorescent adhesion molecules in intercellular contact areas with microfluorimetric techniques.[76]

Shearing methods were also used to compare the strength of concanavalin A-mediated agglutination of two cell lines of different malignant potential. Interestingly, the more tumorigenic line displayed higher ability to form weak adhesions, but this difference disappeared when cell clusters were studied after exposure to a shear of intermediate intensity (33,000 sec^{-1}).[77]

Another study was devoted to the specific binding of target cells by cytotoxic T lymphocytes.[56,78] Under physiological conditions, binding results in the delivery of the so-called "lethal hit" by the lymphocyte, leading to the target disintegration within a few hours. In some experiments, target cells were labeled with radioactive chromium (allowing easy monitoring of cell damage by measuring released radioactivity). The main conclusions of the binding studies were as follows:

1. About half conjugates were disrupted when they were subjected to a shear rate on the order of 90,000 sec^{-1} (corresponding to a force of about 4.3×10^{-8} N.
2. Lymphocytes were made to bind target cells, then subjected to shear of varying intensity, and incubated under physiological conditions for several hours. Target lysis was measured immediately after shear and after the last incubation. It was found that shearing did not decrease the final lysis, but it yielded immediate lysis with a parallel decrease of delayed lysis. Also, free target cells were not damaged by shear. The simplest interpretation of these findings was that the target cells that were bound tightly enough to be damaged during shear were the same as were bound to be lysed during the final incubation (in the absence of shear).

V. GENERAL CONCLUSION AND PERSPECTIVES IN THE NEAR FUTURE

Several conclusions may be drawn from the present studies. First, flow methods were shown to allow quantitative or semiquantitative evaluation of several parameters of adhesion such as the time required for bond formation and mechanical resistance of intercellular bonds. However, it must be borne in mind that adhesion is a continuous process, with progressive formation of an increasing number of bonds and concomitant cell deformation. Hence, no exhaustive description of the adhesive process was achieved.

Second, the significance of binding strength deserves some comments. Indeed, cell separation must also be considered as a time-dependent process, which may account for the difference between the estimates for cell-cell binding strength (10^{-8} to 10^{-7} N) and cell-substrate adhesion (10^{-11} to 10^{-8} N), since cell aggregates were disrupted by very brief exposure to a separating force, say, less than 1 sec (see Section II), whereas substrate-bound cells were subjected to a disruptive force for a period of time on the order of 1 min, allowing progressive cell deformation to occur, with sequential rupture of adhesive bonds. A second point is that binding strength is limited at the same time by the mechanical resistance of cell membranes, the strength of intercellular bonds, and the strength of association between cells and cell-surface molecules.

Third, the wide heterogeneity of the properties of individual cells in any population must always be borne in mind. Indeed, this is a general feature of cell biology, which makes attractive the studies conducted at the single-cell level.

Fourth, the experimental results we reviewed strongly suggest that binding efficiency and binding strength are independent parameters, representative of different cell properties. Several recent studies strongly suggested that the formation of strong adhesion is an active cellular process[73] that may be correlated with the triggering of important cell functions such as phagocytosis[32] and macrophage-[73] or lymphocyte[78]-mediated cytotoxicity.

Several points should be clarified in the near future:

1. Whereas recent theoretical and experimental studies clarified the problem of hydrodynamic interaction between smooth spheres in simple shear flows, there is a need for more information on the influence of cell-surface asperities and glycocalyx on hydrodynamic forces. This is particularly important since some results may be critically dependent on intercellular distances at the nanometer level.[31] It is hoped that the wide availability of computer facilities will allow rapid gathering of numerical data relevant to reasonable cellular models within the next few years.

2. It seems that systematic variation of the medium viscosity should allow independent variation of the intensity and duration of hydrodynamic forces used to break intercellular aggregates, thus allowing quantitative study of rapid cellular responses to mechanical stimuli.

3. Since centrifugation methods seems an attractive alternative to flow methods, it is hoped that they will be adapted to the measurement of cell-cell binding strength (by using ultracentrifugation in high-density media), in order to complete the information yielded by shearing techniques.

The information made available by these techniques may allow experimental testing of quantitative models of cell adhesion.

APPENDIX:
SELECTED DEFINITIONS AND RESULTS IN FLUID MECHANICS

This appendix is devoted to the description of basic concepts of fluid mechanics in order to allow easier understanding and use of the results and equations reviewed in the present chapter. Also, underlying assumptions will be stated more explicitly.

Provided we are interested in systems much larger than fluid molecules, a macroscopic point of view is warranted. The fluid flow may then be described by indicating the mean molecular velocity $\vec{u}(M,t)$ at any point M and time t. Further, using a Cartesian reference frame with coordinates x_1, x_2, x_3 and unit vectors $\vec{e}_1$, $\vec{e}_2$, $\vec{e}_3$, the first order variation of $\vec{u}$ with respect to the position of M may be expressed as a combination of nine first-order derivatives $u_{i,j} = \partial u_i/\partial u_j$, where u_i is the ith component of u:

$$d\vec{u} = \sum_{i=1}^{3} \sum_{j=1}^{3} u_{i,j}\, dx_j\, \vec{e}_i \tag{A1}$$

Equation A1 is equivalent to the more compact matricial form:

$$dU = G \cdot dM \tag{A2}$$

where U and dM are 3×1 matrices and G is a 3×3 matrix with $u_{i,j}$ on line i and row j. It is usual to split G into an antisymmetric and a symmetric component:

$$G = A + S, \quad A = (G - G^+)/2 \quad S = (G + G^+)/2 \tag{A3}$$

where G^+ is the transpose of G. The interest of this procedure is that A represents a rotation, since Equation A2 yields, using a tensorial formalism:

$$d\vec{u} = \vec{\omega} \times d\vec{M} + \overleftrightarrow{S} \cdot d\vec{M}$$

$$\vec{\omega} = (1/2)\,\mathrm{rot}\,\vec{u}$$

$\overleftrightarrow{S}$ is a second rank tensor called the rate of strain. It is represented by matrix S.

Now, the fundamental dynamic equations express the variation of the velocity of a fluid element as a sum of applied forces:

$$\rho(\partial\vec{u}/\partial t + u_1\partial\vec{u}/\partial x_1 + u_2\partial\vec{u}/\partial x_2 + u_3\partial\vec{u}/\partial x_3) = \vec{F}_i + \vec{F}_e \tag{A4}$$

$\vec{F}_i$ and $\vec{F}_e$ are the volume densities of forces exerted by the fluid and external systems, respectively (that interactions may be represented by volume forces is not obvious.) Now, we shall restrict our analysis to the study of incompressible Newtonian fluids, a suitable model for aqueous solutions.

Incompressibility requires that the net fluid flow through any closed surface be zero. It is readily shown with elementary vector analysis that this is equivalent to:

$$\mathrm{div}\,\vec{u} = u_{1,1} + u_{2,2} + u_{3,3} = 0 \tag{A5}$$

The second conditions leads to a simple expression for $\vec{F}_i$:

$$\vec{F}_i = -\vec{\mathrm{grad}}\,p + \mu\Delta\vec{u} \tag{A6}$$

where p is the pressure field and μ the fluid viscosity. Δ is the usual notation for the Laplacian operator. Finally, $\vec{F}_e$ is usually $\rho\vec{g}$, where g is the gravitational acceleration. Another simplification is the common restriction to stationary flows. In this case, all partial derivatives with respect to time are zero and Equation A4 yields:

$$\rho(u_1 \partial \vec{u}/\partial x_1 + u_2 \partial \vec{u}/\partial x_2 + u_3 \partial \vec{u}/\partial x_3) = -\overrightarrow{\text{grad}}p + \mu \Delta \vec{u} + \rho \vec{g} \qquad (A7)$$

Now, it is instructive to examine the order of magnitude of the different terms of the above equation. Let d be a characteristic length of the system and G an order of magnitude for the rate of variation of velocities ($\partial u/\partial x$, expressed in $\sec^{-1}$). Since the flow problem is not altered by subtracting a constant velocity to all fluid elements, the significant part of u is on the order of Gd. The left side of Equation A7 is thus on the order of $\rho G^2 d$ and μ u is on the order of $\mu G/d$. The ratio between these quantities is therefore:

$$R = \rho G D^2 / \mu$$

This dimensionless quantity is the well-known Reynolds' number. Considering the flow around a cell, we take ρ as 10^3 kg/m^3, μ as 10^{-3} (i.e., kg.m^2.sec^{-1}) and d as about 10^{-5} m, which yields:

$$R = 10^{-4} G$$

It is concluded that the inertial term is negligible when the shear rate G is much smaller than 10,000. A similar analysis shows that the gravitational term is negligible when u is much higher than 10^{-4} m/sec (recall than the difference between the cell and the medium density is less than 10% of the medium density). When the above conditions hold, Equation A7 yields:

$$\overrightarrow{\text{grad}}p = \mu \Delta \vec{u} \qquad (A8)$$

This is the Navier-Stokes equation for quasistatic motion. This determines the flow together with the incompressibility Equation A5 and boundary conditions. The usual "nonslip" condition requires that the flow velocity be equal to the solid velocity at any point of the surfaces of solid particles embedded in the flowing medium.

The advantage of Equations A5 and A8 is their linearity. Hence, any linear combination of solutions of the flow equations is also a solution of these equations.

A general formalism elaborated by Brenner and O'Neill[14] allowed optimal use of this property to combine the results obtained by different authors and gain a thorough understanding of the behavior of interacting spheres exposed to simple flows. Since this formalism is now widely used, we shall give a brief account of the main points.

BEHAVIOR OF PARTICULATE SYSTEMS IN SHEAR FLOWS

Single-Solid Particle

The interaction between a sphere and a flow that would be linear (i.e., coefficients $u_{i,j}$ defined in Equation A1 would be constant) in its absence is considered. The basic idea is to take advantage of the linearity of Equations A5 and A8 and represent the flow as a sum of three simpler flows:

1. The undisturbed flow velocity is considered in the absence of the particle. The flow velocity at any point M may be expressed (following Equation A3) as:

$$\vec{u}(M) = \vec{u}(O) + \vec{\omega}_f \wedge \vec{OM} + \overleftrightarrow{S}.\vec{OM}$$

where O is an arbitrary point, $\vec{u}(O)$ is the flow velocity at 0, and $\vec{\omega}_f$ and $\overleftrightarrow{S}$ are a constant vector and a constant second-rank tensor, respectively. These are independent of the choice of O. The pressure is zero everywhere.

2. The flow velocity is $-\overleftrightarrow{S}.\vec{O}M$ on every point M of the particle surface and zero at infinity.

3. The flow velocity is $\vec{U}(O) - \vec{u}(O) + (\vec{\omega} - \vec{\omega}_r)\wedge\vec{O}M$ on the particle surface and zero at infinity. $\vec{U}(O) + \vec{\omega}\wedge\vec{O}M$ is the velocity of any point M of the particle during the actual flow.

Now, Brenner and O'Neill defined the following quantities: the force-torque vector $\vec{F}$ is a six-component vector. Its components $(F_1,F_2,F_3,To_1,To_2,To_3)$ are the components of the resultant $\vec{F}$ of external forces acting on the solid and their torque at point O $\vec{T}o$.

The velocity-spin vector U is a six-component vector consisting of the components of U(O) and ω (U$_1$[O], U$_2$[O], U$_3$[O], ω_1, ω_2, ω_3). The grand resistance matrix R is a 6 × 6 matrix that is dependent on O and the particle shape, but is is independent of the flow and medium viscosity. R is nonsingular (as demonstrated by noticing that energy dissipation by the flow must be (positive) and possesses an inverse. The shear-resistance matrix ϕ is a 6 × 6 matrix that is independent of the flow. The shear vector S is a six-component vector derived from the shear tensor $\overleftrightarrow{S}$ by ordering its components as follows: S_{11}, S_{22}, S_{33}, S_{23}, S_{13}, S_{12} where S_{ij} denotes the component on the ith line and jth row of matrix S. Now, the above quantities are related by the following equation:

$$F = -\mu(R\,U + \phi\,S) \qquad (A9)$$

Systems of Particles

A similar formalism was found to hold for a system of n particles. In this case, F and U are 6n × 1 matrices, ϕ is a 6n × 6 matrix, and R is a 6n × 6n matrix. For example, F is $(F_1, F_2, \ldots F_n, To_1, \ldots To_n)$, where F_i and To_i are the resultant external force and torque at point O_i of particle i.

The interest of this general formalism stems in the following two points:

1. Since R and ϕ are flow-independent, the grand resistance and shear resistance matrix relevant to a particular system (e.g., two spheres) may be obtained by combining solutions of restricted problems obtained by different authors (e.g., the movement of two spheres with identical velocity in a resting medium[79] or the movement of two spheres in contact in shear flow[80]).

2. Since R is regular, it may be inverted and the particle motion in response to external forces is readily obtained. This was especially useful when electrostatic repulsion or electrodynamic attraction were accounted for in different studies on the coagulation of spherical particles in shear flows.

Force Exerted on a Sphere Doubled in Simple Shear Flow

Arp and Mason[16] reported on the interaction between two identical spheres in single shear flows. In view of the symmetry of the systems, consideration of a single sphere allowed a complete description of the dynamics of the system. Now, in the absence of external forces, Equation A9 yields:

$$0 = \mu\,(R\,U + \phi\,S) \qquad (A10)$$

where U represents the velocity of sphere 1 in contact with sphere 2 in the absence of any bond. Now, if cells are linked by adhesive bonds and move as a single entity, the velocity of the system is entirely known when the rotation $\vec{\omega}^D = (\omega_1{}^D, \omega_2{}^D, \omega_3{}^D)$ has been calculated (recall that the point of contact between spheres has zero relative relocity with respect to undisturbed fluid flow). This rotation requires that a force $\vec{F}$ and torque $\vec{T}$ be applied on sphere 1. For reasons of symmetry, a force $-\vec{F}$ and torque $\vec{T}$ must be applied on sphere 2. Since the cell doublet is subjected to no external force, $\vec{T}$ must be zero. We may thus write:

$$F = \mu(R\ U^D + \phi\ S) \qquad (A11)$$

where $F = (F_1, F_2, F_3, O, O, O,)$ and U^D depends on three unknown parameters $\omega_1{}^D$, $\omega_2{}^D$, and $\omega_3{}^D$. Subtracting Equation A10 from Equation A11, we obtain:

$$F = \mu\ R\ (U^D - U) \qquad (A12)$$

This yields six equations allowing determination of six unknown parameters, namely the components of F and ω^D. This may be readily done using a reference frame $Ox_1x_2x_3$ linked to the cell doublet with the sphere centers O_1 and O_2 on axis Ox_3. Using equation 11 (for R) and 26 (for U) from Reference 16 and taking the limit at zero distance between the sphere surfaces, we obtain:

$$F = 19.2\mu a^2 G\ \cos\theta_2\ \sin2\theta_2 \qquad (A13)$$

$$T = 3.40\mu a^3 G(\cos^2\phi_2\ \cos^2 2\theta_2 + \sin^2\phi_2\ \cos^2\theta_2) \qquad (A14)$$

where F is the component of the hydrodynamic force that is parallel to the line-bearing centers, T is the hydrodynamic torque and ϕ_2 and θ_2 are the polar angles of Ox_3 relative to axis O_y of Figure 4. The maximum values of F and T are readily obtained:

$$Fm = 19.2\ \mu a^2 G \qquad (A15)$$

$$T_m = 3.4\ \mu a^3 G \qquad (A16)$$

where μ is the medium viscosity, a the sphere radius and G the shear rate.

Equations similar to A15 (with sometimes slightly different numerical coefficients) were used in different reports.[37,61,65] It may be noticed that depending on the size of the interaction area, the torque may play a dominant role in cell-cell separation.

REFERENCES

1. **Curtis, A. S. G. and Hocking, L. M.,** Collision efficiency of equal spherical particles in a shear flow, *Trans. Faraday Soc.*, 66, 1381, 1970.
2. **Van de Ven, T. G. M. and Mason, S. G.,** The microrheology of colloidal dispersions. VII. Orthokinetic doublet formation of spheres, *Colloid Polym. Sci.*, 255, 468, 1977.
3. **Von Fuchs, N.,** Uber die Stabilitat und Aufladund der Aerosole, *Z. Phys.*, 89, 736, 1934.
4. **Creighton, T. E.,** *Proteins — Structures and Molecular Properties*, W. H. Freeman, San Francisco, 1983, 182.
5. **Schlessinger, J., Koppel, D. E., Axelrod, D., Jacobson, K., Webb, W. W., and Elson, E. L.,** Lateral transport on cell membranes: mobility of concanavalin A receptors on myoblasts, *Proc. Natl. Acad. Sci. U.S.A.*, 75, 2409, 1976.

6. **Bell, G. I.**, Models for the specific adhesion of cells to cells, *Science*, 200, 618, 1978.

7. **Schmid-Schönbein, G. W., Sung, K. L. P., Tozeren, H., Skalak, R., and Chien, S.**, Passive mechanical properties of human leukocytes, *Biophys. J.*, 36, 243, 1981.

8. **Evans, E. A.**, Structural model for passive granulocyte behavior based on mechanical deformation and recovery after deformation tests, in *White Cell Mechanics: Basic Science and Clinical Aspects*, Meiselman, H. J., Lichtman, M. A., and LaCelle, P. L., Eds., Alan R. Liss, New York, 1984, 19.

9. **Capo, C., Garrouste, F., Benoliel, A. M., Bongrand, P., Ryter, A., and Bell, G. I.**, Concanavalin A-mediated thymocyte agglutination: a model for a quantitative study of cell adhesion, *J. Cell Sci.*, 56, 21, 1982.

10. **Weiss, L.**, Cell movement and cell surfaces: a working hypothesis, *J. Theor. Biol.*, 2, 236, 1962.

11. **Einstein, A.**, *Investigations on the Theory of the Brownian Movement*, Dover Press, New York, 1956, 1.

12. **Sommerfeld, A.**, *Mechanics of Deformable Bodies*, Academic Press, New York, 1964, 246.

13. **Reif, F.**, *Fundamentals of Statistical and Thermal Physics*, McGraw-Hill, New York, 1965, 487.

14. **Brenner, H. and O'Neill, M. E.**, On the Stokes resistance of multiparticle systems in a linear shear field, *Chem. Eng. Sci.*, 27, 1421, 1972.

15. **Bossis, G. and Brady, J. F.**, Dynamic simulation of sheared suspensions. I. General methods, *J. Chem. Phys.*, 80, 5141, 1984.

16. **Arp, P. A. and Mason, S. G.**, The kinetics of flowing dispersions. VIII. Doublets of rigid spheres (theoretical), *J. Colloid Interface Sci.*, 61, 21, 1977.

17. **Lichtman, M. A., Santillo, P. A., Kearney, E. A., Roberts, G. W., and Weed, R. I.**, The shape and surface morphology of human leukocytes. In vitro effect of temperature, metabolic inhibitors and agents that influence membrane structure, *Blood Cells*, 2, 507, 1976.

18. **Erickson, L. A. and Trinkaus, J. P. S.**, Microvilli and blebs as sources of reserve surface membrane during cell spreading, *Exp. Cell Res.*, 99, 375, 1976.

19. **Loor, F.**, Structure and dynamics of the lymphocyte surface, in *B and T Cells in Immune Recognition*, Loor, F. and Roelants, G. E., Eds., John Wiley & Sons, New York, 1977, 153.

20. **Wiegel, F. W.**, *Fluid Flow Through Porous Macromolecular Systems*, Springer-Verlag, Berlin, 1980, 1.

21. **Pumphrey, R.**, Computer models of the human immunoglobulins. I. Shape and flexibility, *Immunol. Today*, 7, 174, 1986.

22. **Hynes, R. O. and Yamada, K. M.**, Fibronectins: multifunctional modular glycoproteins, *J. Cell. Biol.*, 95, 369, 1982.

23. **Roseman, S.**, Complex carbohydrates and intercellular adhesion, in *The Cell Surface in Development*, Moscona, A. A., Ed., John Wiley & Sons, New York, 1974, 255.

24. **Bongrand, P., Capo, C., and Depieds, R.**, Physics of cell adhesion, *Prog. Surf. Sci.*, 12, 217, 1982.

25. **Ferrante, A. and Thong, Y. H.**, A rapid one-step procedure for purification of mononuclear and polymorphonuclear leukocytes from human blood using a modification of the Hypaque-Ficoll technique, *J. Immunol. Methods*, 24, 389, 1978.

26. **Von Smoluchowski, M.**, Versuch einen Mathematischen Theorie der Koagulationskinetic Kolloider Losungen, *Z. Phys. Chem.*, 92, 129, 1917.

27. **Swift, D. L. and Friedlander, S. K.**, The coagulation of hydrols by brownian motion and laminar shear flow, *J. Colloid Sci.*, 19, 621, 1964.

28. **Spielman, L. A.**, Viscous interactions in brownian coagulation, *J. Colloid Interface Sci.*, 33, 562, 1970.

29. **Sampson, K. J. and Ramkrishna, D.**, Particle size correlations and the effects of limited mixing on agglomerating particulate systems, *J. Colloid Interface Sci.*, 104, 269, 1985.

30. **Naishuman, G. and Ruckenstein, E.**, The brownian coagulation of aerosols over the entire range of Knudsen numbers: connection between the sticking probability and the interaction forces, *J. Colloid Interface Sci.*, 104, 344, 1985.

31. **Goldman, A. J., Cox, R. G., and Brenner, H.**, Slow viscous motion of a sphere parallel to a plane wall. II. Couette flow, *Chem. Eng. Sci.*, 22, 653, 1967.

32. **Capo, C., Bongrand, P., Benoliel, A. M., and Depieds, R.**, Dependence of phagocytosis on strength of phagocyte-particle interaction, *Immunology*, 35, 177, 1978.

33. **Mège, J. L., Capo, C., Benoliel, A. M., and Bongrand, P.**, Determination of binding strength and kinetics of binding initiation — a model study made on the adhesive properties of P388D1 macrophage-like cells, *Cell Biophys.*, 8, 141, 1986.

34. **Rabinovitch, M.**, Macrophage spreading in vitro, in *Mononuclear Phagocytes in Immunity, Infection and Pathology*, Van Furth, R., Ed., Blackwell Scientific, Oxford, 1975, 369.

35. **Folkman, J. and Moscona, A.**, Role of cell shape in growth control, *Nature (London)*, 273, 345, 1978.

36. **Jeffery, G. B.**, The motion of ellipsoidal particles immersed in a viscous fluid, *Proc. R. Soc. London Ser. A*, 102, 161, 1922.

37. **Albers, W. and Overbeek, J. T. G.**, Stability of emulsions of water in oil. III. Flocculation and redispersion of water droplets covered by amphipolar monolayers, *J. Colloid Sci.*, 15, 489, 1960.

38. **Derjaguin, B. V. and Landau, L. D.**, Theory of the stability of strongly charged sols and of the adhesion of strongly charged particles in solutions of electrolytes, *Acta Physicochim. URSS*, 14, 633, 1941.

39. **Verwey, E. J. W. and Overbeek, J. T. G.**, *Theory of the Stability of Lyophobic Colloids*, Elsevier, Amsterdam, 1948.

40. **Adler, P. M.**, Interaction of unequal spheres. I. Hydrodynamic interactions: colloidal forces, *J. Colloid Interface Sci.*, 84, 461, 1981.

41. **Adler, P. M., Takamura, K., Goldsmith, H. L., and Mason, S. G.**, Particle motions in sheared suspensions. XXX. Rotations of rigid and flexible dumbbells (theoretical), *J. Colloid Interface Sci.*, 83, 502, 1981.

42. **Arp, P. A. and Mason, S. G.**, Orthokinetic collisions of hard spheres in simple shear flow, *Can. J. Chem.*, 54, 3769, 1976.

43. **Takamura, K., Goldsmith, H. L., and Mason, S. G.**, The microrheology of colloidal dispersions. XII. Trajectories of orthokinetic pair-collisions of latex spheres in a simple electrolyte, *J. Colloid Interface Sci.*, 82, 175, 1981.

44. **Takamura, K., Adler, P. M., Goldsmith, H. L., and Mason, S. G.**, Particle motions in sheared suspensions. XXXI. Rotation of rigid and flexible dumbbells (experimental), *J. Colloid Interface Sci.*, 83, 516, 1981.

45. **Hochmuth, R. M., Mohandas, N., Spaeth, E. E., Williamson, J. R., Blackshear, P. L., Jr., and Johnson, D. W.**, Surface adhesion, deformation and detachment at low shear of red and white cells, *Trans. Am. Soc. Artif. Intern. Organs*, 18, 325, 1972.

46. **Doroszewski, J., Skierski, J., and Przadka, L.**, Interaction of neoplastic cells with glass under flow conditions, *Exp. Cell Res.*, 104, 335, 1977.

47. **Doroszewski, J.**, Short term and incomplete cell-substrate adhesion, in *Cell Adhesion and Motility*, Curtis, A. S. G. and Pitts, J. D., Eds., Cambridge University Press, London, 1980, 171.

48. **Charo, I. F., Yuen, C., and Goldstein, I. M.**, Adherence of human polymorphonuclear leukocytes to endothelial monolayers: effects of temperature, divalent cations and chemotactic factors on the strength of adherence measured with a new centrifugation assay, *Blood*, 65, 473, 1985.

49. **Weiss, L.**, The measurement of cell adhesion, *Exp. Cell Res. Suppl.*, 8, 141, 1961.

50. **Weiss, L.**, Studies on cellular adhesion in tissue culture. IV. The alteration of substrata by cell surfaces, *Exp. Cell Res.*, 25, 504, 1961.

51. **Bongrand, P., Capo, C., Benoliel, A. M., and Depieds, R.**, Evaluation of intercellular adhesion with a very simple technique, *J. Immunol. Methods*, 28, 133, 1979.

52. **Curtis, A. S. G.**, The measurement of cell adhesiveness by an absolute method, *J. Embryol. Exp. Morphol.*, 23, 305, 1969.

53. **Hoover, R. L.**, The effect of folic acid on glycosyl transferase activity and cell adhesion, *Exp. Cell Res.*, 106, 185, 1977.

54. **Evans, C. W. and Proctor, J.**, A collision analysis of lymphoid cell aggregation, *J. Cell Sci.*, 33, 17, 1978.

55. **Goldsmith, H. L. and Mason, S. G.**, Some model experiments in hemodynamics. V. Microrheological techniques, *Biorheology*, 12, 181, 1975.

56. **Bongrand, P. and Golstein, P.**, Reproducible dissociation of cellular aggregates with a wide range of calibrated shear forces: application to cytolytic lymphocyte-target cell conjugates, *J. Immunol. Methods*, 58, 209, 1983.

57. **Turitto, V. T. and Baumgartner, H. R.**, Platelet deposition on subendothelium exposed to flowing blood: mathematical analysis of physical parameters, *Trans. Am. Soc. Artif. Intern. Organs*, 21, 593, 1975.

58. **Turrito, V. T., Muggli, R., and Baumgartner, H. R.**, Physical factors influencing platelet deposition on subendothelium: importance of blood shear rate, *Ann. N.Y. Acad. Sci.*, 283, 284, 1977.

59. **Hochmuth, R. M. and Mohandas, N.**, Uniaxial loading of the red cell membrane, *J. Biomech.*, 5, 501, 1972.

60. **Houdjik, W. P. M., Sakariassen, K. S., Nievelstein, P. F. E. M., and Sixma, J. J.**, Role of factor VIII-von Willebrand factor and fibronectin in the interaction of platelets in flowing blood with monomeric and fibrillar human collagen types I and III, *J. Clin. Invest.*, 75, 531, 1975.

61. **Goldsmith, H. L., Lichtarge, O., Tessier-Lavigne, M., and Spain, S.**, Some model experiments in hemodynamics. VI. Two-body collisions between blood cells, *Biorheology*, 18, 531, 1981.

62. **Goldsmith, H. L., Gold, P., Schuster, J., and Takamura, K.**, Interactions between sphered human red cells in tube flow. Technique for measuring the strength of antigen-antibody bonds, *Microvasc. Res.*, 23, 231, 1982.

63. **Bell, D. N., Teirlinck, H. C., and Goldsmith, H. L.**, Platelet aggregation in Poiseuille flow. I. A double infusion technique, *Microvasc. Res.*, 27, 297, 1984.

64. **Bell, D. N. and Goldsmith, H. L.**, Platelet aggregation in Poiseuille flow. II. Effects of shear rate, *Microvasc. Res.*, 27, 316, 1984.

65. **Duszyk, M., Karvalec, M., and Doroszewski, J.,** Specific cell-to-cell adhesion under flow conditions, *Cell Biophys.,* 8, 131, 1986.

66. **Duszyk, M. and Doroszewski, J.,** Poiseuille flow method for measuring cell-to-cell adhesion, *Cell Biophys.,* 8, 119, 1986.

67. **Weiss, L. and Coombs, R. R. A.,** The demonstration of rupture of cell surfaces by an immunological technique, *Exp. Cell Res.,* 30, 331, 1963.

68. **Schmid-Schöenbein, G. W., Fung, Y., and Zweifach, B. W.,** Vascular endothelium-leukocyte interaction. Sticking shear forces in venules, *Circ. Res.,* 36, 173, 1975.

69. **Coman, D. R.,** Decreased mutual adhesiveness, a property of cells from squamous cell carcinomas, *Cancer Res.,* 4, 625, 1944.

70. **Zeidman, I.,** Chemical factors in the mutual adhesiveness of epithelial cells, *Cancer Res.,* 7, 386, 1947.

71. **McKeever, P. E.,** Methods to study pulmonary alveolar macrophage adherence. Micromanipulation and quantitation, *J. Reticuloendothelial Soc.,* 16, 313, 1974.

72. **Easty, G. C., Easty, D. M., and Ambrose, E. J.,** Studies of cellular adhesiveness, *Exp. Cell Res.,* 19, 539, 1960.

73. **McClay, D. R., Wessel, G. M., and Marchase, R. B.,** Intercellular recognition: quantitation of initial binding events, *Proc. Natl. Acad. Sci. U.S.A.,* 78, 4975, 1981.

74. **Sommers, S. D., Whisnant, C. C., and Adams, D. O.,** Quantification of the strength of cell-cell adhesion: the capture of tumor cells by activated murine macrophages proceeds through two distinct stages, *J. Immunol.,* 136, 1490, 1986.

75. **Lepoivre, M. and Lemaire, G.,** Quantitation of intercellular binding strength by disruptive centrifugation: application to the analysis of adhesive interactions between P815 tumor targets and activated macrophages, *Ann. Immunol. (Inst. Pasteur),* 137C, 329, 1986.

76. **McCloskey, M. A. and Poo, M. M.,** Contact induced redistribution of specific membrane components: local accumulation and development of adhesion, *J. Cell Biol.,* 102, 2185, 1986.

77. **Capo, C., Benoliel, A. M., Bongrand, P., Mishal, Z., and Berebbi, M.,** T-cell-fibroblast hybridoma deformability and concanavalin A-induced agglutination, *Immunol. Invest.,* 14, 27, 1985.

78. **Bongrand, P., Pierres, M., and Golstein, P.,** T-cell mediated cytolysis: on the strength of effector-target cell interaction, *Eur. J. Immunol.,* 13, 424, 1983.

79. **Stimson, M. and Jeffery, J. B.,** The motion of two spheres in a viscous fluid, *Proc. Roy. Soc. London Ser. A,* 111, 110, 1926.

80. **Nir, A. and Acrivos, A.,** On the creeping motion of two arbitrary-sized touching spheres in a linear shear field, *J. Fluid Mech.,* 59, 209, 1973.

Chapter 6

THE USE OF FLOW CYTOMETRY IN THE STUDY OF INTERCELLULAR AGGREGATION

David M. Segal

TABLE OF CONTENTS

I. INTRODUCTION

The simplest and perhaps most commonly used method for measuring intercellular aggregation is direct observation with a microscope. This method is extremely laborious and suffers from lack of objectivity and inaccuracies inherent in the measurement of relatively small numbers of cells. Flow techniques offer the possibility of objectively and rapidly scanning large numbers of cells for aggregation, and Orr and Roseman[1] were the first to apply flow techniques to measure cell-cell interactions. They followed the loss of single cells in suspension by using a Coulter Counter, a loss which occurred as single cells went into aggregates. This allowed a rapid, objective measurement, but did not directly detect multicellular aggregates.

Dual-fluorescence flow cytometry offers major improvements in the technology for measuring intercellular aggregation, especially in cases where conjugates between different types of cells are formed. To illustrate the technique conceptually, consider the system where cell A aggregates with cell B. In order to detect intercellular aggregation, cell A is labeled with a red fluorophore and cell B with a green one. Aggregation is allowed to proceed, and the cells are analyzed for red and green fluorescence in a flow cytometer as shown in Figure 1. In the cytometer, individual particles (single cells or conjugates) are channeled into a flow stream and are sequentially interrogated for red and green fluorescence at a rate of several hundred cells per second. The cytometer shown in Figure 1 contains two lasers; cells first pass the argon laser which excites the green fluorophore (usually fluorescein), but not the red, and emission is measured by detector 1. After traversing a distance of about 350 μm, the cell is next illuminated by a dye laser which excites the red fluorophore, but not the green, and emission is measured in detector 2. It is easy to see that single cells will give positive emission signals in either detector 1 or detector 2, but not in both detectors, whereas a conjugate which contains both types of cells will give positive signals in both detectors. Moreover, the intensity of each signal from a conjugate will be proportional to the number of cells of each type within the conjugate. Therefore, by using dual-fluorescence flow cytometry, samples containing 10^4 to 10^5 particles can be rapidly analyzed for relative contents of single cells and multicellular conjugates, and the cellular compositions of conjugates can be estimated from their fluorescence intensities relative to those of single (unconjugated) cells.

II. STAINING OF CELLS

In order to measure cell-cell interactions by dual-fluorescence flow cytometry, the interaction cell types must first be stained with fluorophores which are readily distinguishable from one another by the cytometer. For dual-laser instruments, dyes with widely separated excitation maximums, such as fluorescein (excitation maximum 492 nM) and various derivatives of rhodamine (e.g., rhodamine X-isothiocyanate [X-RITC], excitation maximum 580 to 590 nM, or Texas Red, excitation maximum 596 nM, can be used[2]). Figure 2 shows that cells stained with fluorescein isothiocyanate (FITC) give a strong positive signal when excited at 488 nM with an argon laser, but a very weak signal when excited with a krypton laser at 568 nM. Conversely, cells stained with X-RITC emit brightly when excited at 568 nM, but very weakly when excited at 488 nM. Mixtures of the two stained-cell populations give two easily distinguishable peaks.

Single-beam instruments, which are considerably less expensive than dual-laser cytometers, may also be used to measure intercellular aggregation. In this case, two dyes with similar excitation maximums but different emission maximums — e.g., fluorescein (emission maximum 517 nM) and phycoerythrin ([PE], excitation 480 nM; emission 578 nM) — can be used. Cells are excited near the excitation maximums and the emitted beams from the

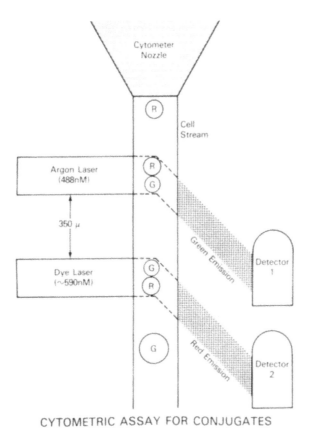

CYTOMETRIC ASSAY FOR CONJUGATES

FIGURE 1. Schematic representation of the dual color flow cytometric assay for cell-cell interactions. The cytometer measures interactions between two types of cells, one labeled with a red-emitting fluorophore (R), the other with a green (G) fluorophore. As particles pass through the cytometer they are first interrogated for green emission (by exciting with the argon laser and measuring emission in detector 1), and then for red emission (using the dye laser and detector 2). Single cells give positive signals in either detector 1 or detector 2, while conjugates give positive signals in both detectors.

two fluorophores are split, based on wavelength, by a dichroic mirror, each beam going to a separate detector. Because of some overlap in emission spectra, corrections must be applied for spillover of fluorescein emission into the PE detector and vice versa; this correction is usually applied electronically by the cytometer.

Dyes can be attached to cells in various ways. Direct surface labeling using FITC and X-RITC has been successfully employed for measuring multicellular conjugates at low temperature.[3,4] At higher temperatures, X-RITC-labeled cells lose some of their stain, which is then picked up by the second, FITC-labeled, population. With X-RITC, this fluorophore transfer leads to uninterpretable results, so that a different stain must be used. To overcome this problem, Perez et al.[5] surface-labeled cells with biotin, followed by Texas Red avidin. When used with FITC-labeled cells, transfer of the red dye was reduced to the point where meaningful results could be obtained at 37°C.

Surface labeling can affect intercellular aggregation, especially when the cells are labeled with high concentration of fluorophore. The extent of aggregation as measured cytometrically should be compared with the amount of aggregation measured microscopically using unlabeled cells. If the two measurements give widely differing estimates of aggregation, then

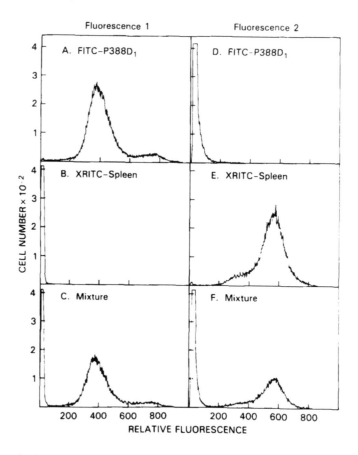

FIGURE 2. Fluorescence profiles of FITC-labeled P388D₁ cells (A,D) X-RITC-labeled, antibody-coated spleen cells (B,E), and a mixture of the two (C,F). Fluorescence 1 (A to C) is emission resulting from excitation with an argon laser, fluorescence 2 (D to F) is emission resulting from excitation with a krypton laser. The cytometer configuration is similar to that shown in Figure 1, except that a krypton laser is used in place of the dye laser.[3]

the surface stain is artifactually affecting the results and should not be used. In order to circumvent problems inherent in surface labeling, Luce et al.[6] have successfully used internal stains to measure conjugates. The dyes which they have found satisfactory for measuring T-cell conjugates are sulfo-fluorescein diacetate (SFDA) and hydroethidine (HE). In their precursor forms, both dyes are hydrophobic and readily pass through plasma membranes. Inside cells, they are chemically modified, becoming charged, fluorescent molecules which cannot excape from the cellular interior. The excitation maximums of both SFDA and HE are in the same range (around 490 to 500 nM), but the emission maximums are widely separated (530 nM for SFDA and 610 nM for HE). Therefore, a single-beam cytometer can be used with these dyes. More recently, Storkus et al.[7] have measured conjugates between NK cells, internally labeled with fluorescein diacetate, and tumor cells internally labeled with Hoechst 33342. With these two dyes, a dual-beam instrument is required.

Specific labeling, e.g., with fluorescent antibodies or ligands, offers another approach for providing fluorescently labeled cells. While studies using specific labels in the measurement of intercellular aggregations have not yet been published, the possibility could be extremely useful when heterogeneous cell populations are examined for conjugate formation. For examp e, in an experiment described below, studies were done to determine the phenotypes o eripheral blood leukocyte (PBL) subsets which form conjugates with antibody-coated target cells. The target cells were labeled directly with a red fluorophore, while the

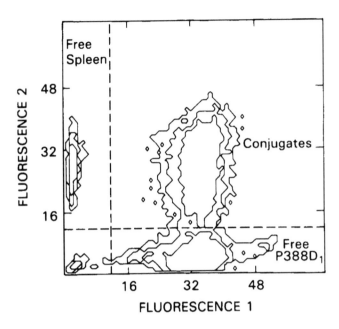

FIGURE 3. Contour plot of antibody-coated spleen cells labeled with a red fluorophore (fluorescence 2) mixed with P388D₁ cells labeled with a green fluorophore (fluorescence 1). 5×10^4 particles were analyzed and contours were drawn to indicate the major peaks. P388D₁ cells were considered to be unconjugated if their red fluorescence lay below the horizontal line, and spleen cells were unconjugated if their green fluorescence lay to the left of the vertical line. Conjugates appear in the center of the plot, as particles which are positive in both the red and green.[3]

PBL were labeled with fluorescein-conjugated monoclonal antibodies. Red-green particles appeared only when the antibody bound to the subset of PBL which formed conjugates, thus defining the phenotype of the conjugate-forming cells. Fluorescent probes which measure cellular properties,[8] such as cytoplasmic calcium concentration or membrane potential, might also be useful in measuring intercellular aggregation. Thus, it might be possible to detect a calcium influx or a change in membrane potential when two or more cells aggregate.

III. MEASUREMENT OF CONJUGATES BY FLOW CYTOMETRY

Once the cells are stained, intercellular aggregation is allowed to proceed according to the requirements of the system. The cells are diluted to about $10^6/m\ell$ in a medium which does not disaggregate conjugates, and are analyzed in the cytometer. The most convenient output is a contour diagram such as that shown in Figure 3 for the Fc receptor-mediated aggregation of antibody-coated splenocytes (stained with X-RITC) with the P388D₁ mouse macrophage line (stained with FITC). In Figure 3, fluorescence 1 and fluorescence 2 refer to green (FITC) and red (X-RITC) fluorescence emissions, respectively, and fluorescence is measured on a logarithmic scale, six units corresponding to a factor of two increase in fluorescence. Contour lines indicate the numbers of particles emitting with various red and green intensities. Thus, some particles are red but not green (free spleen cells), others are green but not red (free P388D₁ cells), and others are both red and green (conjugates). The percentages of particles lying within each group can be rapidly calculated, using a digital computer, by summing the particles within each area and dividing by the total number of particles analyzed (5×10^4 in Figure 3).

An important question concerning the measurement of multicellular aggregates by flow

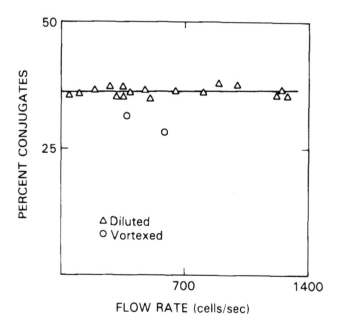

FIGURE 4.   The detection of conjugates at varying flow rates. Conjugates between antibody-coated lymphocytes and P388D$_1$ cells were measured at varying flow rates. Triangles represent multiple measurements from a single sample; the circles represent measurements from samples which had been vortexed prior to assay.[3]

cytometry is whether the shear forces generated by the flow system of the cytometer are sufficient to disaggregate conjugates. This should be investigated experimentally for each system, e.g., by comparing microscopic with cytometric results. Since shear forces within the cytometer will vary with flow rate, another method of detecting disaggregation is to measure conjugates at differing flow rates. Figure 4 shows that the number of conjugates measured between antibody-coated lymphocytes and P388D$_1$ cells is independent of flow rate, whereas if the cells were vortexed prior to measurement, the number of conjugates decreased. In other controls,[3] the number of conjugates measured did not vary with nozzle diameter. Therefore, shear forces do not seem to affect the flow cytometric measurement of conjugates between P388D$_1$ cells and antibody-coated lymphocytes. These results agree with the theoretical considerations of Bell,[9] who estimated that even a few relatively weak bonds between two cells should lead to stable conjugates.

Data such as those presented in Figure 3 can be used to estimate the composition of multicellular conjugates. If a conjugate contains multiple red and green cells, then the red and green fluorescence-emission intensities from the conjugate will be multiples of those arising from single, unconjugated cells. For example, in Figure 5 (A to C), the green fluorescence of P388D$_1$ cells in conjugates is compared with the green fluorescence of free P388D$_1$ cells. At a variety of P388D$_1$-to-lymphocyte ratios, the fluorescence profiles of conjugates overlap those of free P388D$_1$ cells, demonstrating that the great majority of conjugates contain one P388D$_1$ cell. By contrast, the red fluorescence of conjugates is greater than that of free spleen cells (Figure 5 D to F), and the difference increases as the spleen-to-P388D$_1$ ratio increases. Because of the relatively broad staining distribution of the unconjugated cells, it is impossible to accurately determine the numbers of conjugates containing 1, 2, 3, or more spleen cells. However, the peak fluorescence of the red cells within the conjugate peak in panel D is 2.7 times that of the single cell (when logarithmic data is converted to linear), and the fluorescence of the major peak in panel E is 1.9 that of the unconjugated peak.

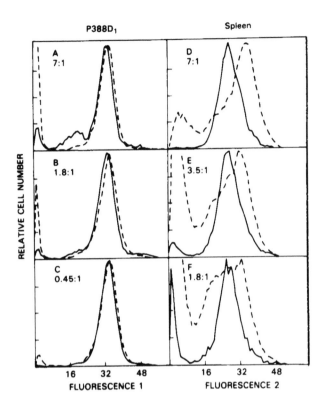

FIGURE 5. Comparison of fluorescence distributions of conjugated and unconjugated cells. In A to C the (green) fluorescence profiles of unconjugated P388D₁ cells (solid lines) are superimposed upon those of the conjugated cells (dashed lines). In D to F the (red) profiles of conjugated and unconjugated spleen cells are compared. Ratios of spleen to P388D₁ cells in the original mix are indicated on each panel. This figure shows that conjugates contain one P388D₁ cell and several spleen cells, depending upon the initial cell ratios.[3]

The numbers of particles which are conjugates or single cells can be determined directly from the cytometer data by integrating the appropriate areas of the contour plot as shown, for example, in Figure 3. The percentage of particles lying within a given region of the contour plot can then be calculated by dividing the number of particles within that region by the total number of particles analyzed and multiplying by 100. We designate these percentages as P_1, P_2, or P_c, where P_1 and P_2 are the percentages of particles of type 1 or 2 lying within the single cell regions, and P_c is the percentage of particles falling within the conjugate region. If the conjugates contain, on the average, N_i cells of type i, then the fraction of cells of type i (f_i) which are in conjugates can be calculated from Equation 1:

$$f_i = \frac{N_i\, P_c}{P_i + N_i\, P_c} \tag{1}$$

For example, if in Figure 3, 30% of the particles were free spleen cells and 10% of the particles were conjugates with an average of two spleen cells per conjugate, the fraction of spleen cells in conjugates would be 0.4.

Because conjugates contain multiple numbers of cells, the number of particles measured by the cytometer is less than the total number of cells within a given sample. The percentages of particles P_1, P_2, or P_c, as determined directly from the contour plots, can be converted to fractions of total cells F_1, F_2, or F_c using Equations 2 to 4:

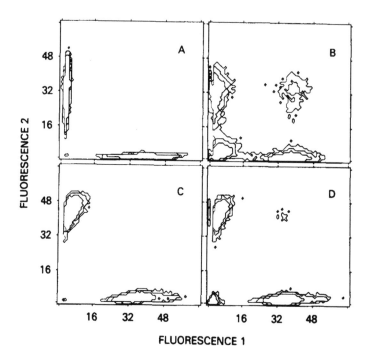

FIGURE 6.   Test for homoaggregation between antibody-coated splenocytes (A) or P388D$_1$ cells (C). In A, one aliquot of spleen cells was labeled red and a second aliquot green. When the two were mixed, no conjugates were observed. Panel C, same as A except P388D$_1$ cells were tested. In panel B, unlabeled P388D$_1$ cells were added to the mixture of red- and green-labeled antibody-coated spleen cells. Red-green particles are formed when a P388D$_1$ cell binds both red and green spleen cells. In panel D, unlabled, antibody-coated spleen cells are added to red- and green-labeled P388D$_1$ cells. Very few red-green particles are observed because the great majority of conjugates contain one P388D$_1$ cell.[3]

$$F_1 = \frac{P_1}{100 + P_c (N_1 + N_2 - 1)} \tag{2}$$

$$F_2 = \frac{P_2}{100 + P_c (N_1 + N_2 - 1)} \tag{3}$$

$$F_c = \frac{P_c (N_1 + N_2)}{100 + P_c (N_1 + N_2 - 1)} \tag{4}$$

For example, consider a sample containing 100 cells of type 1, 100 cells of type 2, and 100 conjugates, each with one type 1 cell and one type 2 cell. Then P_1, P_2, and P_c, as measured by the cytometer, would each be 33.3%. However, since the sample contains a total of 400 cells, $F_1 = F_2 = 0.25$, and $F_c = 0.5$; that is, half of the cells would be in conjugates.

IV. SPECIFICITY OF CONJUGATES

The flow cytometric technique offers a sensitive and accurate means for assessing the specificity of intercellular aggregation. The first question to be asked about specificity is whether a particular cell type binds to itself or whether it will only aggregate with a different type of cell. In Figure 6, antibody-coated spleen cells were separated into two fractions,

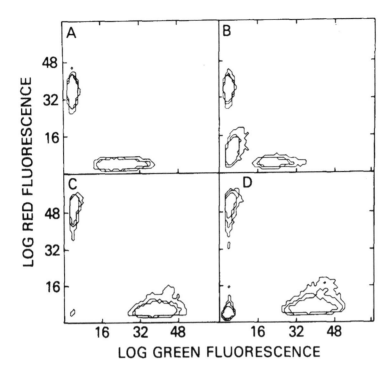

FIGURE 7. Test for homoaggregation of B6 spleen cells (panel A) or CTL clone 5.6 cells (panel C). No homoaggregation is observed. When unlabeled 5.6 cells are added to a mixture of red and green B6 lymphocytes (their natural target, compare Figure 8) (panel B), or when the reverse experiment is done (panel D), no multicolor particles are detected, confirming that each conjugate contains 1 cytotoxic cell (5.6) and 1 target cell.[5]

one labeled with a red fluorophore, the other with a green one. When mixed (panel A), no red to green particles were detected, showing that these cells do not aggregate with themselves. Similarly, P388D$_1$ cells (panel D) also do not form homoconjugates. Taken together with the data of Figure 3, it is clear that when P388D$_1$ cells are mixed with antibody-coated lymphocytes, only heteroconjugates are formed. In a similar experiment (Figure 7), neither cloned cytotoxic T lymphocytes (CTL) nor a target cell with which they form conjugates (see below) show any homoaggregate formation.

Also shown in Figures 6 and 7 are contour plots of experiments in which red and green cells of one type are mixed with unstained cells of the second type. Because multiple spleen cells bind to a single P388D$_1$, particles which are both red and green appear when unstained P388D$_1$ cells are mixed with red and green antibody-coated lymphocytes (Figure 6B). By contrast, since most conjugates contain a single P388D$_1$ cell, few red to green particles are formed when unstained lymphocytes are added to red and green P388D$_1$ cells (Figure 6D). The great majority of CTL conjugates contain one CTL and one target cell, and therefore no red to green particles are seen when similar experiments are done with this system (Figure 7, panels B and D).

In systems where heteroconjugates form, specificity of aggregation is conveniently measured by changing cell types. For example, in Figure 8, conjugation is examined between CTL clones and lymphocytes from mice of varying haplotypes. Clones 5.6 and 37 are of Bm.10 origin and specifically lyse cells of the b haplotype. Panels A to F show that the lytic specificities are reflected in their ability to form conjugates. Thus, in panels A and C, 3 and 7% of clone 5.6 cells form conjugates with the B10.D2 and Bm.10 lymphocytes, respectively, while 28% of the 5.6 cells form conjugates with the correct (B6) target. Similar

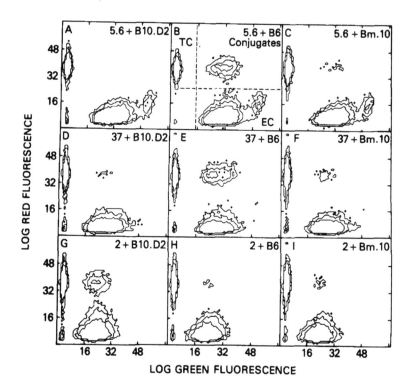

FIGURE 8.   Specificity of conjugate formation between cloned CTL and spleen cells. FITC (green)-labeled cells from murine CTL clones 5.6 (A to C), 37 (D to F), and 2 (G to I) were mixed with equal numbers of Texas Red-labeled B10.D2 (A, D, G), B6 (B, E, H), and Bm.10 (C, F, I) spleen cells. The cells were centrifuged, incubated for 30 min at 24°C and analyzed by flow cytometry. 5×10^4 particles were analyzed in each sample. Cones 5.6 and 37 specifically lyse H-2^b but not H-2^{bm10} nor H-2^d targets, while clone 2 is specific for H-2^d targets.[5]

results are obtained with clone 37 (6 and 10% nonspecific, 26% specific, panels D to F). Another clone (2, panels G to I) specifically lyses cells from H-2^d mice, and conjugate formation again parallels lysis. In the Fc receptor-dependent system (Figure 9), specificity is demonstrated either by using lymphocytes which have no antibody on their surfaces or by blocking the Fc receptors on the P388D$_1$ cells with an anti-Fc receptor monoclonal antibody. In this system, specificity of conjugate formation is greater than for CTL. In panels A and B of Figure 9, the percentages of particles in the conjugate region are less than 1%, while in panel C, where antibody-coated splenocytes form conjugates with unblocked P388D$_1$ cells, over 25% of the particles are conjugates.

V. APPLICATIONS

To date, dual fluorescence-flow cytometry has been used mainly to measure cell-cell interactions in the immune system, especially those involving the interaction of cytotoxic cells with various types of target cells.[3-7,10-12] Flow cytometry is a particularly powerful technique for measuring the kinetics of intercellular aggregation, as for example in Figure 10, where CTL clone 5.6 cells are forming conjugates with B6 lymphocytes. Fluorescently stained effector and target cells were mixed and sedimented at 0 time and analyzed by flow cytometry at varying time intervals and at three different temperatures. The cytometer data were then corrected using Equations 2 to 4 with $N_1 = N_2 = 1$ (Figure 7), and fit to first-order rate equations using the rate constants indicated in Table 1. These data show that the

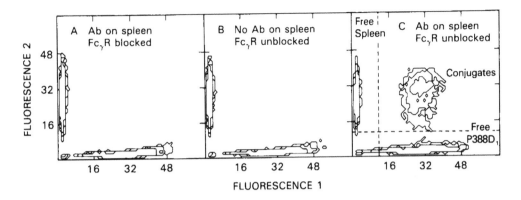

FIGURE 9. Specificity of conjugates formation between antibody-coated spleen cells and P388D₁ cells. In panel A, P388D₁ cells were coated with the anti-FcR monoclonal antibody 2.4G2, and then mixed with antibody-coated splenocytes. In B, untreated P388D₁ cells were incubated with spleen cells not coated with antibody, and in C, untreated P388D₁ cells were incubated with antibody-coated spleen cells. Conjugates are detected only in panel C, showing that both free Fc receptors on the P388D₁ cells, and IgG antibodies on the spleen cells are required for conjugate formation.[3]

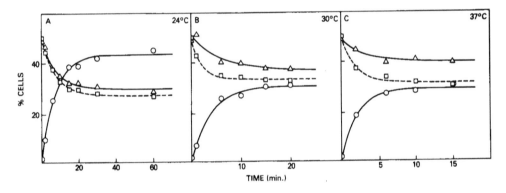

FIGURE 10. Kinetics of conjugate formation between CTL clone 5.6 and B6 spleen cells at varying temperatures. (○), percentages of cells in conjugates, (△) percentage of free 5.6 cells, and (□) percentage of free B6 spleen cells. Cells were mixed, pelleted, gently resuspended, and were assayed by flow cytometry. Data were corrected using Equations 2 to 4, with $N_1 = N_2 = 1$. Single first-order fits were made to the data using the rate constants given in Table 1.[5]

rates of conjugate formation or single-cell depletion increase by factors of 3 to 4 when the temperature increases from 24 to 37°C. By contrast, the number of cells in conjugates after the reaction has reached a plateau decreases with increasing temperature, from 43% at 24° to 29% at 37°C.

If the total number of effector cells and target cells are measured at each time point, one finds that the effector-target (E:T) ratios remain constant with time at 24° and 30°C (temperatures where lysis does not occur in ^{51}Cr release assays), but increase after about 15 min at 37°C (Figure 11A). Presumably this increase in E:T ratio is a result of target-cell lysis. Therefore, the cytometric data can be used to follow target-cell lysis in addition to conjugate formation. Figure 11B shows that after a 10 to 15 min lag phase, target-cell lysis commences, with 30% of the targets being lysed after 1 hr incubation. Measurements of conjugates at time points later than those shown in Figure 10C revealed that no change in the percentage of cells in conjugates occurred for the first 30 min, while the percentage of cells in conjugates decreased gradually from 31 to 24% over the next 30 min. This low-level decrease of conjugates suggests that new conjugates constantly form as target cells are lysed and released

Table 1
KINETIC PARAMETERS
FOR THE FORMATION OF
CONJUGATES BETWEEN
CTL CLONE 5.6 AND B6
SPLEEN CELLS[5]

Kinetic constant[a]	24°C	30°C	37°C
CC^{max}	43.3	32.6	29.0
k_1	0.14	0.20	0.48
k_2	0.13	0.16	0.35
k_3	0.14	0.26	0.56

[a] CC^{max}, the percentage of cells in conjugates at infinite time k_1, k_2, and k_3, first-order rate constants (per minute) for conjugate formation, free 5.6 depletion, and free B6 spleen cell depletion, respectively.

from effectors. Recycling of CTL during target-cell lysis is a well-known phenomenon in this system.

Target-cell lysis can, in some cases, be followed more directly by internally labeling the target cell and then monitoring dye release. Figure 12 shows the results of an experiment where antibody-coated chicken red blood cells (CRBC) are lysed by human PBL. The PBL are labeled with a red fluorophore (Texas Red avidin on biotinylated PBL) and the CRBC are internally labeled with carboxyfluorescein. Internal labeling with carboxyfluorescein is done by incubating the red cells with 10 to 15 μM carboxyfluorescein diacetate for 15 min at 37°, washing, and incubating 30 min at 37°. The carboxyfluorescein diacetate, being an apolar molecule, passes through the plasma membrane into the cytoplasm, where it is hydrolyzed by esterases. Once hydrolyzed, the resulting carboxyfluorescein is a charged, polar, highly fluorescent molecule which escapes very slowly from the interior of the cell, unless, of course, the cell is lysed, in which case it is rapidly released. In Figure 12A, the antibody-coated CRBC form clearly discernible conjugates with PBL at 0°C. Upon warming to 37°C, much of the carboxyfluorescein is lost from the conjugated CRBC, but not from the free CRBC, by 30 min, and most is gone by 108 min. The loss of dye from conjugated CRBC can be seen more clearly in single-parameter histograms (Figure 13). The conjugated target cells vanish at 30 min in the contour plots (Figure 12B) because the distribution of fluorescence intensities of conjugated target cells broadens (Figure 13, middle panel); therefore, the conjugates fall below the contour levels and disappear. The target cells then reappear (Figure 12C) at 108 min because their fluorescence intensities sharpen around a low-average green fluorescence value (Figure 13, lower panel).

In the experiment shown in Figures 12 and 13, the cytotoxic cells are a heterogeneous mixture of lymphocytes and monocytes. In order to determine the types of cells in conjugates, the CRBC were labeled with a red fluorophore (Texas Red avidin on biotinylated CRBC), while the PBL were labeled with various monoclonal antibodies conjugated with the green fluorophore, FITC. Figure 14 shows the results of an experiment in which the PBL were stained with FITC-OKM1, an antibody which binds to monocytes, and to a lesser extent to K cells (cells which express E rosette and Fcγ receptors and mediate ADCC). Control experiments showed that OKM1 did not bind to CRBC. The contour plot of antibody-coated,

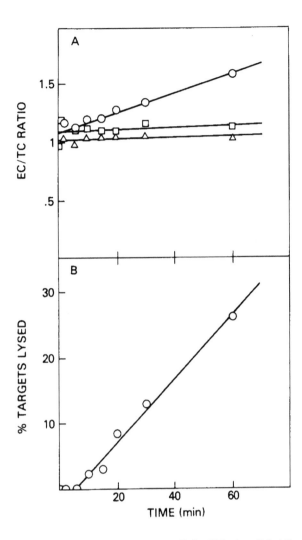

FIGURE 11. Lysis of B6 spleen cells by CTL clone 5.6. (A) In the experiment shown in Figure 10, the percentages of free plus conjugated effector cells (EC) and free plus conjugated target cells (TC) were calculated at each time point, and the EC/TC ratios plotted at 24°C ($\triangle$), 30°C ($\square$), and 37°C ($\bigcirc$). Lysis is detected at 37° by the time-dependent increase in the EC/TC ratio; (B) the percentages of target cells (B6 spleen) lysed at 37° vs. time, calculated from the cytometer data as described in Reference 5.

Texas-Red-labeled CRBC mixed with FITC-OKM1-labeled PBL is shown in Figure 14, panel A. The cells which are negative in the red (free PBL) have three distinct peaks, negative, dull, and bright (panels C and F). Those which are positive in the red, free CRBC and conjugates also show a complex staining pattern. In order to distinguish free CRBC from OKM1-negative conjugates, we made use of the fact that CRBC are smaller by low-angle light scatter than PBL and conjugates. Figure 14B shows the OKM1 distribution on large, red positive particles, i.e., conjugates. The distribution falls into three clearly discernible peaks. The lowest peak probably results from a small amount of agglutination of the red cells which occurred in this experiment, while the intermediate and high OKM1 peaks correspond to cells which were previously identified as K cells and monocytes,

LYSIS OF Ab-CRBC BY HUMAN PBL

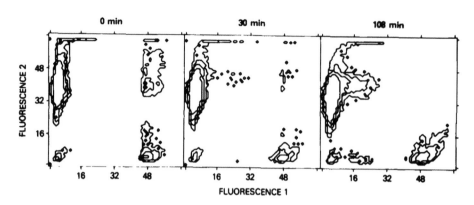

FIGURE 12.   Lysis of antibody-coated chicken red blood cells by human peripheral blood mononuclear cells. Mononuclear cells were labeled with biotin followed by Texas Red avidin, and chicken erythrocytes were internally labeled by incubating them with carboxyfluorescein diacetate. They were then modified with trinitrobenzene sulfonate and coated with anti-DNP IgG antibodies. Conjugates were formed by incubating red cells and mononuclear cells together at 0°. At 0 time the cells were warmed to 37° and analyzed periodically by flow cytometry.

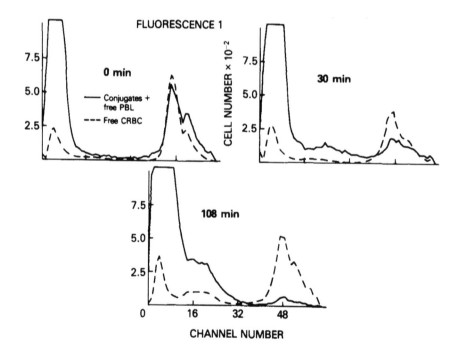

FIGURE 13.   The loss of green fluorescence from conjugated CRBC. The distributions of green fluorescence in cells which are negative (free CRBC) and positive (conjugates + free PBL) in the red are indicated at various incubation times. Histograms are generated from the data shown in Figure 12, and show the loss of carboxyfluorescein from conjugated, but not from free CRBC.

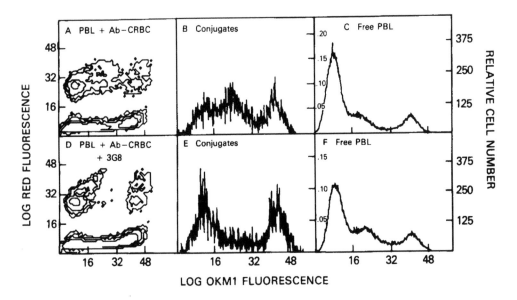

FIGURE 14.   Distribution of OKM1 positive cells in conjugates containing antibody-coated CRBC (labeled with a red fluorophore) and human PBL. Conjugates were formed by incubating CRBC and PBL at 0°, and then staining the mixture with FITC-OKM1. In A to C, conjugates were formed in medium alone. in D to F, they were formed in medium containing monoclonal antibody 3G8, which binds to K cell but not to monocyte Fc receptors. A and D are the contour plots of all the cells. B and E show the OKM1 distributions on conjugates (red positive particles with high light scatter), and C and F show the OKM1 distributions on free PBL (cells negative in the red).

respectively.[13] When conjugates were formed in the presence of 3G8, a monoclonal antibody against Fcγ receptors on K cells, but not on monocytes,[14,15] the intermediate peak disappears (Figure 14E), thus confirming that conjugates between K cells and antibody-coated CRBC comprise the intermediate peak.

VI. FUTURE POSSIBILITIES

Multicolor-flow cytometric analyses of cell-cell interactions should be adaptable to many different systems and applications. Many cytometers have the capacity to sort cells on the basis of fluorescence emission, and Storkus et al.[7] have recently shown that NK conjugates could be isolated from unconjugated cells by cell sorting. This might be particularly useful for cloning conjugate-forming cells. Although flow cytometry has been used mainly to probe cell-cell interactions in the immune system, it should also be suitable for measuring the initial events in organogenesis and fertilization, and for following more complex interactions such as the specific binding of one nerve cell to another. The use of fluorescently labeled monoclonal antibodies to stain the interacting partners (as in Figure 14) might allow one to follow the progression of interactions which result in the formation of complex structures. Many of the new cytometers can detect three colors simultaneously; by using two colors to measure conjugates and the third to measure antibody binding, the sensitivity of the method should be improved. Finally, several dyes are now available which measure cellular properties such as cytoplasmic calcium concentration and membrane potential.[8] By using such dyes, it may be possible to measure the immediate effects of conjugate formation on the physiology of the interacting cells.

REFERENCES

1. **Orr, C. W. and Roseman, S.,** Intercellular adhesion. I. A quantitative assay for measuring the rate of adhesion, *J. Membr. Biol.,* 1, 109, 1969.
2. **Segal, D. M., Titus, J. A., and Stephany, D. A.,** The use of fluorescence flow cytometry for the study of lymphoid cell receptors, *Methods Enzymol.,* in press.
3. **Segal, D. M. and Stephany, D. A.,** The measurement of specific cell:cell interaction by dual-parameter flow cytometry, *Cytometry,* 5, 169, 1984.
4. **Segal, D. M. and Stephany, D. A.,** The mechanism of intercellular aggregation. I. The kinetics of $Fc\gamma$ receptor-mediated aggregation of $P388D_1$ cells with antibody-coated lymphocytes at 4°C, *J. Immunol.,* 132, 1924, 1984.
5. **Perez, P., Bluestone, J. A., Stephany, D. A., and Segal, D. M.,** Quantitative measurements of the specificity and kinetics of conjugate formation between cloned cytotoxic T lymphocytes and splenic target cells by dual parameter flow cytometry, *J. Immunol.,* 134, 478, 1985.
6. **Luce, G. G., Sharrow, S. O., and Shaw, S.,** Enumeration of cytotoxic cell-target cell conjugates by flow cytometry using internal fluorescent stains, *Biol. Tech.,* 3, 270, 1985.
7. **Storkus, W. J., Balber, A. E., and Dawson, J. R.,** Quantitation and sorting of vitally stained natural killer cell-target cell conjugates by dual beam flow cytometry, *Cytometry,* 7, 163, 1986.
8. **Shapiro, H. M.,** *Practical Flow Cytometry,* Alan R. Liss, New York, 1985, chap. 7.
9. **Bell, G. I.,** Models for the specific adhesion of cells to cells, *Science,* 200, 618, 1978.
10. **Berke, G.,** Enumeration of lymphocyte-target cell conjugates by cytofluorometry, *Eur. J. Immunol.,* 15, 337, 1985.
11. **Perez, P., Hoffman, R. W., Shaw, S., Bluestone, J. A., and Segal, D. M.,** Specific targeting of cytotoxic T cells by anti-T3 linked to anti-target cell antibody, *Nature (London),* 316, 354, 1985.
12. **Karpovsky, B., Titus, J. A., Stephany, D. A., and Segal, D. M.,** Production of target-specific effector cells using hetero-cross-linked aggregates containing anti-target cell and anti-$Fc\gamma$ receptor antibodies, *J. Exp. Med.,* 160, 1686, 1984.
13. **Titus, J. A., Sharrow, S. O., and Segal, D. M.,** Analysis of Fc (IgG) receptors on human peripheral blood leukocytes by dual fluorescence flow microfluorometry. II. Quantitation of receptors on cells that express the OKM1, OKT3, OKT4 and OKT8 antigens, *J. Immunol.,* 130, 1152, 1983.
14. **Fleit, H. B., Wright, S. D., and Unkeless, J. C.,** Human neutrophil Fc receptor distribution and structure, *Proc. Natl. Acad. Sci. U.S.A.,* 79, 3275, 1982.
15. **Perussia, B., Trinchieri, G., Jackson, A., Warner, N. L., Faust, J., Rumpold, H., Kraft, D., and Lanier, L. L.,** The Fc receptor for IgG on human natural killer cells: phenotypic, functional, and comparative studies with monoclonal antibodies, *J. Immunol.,* 133, 180, 1984.

Chapter 7

MICROMETHODS FOR MEASUREMENT OF DEFORMABILITY AND ADHESIVITY PROPERTIES OF BLOOD CELLS AND SYNTHETIC MEMBRANE VESICLES

Evan Evans

TABLE OF CONTENTS

I. INTRODUCTION

In Chapter 4, an outline was presented for the mechanics of cell and membrane capsule deformation and the mechanics of membrane-surface adhesion. Two separate aspects of adhesion were shown to be important: contact formation and subsequent separation. Formation of adhesive contact is promoted by the long-range attraction between surfaces, and is represented by a reversible free-energy potential (affinity) per unit area. On the other hand, contact separation may be opposed by additional forces due to specific molecular bonds (see the discussion in Chapter 10) that greatly augment the adhesion energy. Consequently, the stresses induced in cells and membrane surfaces by spontaneous spreading (formation) of an adhesive contact may be much lower than the stresses created by external forces applied to separate the contact. Distinct experiments must be designed to probe these two aspects of adhesion. In all experiments, a cell or membrane capsule acts as an intermediate body (coupler) between the contact zone and the transducer that is used to measure and apply the force. Therefore, the mechanical properties of the cell or membrane are requisite parameters in the analysis of data for forces applied to the body, either to control contact formation or to separate an adherent contact. The goal is to derive the stresses (e.g., membrane tension) proximal to the contact zone from the forces measured by the transducer. Thus, it is necessary to briefly consider methods for measurement of mechanical properties of cells and membrane vesicles as well as methods for testing adhesive properties.

Direct mechanical experiments on single cells and large lipid vesicles offer distinct advantages over the study of these capsules in suspension because the forces and associated deformations of individual bodies can be controlled and measured. There are two significant difficulties in the design of mechanical tests for single cells and lipid vesicles. First, the size of these capsules and regions of interest are microscopic, with dimensions that are on the order of 10^{-4} cm or greater and as such are difficult to observe accurately because of optical diffraction. Second, the forces required to deform (or even destroy!) elements of a cell or vesicle may be 10^{-6} dyn (10^{-9} g) or less when applied to regions on the scale of 10^{-4} cm. Thus, accurate measurement and control of these forces and deformations are extremely difficult. To circumvent these difficulties, amplification methods are required to facilitate control and measurement at the laboratory level. In the discussion to follow, methods and equipment are described that have been developed to greatly facilitate micromechanical experiments on small cells and vesicles. This is followed by examples of application to red cells, white cells, and vesicle studies to illustrate the approaches.

II. METHODS AND EQUIPMENT

The system which has been developed for micromechanical study of single cells and vesicles is centered around an inverted microscope that has several micromanipulators mounted directly on the microscope stage. The small manipulators are pneumatically controlled by hand-operated "joy sticks". Small glass suction micropipettes or thin platinum beam transducers (described later) are attached to the micromanipulators. Cells or vesicles are placed in very small concentrations into one or more microchambers on the microscope stage. The time course of each experiment as well as the pertinent data (e.g., temperature, pipette suction pressures, time, etc.) are recorded on videotape with the use of video multiplexing. A mercury vapor lamp with narrow-band interference filters provides the best image quality for video recording (e.g., at 436 or 546 nm). It is essential that a high-numerical-aperture (greater than 0.6) objective be used, followed by large, "empty" magnification. A large video image with good optical resolution gives the best video resolution since there are several video scan lines within the diffraction border surrounding the cell. If temperature-controlled stages are used, it is necessary to have a long-working-length objective (working distance greater than 1 mm) in order to avoid heat conduction from the chamber. For bilayer vesicles studies, the Hoffman modulation optical system (Modulation Optics, Inc., Greenvale, N.Y.) provides an excellent interference contrast image of single bilayers.

It is often desirable to compare the properties measured on single cells or vesicles in separate solution environments. The advantage is that the properties can be compared directly without cell-cell variations. The procedure involves the use of adjacent, but separate, chambers on the microscope stage which have open ports for entry through air-liquid interfaces. A large ($\sim 50 \times 10^{-4}$ cm) micropipette traverses one chamber and spans the air gap between chambers to enter the second chamber, where a cell is inserted into the lumen of this pipette. The stage is then translated to displace the large micropipette from one chamber into the second chamber through the air-gap interface. The cell or vesicle that is being transferred is sheltered from exposure to the air in the transfer process. In the second chamber, the cell is simply withdrawn from the interior of the transfer pipette with the use of a second, smaller suction micropipette. Contamination by solution in the transfer pipette is kept to a minimum by backfilling this pipette with oil so that only a small reservoir space remains near the tip of the pipette. Also, the presence of the oil restricts motion of fluid in the pipette. The result is that the small fluid space at the tip of the transfer pipette acts simply as a cavity to shelter the cell from the exposure to air.

The major component in a mechanical test system is the device used to apply force to the body. In addition, this component usually acts as a transducer to measure the applied force. As noted in the introduction, the levels of force required to deform cells can be very small, indeed smaller than the measurable weights on the most sensitive microbalance. Hence, the force transducer must have a large built-in amplification. One of the simplest methods is to use a small micropipette with an internal diameter on the order of 10^{-4} cm coupled by a continuous water system to a micrometer-positioned water manometer. The pipette suction pressures can be controlled and measured with sensitive pressure transducers which have a resolution of 10^{-6} atmospheres (1 dyn/cm^2). Therefore, the suction force can be as low as 10^{-8} dyn (i.e., 10^{-11} g). For the experiments described in the next section, the suction pressures range from 10 to 10^5 dyn/cm^2. Calibration of the suction micropipette is made easy by the use of a micrometer-positioned water manometer coupled to a sensitive pressure transducer. However, it is essential that the system be zeroed accurately before every experiment since any alteration in curvature of the air-water interface through which the micropipette passes can change the suction pressure. The zero pressure adjustment is simply accomplished by observing the movement of small particles within the lumen of the micropipette. When motion of these particles ceases, then the pressure difference can be treated as zero. It is very important that no air gaps or bubbles be present in the water lines from the manometer to the pipette tip.

Micropipette suction is a specific type of force application which may not always be useful. Another sensitive force transducer has been developed[1] that can measure forces in the 10^{-9} g range. This transducer is a thin platinum beam with a thickness in the range of 2 to 4×10^{-4} cm, a width of 10 to 14×10^{-4} cm, and a length on the order of a fraction of a centimeter. The thin beam transducer is mounted with its width dimension vertical to stabilize against seismic vibrations; the force is measured by displacement in the horizontal direction. Figure 1 shows both types of transducers; here, a red blood cell (held with a suction micropipette) is pushed against the beam to produce deflection. This provides a simple method to calibrate the force coefficient of the beam. The red blood cell is pressurized by the suction micropipette and acts as a fluid coupler to apply force to the platinum beam. Given the suction pressure in the micropipette and the observation of the aspirated length of the red cell inside the pipette vs. the deflection of the beam, the force applied to the beam is directly given from the red cell membrane mechanics as a function of beam displacement.[1] Analysis of the data yields the beam constant which gives the applied force simply in proportion to the beam deflection.

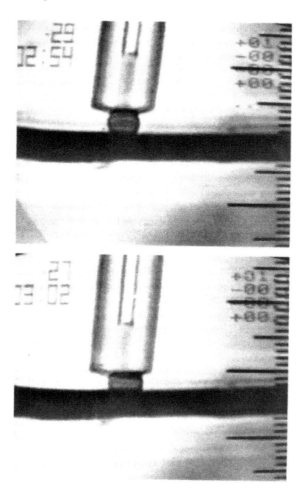

FIGURE 1.   Video micrographs of a beam transducer calibration
test where a human red blood cell is used as a "fluid coupler"
between the suction micropipette and the platinum beam.[1] The scale
at the right side of the video recording is approximately 10^{-4} cm/
division. Note the changes in cell aspiration length and beam position between the two photographs.

The last major element in a micromechanical test is the method for quantitation of deformation and rate of deformation of the capsule. As noted before, it is important to optimize
the optical-video image with the use of a high-numerical aperture, long-working-length
objective, followed by large empty magnification to create an image of the cell which is as
big as possible on the TV screen. Additional enhancement features (e.g., interference contrast) greatly improve the inherent contrast enhancement of the video system. The result is
an image which is not limited by the resolution of the video system, but rather by the
resolution limit of the microscope. Even with this optimization, the uncertainty created by
the diffraction pattern remains around the edges of the body. As such, the dimensions of
cells and vesicles are limited in accuracy to $\pm 0.25 \times 10^{-4}$ cm. To circumvent this problem,
it is recognized that accurate detection of displacements of a position on the cell are possible
even though the absolute position is not defined. Hence, the displacements are used in
conjunction with mathematical analysis of whole cell geometry to quantitate cellular deformation. For example, changes in red cell membrane area on the order of 0.1% can be
detected even though total area cannot be measured more accurately than about 10 to 15%.

Detection of displacements in the video image is facilitated by the use of video image processors and position analyzers. These electronic packages allow point-to-point measurement in the video image with an accuracy of 1% of the total length. Because of the excellent edge detection in the video system, it is possible to evaluate displacements of cell surfaces with an accuracy of better than 10% of the wavelength of light.

In summary, the key elements of the micromechanical test apparatus are the microscope-micromanipulation system, the capability of single cell or vesicle transfer between microchambers, microforce transducers, and optical detection and video recording components. With this general system, it is possible to carry out many types of single cell experiments, such as measurement of viscoelastic properties of blood cells and lipid bilayer vesicles, and measurement of cell-cell and vesicle-vesicle adhesivity properties. In the mechanical tests of cell or vesicle deformability, the deformation of a single cell or vesicle is observed in response to the application of either suction pressure or compression against the platinum beam. Studies of cell-cell or vesicle-vesicle adhesivity involve selection of one capsule as a test surface and the manipulation of the second capsule into close proximity for formation of adhesive contact. Two features are evaluated in the adhesion study: (1) the spreading of one capsule on the test capsule surface to ascertain the affinity between these surfaces and (2) the subsequent strength of adhesive contact as measured by stepwise separation of the two capsules.

III. MEASUREMENTS OF CELL AND VESICLE DEFORMABILITY

In Chapter 4, "constitutive recipes" for the relation of material stresses to deformation and rate of deformation were outlined. Also, the magnitude and significance of specific material properties already identified for cells were discussed vis-à-vis deformability of cells in general. Here, some simple procedures will be described that can be used to establish these properties.

A. Red Cell and Vesicle Membrane Area Rigidities

The area compressibility moduli of cell or vesicle membranes are measured by micropipette aspiration of osmotically preswollen capsules (as illustrated in Figure 6 of Chapter 4).[2,3] The capsule deforms and enters the micropipette easily until the portion outside the pipette becomes a rigid spherical surface. As the pipette suction pressure is increased, the membrane isotropic tension $\bar{\tau}$ is increased proportionally:

$$\bar{\tau} = P \cdot R_p / (2 - 2 \cdot R_p / R_o)$$

Close observation of the aspirated length in the micropipette shows a small but linear increase in this length with pipette suction pressure. Since the outside portion of the capsule is spherical, displacement of the aspirated length is directly proportional to the increase in membrane area or decrease in interior volume. As has been shown previously, decrease in capsule volume contributes less than 20% to the displacement of the aspirated length.[2] If the small volume change is neglected, the area change is approximated by the simple expression:

$$\Delta A \cong 2\pi \cdot R_p \cdot (1 - R_p / R_o) \cdot \Delta L$$

where R_p is the pipette radius and R_o is the diameter of the exterior spherical portion. Hence, the observation of aspirated length vs. suction pressure is simply transformed into membrane tension vs. fractional change in surface area which yields moduli of area rigidity on the order of 10^2 to 10^3 dyn/cm.[2-4] Aspiration of flaccid red cells or vesicles, so that the area is not required to increase, require suction pressures that are 10^3 to 10^4 times smaller (i.e., suction pressures of 1 to 100 dyn/cm^2 or less). These lower pressures are characteristic of membrane extensional and bending rigidities.

Another method to test membrane area rigidity is to compress the capsule against a thin platinum beam transducer and to observe the increase in major (equatorial) dimension of the capsule.[3] The increase in cross-sectional dimension is directly related (but not with a simple linear relation) to the fractional increase in membrane area. This approach has been applied to giant phospholipid bilayer vesicles.[3] The elastic area compressibility moduli derived from these experiments are the same as those measured by micropipette aspiration, which demonstrates that the method used to produce the area dilation does not influence the result.

B. Red Cell Membrane Extensional and Bending Rigidity (Static)

Lipid bilayer membranes (in the liquid state) do not resist extension in the plane of the membrane unless area changes are required. On the other hand, the red cell membrane possesses a small but significant extensional rigidity. For example, observation of flaccid red cell aspiration (Figure 7, Chapter 4) shows a proportional increase in aspirated length with increase in suction pressure; this stepwise increase in length vs. aspiration pressure is reversible in that reduction of the aspiration pressure gives a proportional reduction in length. Here, the aspiration pressures are in the range of 100 to 500 dyn/cm^2 for a 1×10^{-4} cm-caliber suction pipette. Analysis of this experiment has shown that the membrane extensional rigidity can be derived from the derivative of the suction pressure with respect to length:

$$\mu \sim R_p^2 \cdot \frac{dP}{dL}$$

Experimental results yield values on the order of 10^{-2} dyn/cm.[5,6]

Computation of the membrane extensional rigidity from the pipette suction vs. length data presumes that the bending rigidity of the membrane can be neglected. This assumption is valid as long as the surface contour of the red cell remains smooth and the extrapolated intercept of the length vs. suction pressure is zero or less.[7] When the suction pressure is sufficient to produce excessively large aspiration lengths, the cell surface begins to wrinkle (buckle) and eventually large folds occur, enabling the cell to move up the pipette until simply limited by the surface area and volume restrictions. It has been shown that the transition from smooth to "buckled" membrane contour is directly related to the ratio of membrane bending to extensional rigidities.[8] From the analysis, the bending modulus is proportional to the pressure $\hat{P}$ at which the cell buckles or folds:

$$B \cong c \cdot R_p^3 \cdot \hat{P}$$

where the coefficient c depends on the pipette inner radius and is in the range of 0.005 to 0.012. Observations of normal red cells show that pressures sufficient to cause buckling are about 500 dyn/cm^2 for a 2×10^{-4} cm caliber pipette.[8] From these values, the bending modulus is determined to be on the order of 10^{-12} dyn-cm (erg). Deformation of a flaccid lipid bilayer vesicle is resisted only by this small bending rigidity.

C. Dynamic Rigidity of Red and White Blood Cells

The membrane bending and extensional rigidities just described represent the static resistance to deformation of the red cell. When cells are deformed by extension and folding (bending) over a very small time increment, large viscous forces oppose the deformation. The dynamic rigidities of the red cell are derived from time constants for rapid elastic recovery from extension and bending deformations; these time constants are t_e and t_f, respectively. The dynamic rigidities are given by the product of the static rigidity $\times$ the characteristic time constant for the deformation response $\div$ by the time interval over which the deformation occurs. As discussed in Chapter 4, the characteristic time constant for membrane extensional deformation is given by the following equation:

$$t_e \sim (\eta_m + \tilde{\eta}_{Hb} \cdot \delta)/\mu$$

where η_m is the membrane surface viscosity for extensional deformations, $\tilde{\eta}_{Hb}$ is the viscosity of the cell interior contents, and δ is the characteristic dimension or thickness of the cell. By comparison, the characteristic folding time is approximated by the equation

$$t_f \sim \tilde{\eta}_{Hb} \cdot D/(B \cdot k_m^2)$$

where only the cell interior viscosity appears to be significant, D is the characteristic dimension of the cell, and k_m is the curvature of the fold. The following procedures are used to measure response times:

1. Cells are extended end to end by diametrically opposed pipettes so that little buckling or folding of the membrane surface occurs (Figure 8, Chapter 4). The cell is then quickly released and the length-to-width recovery time course is recorded. The time-dependent recovery of the cell shape can be analyzed using a simple visco-elastic model to give the characteristic time for extensional recovery.[5] Typically, these time constants are on the order of 0.1 sec for normal red cells.[9]
2. The dynamic response of the cell to folding is tested by aspiration of red cells with large-caliber pipettes so that little cell extension occurs. The cell simply is folded upon entrance into the pipette. The cell is then rapidly expelled from the end of the pipette with a pressure pulse and the time course of the width recovery is recorded. The representative time for the cell width to recover 62% toward its final width is taken at a measure of the time constant for elastic recovery from folding. Values characteristic of the folding response are on the order of 0.3 sec for normal cells.[9]

Unlike its companion red blood cell, white cells are not simple membrane capsules that enclose a liquid interior. They possess a complicated cytoplasm that includes granules and a nucleus. However, micropipette aspiration tests have shown that passive deformations of the white cell into a micropipette are a continuous, but exceedingly slow flow process (illustrated in Figure 9, Chapter 4).[10] Consequently, white cell rigidity must be treated as a dynamic property in all deformation processes. Aspiration of the white cell into a 5×10^{-4} cm pipette with a suction pressure on the order of 300 dyn/cm^2 will result in total cell aspiration after several minutes of exposure to the suction pressure. On the other hand, aspiration of the white cell with the same caliber pipette but a suction pressure of 1000 dyn/cm^2 will result in total aspiration of the cell in less than 30 sec. These results indicate that the effective viscosity of the cytoplasmic mixture is on the order of 10^3 P (dyn-sec/cm^2) at 25°C. Also, there is a net cortical stress or tension of about 10^{-2} dyn/cm which establishes a threshold pressure below which the cell will not enter the micropipette.

IV. TESTS OF BLOOD CELL AND VESICLE ADHESIVITY

An outline was just presented of how the determinants of cell and vesicle deformability can be measured by micromechanical tests. In these final paragraphs, it is shown how micromechanical tests can be used to quantitate the free-energy potentials (affinities) for formation of adhesive contact, as well as to test the subsequent strength of adhesion between membrane surfaces.

A. Red Cell and Vesicle Aggregation Potentials

Adhesive contacts between membranes are mediated not only by a wide variety of specific agents such as agglutinating antibodies and lectins but also by nonspecific electrostatic and electrodynamic (colloidal-type) interactions and aggregating protein and polymer macromolecules. These surface reactions are really only prominent over short range compared with observable cellular dimensions. Details of the actual attraction forces between surfaces are not observable and are best cumulated into an integral of force × displacement. The integral is the reversible work involved in formation of an adhesive contact between two flat surfaces, and is represented thermodynamically as a free-energy reduction. Stable aggregation of cells or vesicles is promoted by the chemical affinity between outer membrane surfaces, but is opposed by the cell or vesicle rigidity. Mechanical equilibrium is established when small virtual decreases in free energy due to contact formation just balance the small increases in work of deformation required to increase the membrane-membrane contact. Micromechanical tests have been designed which take advantage of this equilibrium statement; these tests have been used to study the low-level affinity associated with aggregation of vesicles in simple salt solutions and aggregation of red blood cells and vesicles in salt solutions plus high-molecular-weight polymer fractions and plasma proteins. Some examples of experiments that have been carried out include red cell-red cell aggregation, red cell-synthetic lipid vesicle aggregation, and synthetic lipid vesicle-vesicle aggregation. These experiments are similar in that each involved micromanipulation of two capsules to control the aggregation process. Specifically, one capsule (either red blood cell or lipid vesicle) was aspirated with sufficient suction pressure so as to form a rigid spherical segment outside the micropipette; this was the "test" surface. The other cell or vesicle was aspirated with a lower pressure so as to leave the capsule deformable, then maneuvered into close proximity of the spherical test surface where adhesion was allowed to occur. The free-energy potentials for aggregate formation were derived from the observation of capsule geometry and predetermined membrane elastic properties or pipette suction pressures.

The aggregation potentials for red cells to red cells — or to synthetic lipid vesicles — in dextran polymer solutions and solutions of plasma proteins were derived from observations of the fractional coverage of the spherical test surface by flaccid red cells as shown in Figure 2.[11-14] Aggregation will not normally occur between red cells or between red cells and phospholipid vesicles in simple salt solutions, but aggregation between these capsules occurs spontaneously in the presence of large macromolecules such as the dextran polymers or plasma proteins. The aggregation can be quantitated by measurement of the fractional coverage of the spherical "test" surface at equilibrium. This measure is given by observation of the height of the polar cap at the contact zone divided by the test surface diameter, i.e., fractional coverage $x = z/2R_s$ of the spherical surface. The free-energy reduction per unit area of contact depends only on the elastic rigidity of the red cell membrane:

$$\gamma/\mu \cong x^2/(1 - x) - x \cdot \ln(1 - x) + 2 \cdot B/(\mu \cdot R_s^2)$$

opposed by a small bending rigidity threshold. Thus, observations of the extent of coverage of a spherical surface by the red cell plus the knowledge of the elastic shear modulus can be used to directly determine the aggregation potential. Figure 3 shows the fractional encapsulation of sphered red cells by flaccid red cells as a function of blood plasma concentration. Figure 4 presents the free-energy potentials for aggregation of red cells in plasma concentrates derived from Figure 3 with the mechanical relation above.[14]

For vesicle-vesicle aggregation, free-energy potentials are determined from direct meas-

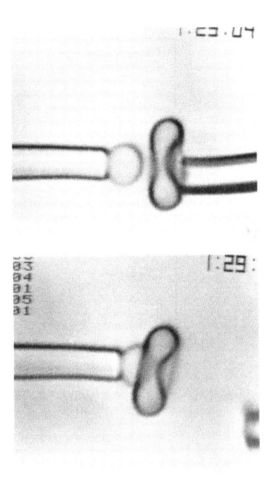

FIGURE 2. Video micrograph of a human red blood cell that has encapsulated (adhered to) a sphered red cell in blood plasma. The equilibrium coverage of the sphere by the red cell is used to determine the membrane-membrane affinity between the two surfaces.[12-14]

urements of the tension induced in an initially flaccid vesicle membrane as it adheres to a rigid "test" surface as shown in Figure 5. Such tests are especially useful because the composition of the lipid vesicle surface can be constituted so that the levels of electric double-layer and steric repulsion (e.g., surface sugars) can be controlled in relation to the specific interactions (e.g., agglutinating antibodies) and nonspecific attractions (e.g., van der Waals' or large nonadsorbing macromolecules).[15-18] Here, the free-energy potentials are directly related to the displacement of the pipette suction force:

$$\gamma = -\pi \cdot R_p^2 \cdot P \cdot \left(\frac{dL}{dA_c}\right)$$

where P is the pipette suction pressures, L is the aspirated length of the vesicle inside the pipette, and R_p is the pipette radius, and A_c is the contact area. The derivative (dL/dA_c) is specified by the vesicle geometry, pipette size, and location. The geometry is too complicated to permit a closed-form analytical solution, but it is a simple matter to solve the problem numerically. Figure 6 shows the experimental observation of fractional coverage of a sphered vesicle by a second vesicle as a function of suction pressure applied to the adherent vesicle.

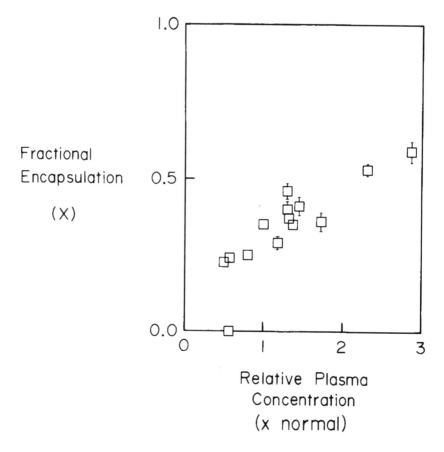

FIGURE 3. Extent (x = z/2R$_s$) of coverage of a sphered red cell surface by an adherent flaccid red cell measured in concentrated plasma fractions (given as ratios of total protein in the concentrate to normal values).

Also plotted is the behavior predicted for a uniform free-energy potential for contact formation. This approach has been used to determine the levels of membrane-membrane affinity for aggregation of neutral and electrically charged vesicles in salt solutions and in solutions containing large dextran polymers and plasma proteins.[15-18] The natural van der Waals attraction between neutral lipid surfaces is characterized by free-energy potentials on the order of 10^{-2} to 10^{-1} erg/cm^2.[17] At first impression, it would seem that these energies are sufficient to cause aggregation of all cells; however, the presence of superficial carbohydrate structures on cell surfaces creates sufficient steric repulsion so as to overcome the van der Waals' attraction. Addition of large polymers augments the free-energy potential for adhesion by factors up to an order of magnitude or more in concentrated solutions ($\geq$10% wt:wt).[15,18]

B. Strength of Adhesion for Specifically Bound Red Cells

In contrast to spontaneous spreading and the reversible adhesion just described, often there is no tendency of one cell or capsule to spread on a test surface; however, when forced together, these capsules subsequently form very strong adhesive contacts. This is the case for cell membranes bound together by lectins or multivalent, agglutinating antibodies. Here, the membrane receptors are present in such low surface densities that the curvature of the cell membrane in the vicinity of the contact zone precludes formation of additional cross bridges. In other words, the mechanical rigidity of the membranes must be overcome by external force in order to make additional membrane-membrane cross bridges. Once contact has been established, large forces may be required to separate or break these molecular

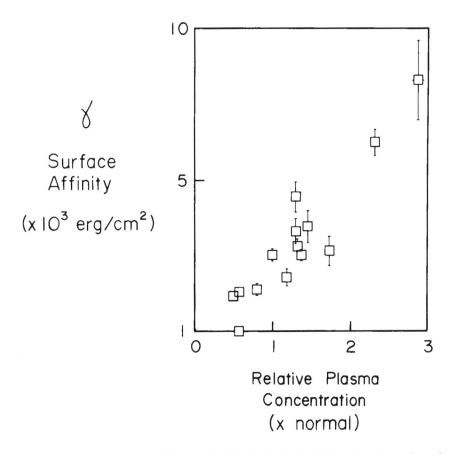

FIGURE 4. Free-energy potentials for red cell-red cell adhesion derived from observations of encapsulation (Figure 3) in concentrated human plasma.

bonds. An example of this is shown in Figure 7, where red cells were forced to make contact. Cross bridges were formed by a snail lectin (helix pomatia, HPA). Here, no tension was induced in the red cell membranes by simple contact (i.e., no spreading was observed). However, separation of the cells required large forces, as illustrated in Figure 8. The work required to separate an area of contact is directly related to the membrane tension at the contact zone. The tension at the contact zone is derived from the pipette suction pressure and the equations of mechanical equilibrium. By using a large-caliber pipette as shown in Figure 7, the tension local to the contact can be expressed as

$$\tau_m = \frac{P \cdot R_p^2 + P_c \cdot R_a^2}{2 \cdot R_a \cdot \sin\theta_m}$$

where P is the pipette suction pressure, R_p is pipette radius, R_a is the radius of the adhesion zone, and θ_m is the angle between the local normal to the membrane and the axis of symmetry (as shown in Figure 7B). The significant unknown is the pressure P_c inside the cell. Additional analysis is required to eliminate this unknown, and implicitly depends on the area, volume, and elastic deformation of the red cell. Limits on the cell pressure are given by:

$$0 < P_c < \frac{P \cdot R_p}{R_a} \left[\frac{(R_p - R_a \cdot \sin\theta_m)}{(R_p \cdot \sin\theta_m - R_a)} \right]$$

When the aspirated cell is highly pressurized as shown in Figure 7, then P_c is close to the

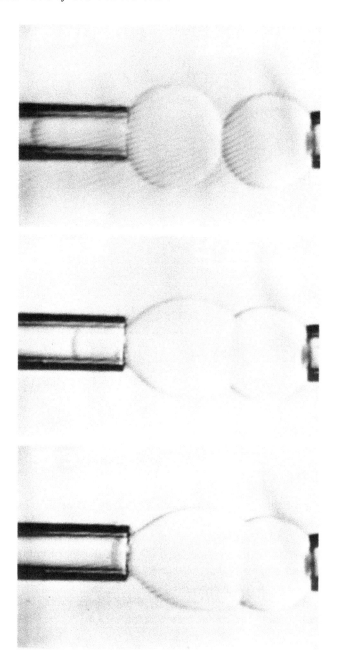

FIGURE 5.   Video micrographs of the equilibrium configuration of two giant lecithin vesicles that were allowed to adhere via van der Waals attraction in saline solution. The suction pressure in the pipette on the left determines the extent of the coverage of the rigid, spherical vesicle surface. This approach is used to measure the membrane-membrane affinity between vesicle surfaces.[15-18]

upper limit. Figure 8 shows the tension vs. contact area calculated for the measured force with the upper-limit estimate for P_c.

Results from these experiments have shown that separation of red cell surfaces subsequent to contact normally results in accumulation of membrane cross bridges and receptors at the

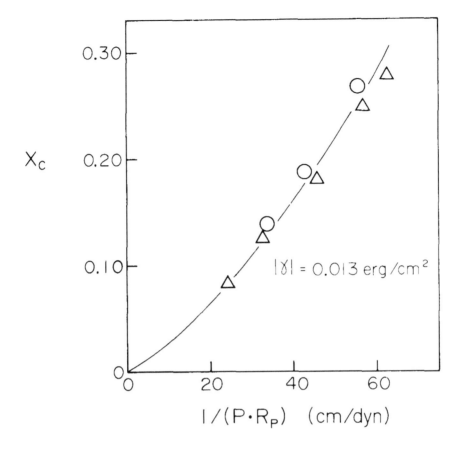

FIGURE 6. Extent of coverage ($x = z/2R_s$) of a pressurized lipid bilayer vesicle by an adherent second vesicle measured as a function of suction pressure applied to this vesicle. Triangles are measurements of contact formation; open circles are measurements of contact separation.

rim of the contact zone.[19] However, with the use of a glutaraldehyde-fixed red cell sphere, receptor accumulation can be avoided (illustrated in Figure 7). Experiments similar to these can be used as sensitive tests for agglutination reactions where the threshold and activity of the ligand can also be determined.

C. Adhesivity of Blood Cells to Vascular Endothelium

Recently, experiments have shown that there is a tendency for red cells from sickle-cell disease patients to adhere to vascular endothelial cells.[20] Here again, micromechanical tests have been useful for sensitive assay of cell-cell adhesivity. Figure 9 shows an example of tests of adhesivity of red cells from healthy individuals and individuals with homozygous SS disease to vascular endothelial cells which were cultured on sepharose beads. As is shown in Figure 9, sickle cells often adhere quite strongly to the endothelial cells. The results show that not only receptors on the sickle cell are present, but a factor in the plasma of sickle-cell disease patients is also involved.[20] Experiments such as these can also be used for sensitive study of the adhesion of other cells (white and tumor cells) to endothelium.

D. White Cell Phagocytosis

To test the adherence and phagocytic properties of blood granulocytes, methods have been developed where the exposure of the white cell to stimuli in solution and on particle surfaces can be controlled. The approach utilizes the cell-transfer methodology described earlier. For

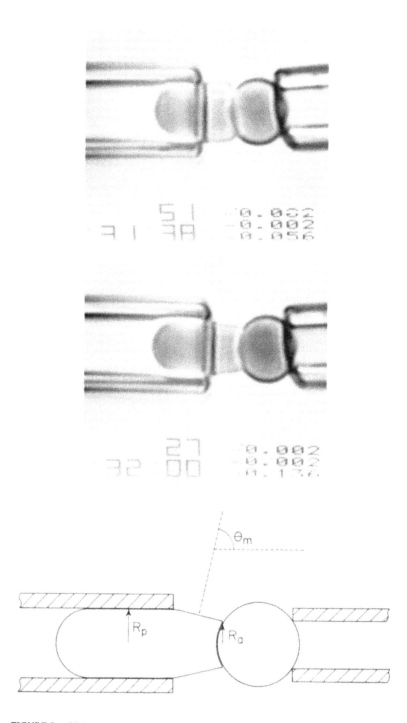

FIGURE 7.   Video micrographs of the stepwise separation of red cells that were bound by
HPA (helix pomatia) lectin cross bridges. The cell on the right was first glutaraldehyde-fixed
as a sphere, coated with HPA, and then transferred to the second chamber that contained
only flaccid red cells in HPA-free buffer. The second (uncoated) cell was then maneuvered
into position and an adhesive contact was made as shown in the top panel. The flaccid cell
was then separated in steps at which the force of attachment was measured from the pipette
suction pressure and the cell geometry as shown in the subsequent panels.

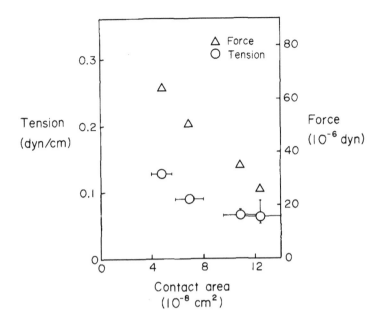

FIGURE 8. Total force and estimated membrane tension required to separate adherent red cells (shown in Figure 7) bound by HPA-lectin cross bridges. The values are derived from pipette suction as a function of contact area.

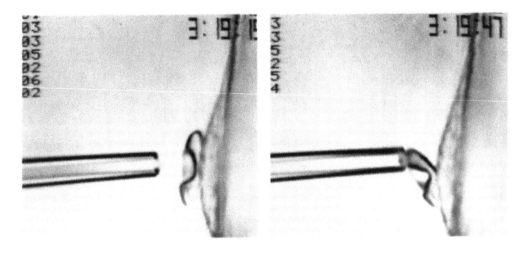

FIGURE 9. Video micrographs of adherence of a sickle red blood cell to a vascular endothelial cell. The sequence illustrates the approach used to test the tendency of cells to adhere to endothelial surfaces.

example, zymosan (yeast) granules are placed in one chamber on the microscope stage and a small blood sample is placed in an opsonating medium in an adjacent microchamber. One micropipette is used to aspirate a single yeast particle and then to transfer it into the chamber which contains the blood granulocytes. A second pipette is used to aspirate a granulocyte and to manuever it into proximity of the yeast particle surface. The yeast particle and white cell are brought into contact, and the kinetics of the phagocytosis process are then observed (shown in Figure 10). This experiment can be used as a sensitive test of opsonation and activity of white cells. With the use of a second micropipette to retard the engulfment of the opsonated particle, the active contractility of the phagocyte can be directly quantitated by measurement of the suction pressure required to arrest the engulfment.

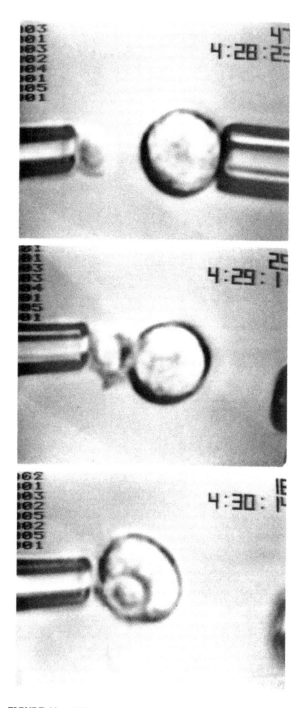

FIGURE 10.   Video micrographs of granulocyte phagocytosis of
a zymosan (yeast) granule. The single zymosan granule was transferred to the chamber which contained the granulocyte; the cells
were then maneuvered into position to permit adhesion and phagocytosis. The time course of the event is demonstrated by the time
recordings in the upper right corner of each image.

REFERENCES

1. **Evans, E. A., Kwok, R., and McCown, T.,** Calibration of beam deflection produced by cellular forces in the 10^{-9} - 10^{-6} gram range, *Cell Biophys.,* 2, 99, 1980.

2a. **Evans, E. A., Waugh, R., and Melnik, L.,** Elastic area compressibility modulus of red cell membrane, *Biophys. J.,* 16, 585, 1976.

2b. **Evans, E. A. and Waugh, R.,** Osmotic correction to elastic area compressibility measurements on red cell membrane, *Biophys. J.,* 20, 307, 1977.

3. **Kwok, R. and Evans, E. A.,** Thermoelasticity of large lecithin bilayer vesicles, *Biophys. J.,* 35, 637, 1981.

4. **Evans, E. A. and Kwok, R.,** Mechanical calorimetry of large DMPC vesicles in the phase transition region, *Biochemistry,* 21, 4874, 1982.

5. **Evans, E. A. and Skalak, R.,** *Mechanics and Thermodynamics of Biomembranes,* CRC Press, Boca Raton, Fla., 1980, 254.

6. **Waugh, R. and Evans, E. A.,** Thermoelasticity of red blood cell membrane, *Biophys. J.,* 26, 115, 1979.

7. **Evans, E. A.,** Minimum energy analysis of membrane deformation applied to pipet aspiration and surface adhesion of red blood cells, *Biophys. J.,* 30, 265, 1980.

8. **Evans, E. A.,** Bending versus shear rigidity of red blood cell membrane, *Biophys. J.,* 43, 27, 1982.

9. **Evans, E. A., Mohandas, N., and Leung, A.,** Static and dynamic rigidities of normal and sickle erythrocytes: major influence of cell hemoglobin concentration, *J. Clin. Invest.,* 73, 477, 1984.

10. **Evans, E. A.,** Structural model for passive granulocyte behaviour based on mechanical deformation and recovery after deformation tests, in *White Blood Cell Mechanics: Basic Science and Clinical Aspects,* Lichtman and Meiselman, Eds., Alan R. Liss, New York, 1983, 53.

11. **Evans, E. A. and Parsegian, V. A.,** Energetics of membrane deformation and adhesion in cell and vesicle aggregation, *Ann. N.Y. Acad. Sci.,* 416, 13, 1983.

12. **Buxbaum, K., Evans, E. A., and Brooks, D. E.,** Quantitation of surface affinities of red blood cells in dextran solutions and plasma, *Biochemistry,* 21, 3235, 1982.

13. **Evans, E. A. and Kukan, B.,** Free energy potential for aggregation of erythrocytes and PC:PS vesicles in dextran solutions and in plasma, *Biophys. J.,* 44, 255, 1984.

14. **Janzen, J., Kukan, B., Brooks, D. E., and Evans, E.,** Free energy potential for red cell aggregation in plasma, serum, and solutions of isolated plasma proteins determined from two-cell adhesion tests, *Biochemistry,* (submitted).

15. **Evans, E. A. and Metcalfe, M.,** Free energy potential for aggregation of mixed PC:PS lipid vesicles in glucose polymer (dextran) solutions, *Biophys. J.,* 45, 715, 1984.

16. **Evans, E. and Metcalfe, M.,** Free energy potential for aggregation of large, neutral lipid bilayer vesicles by van der Waals' attraction, *Biophys. J.,* 46, 423, 1984.

17. **Evans, E. and Needham, D.,** Giant vesicle bilayers composed of mixtures of lipids, cholesterol, and polypeptides: thermo-mechanical and (mutual) adherence properties, *Faraday Discuss. Chem. Soc.,* 81, 267, 1986.

18. **Evans, E., Needham, D., and Janzen, J.,** Non-specific adhesion of phospholipid bilayer membranes in solutions of plasma proteins: measurement of free energy potentials and theoretical concepts, in *Proteins at Interfaces,* American Chemical Society Symp. Ser. No. 343, Brash and Horbett, Eds., ACS, Washington, D.C., 1987, 88.

19. **Evans, E. and Leung, A.,** Adhesivity and rigidity of red blood cell membrane in relation to WGA binding, *J. Cell Biol.,* 98, 1201, 1984.

20. **Mohandas, N. and Evans, E.,** Sickle erythrocyte adherence to vascular endothelium: morphological correlates and the requirement for divalent cations and collagen-binding plasma proteins, *J. Clin. Invest.,* 76, 1605, 1985.

Chapter 8

ELECTRON MICROSCOPIC STUDY OF THE SURFACE TOPOGRAPHY OF ISOLATED AND ADHERENT CELLS

Colette Foa, Jean-Louis Mege, Christian Capo, Anne-Marie Benoliel, Jean-Remy Galindo, and Pierre Bongrand

TABLE OF CONTENTS

I. INTRODUCTION

The surface bumps of nucleated cells play a preeminent role in cell-substrate and cell-cell interactions, and the organization of the various rough patches is directly related to cellular cytoskeletal elements.

First, the role of the microvilli in initiating the intercellular contact has been investigated with electron microscopy. Jones et al.[1] suggested that intercellular contacts elicit a rapid response from the organelles concerned with cell motility. Willingham and Pastan[2] proposed that agglutinability of mouse and rat fibroblasts is regulated through the modulation of cell surface microvilli by cyclic adenosine 3'-5'-monophosphate (cAMP), and that transformed cells are highly agglutinable because their low cAMP levels result in the formation of numerous microvilli. In neural retinal cells, Ben-Shaul and Moscona[3] related cell deformations and cell reaggregation to the presence of microvilli.

Second, there may be marked differences between the molecules located on the tip of microvilli as compared to different areas of the cell membrane. Using scanning electron microscopy (SEM), Weller[4] demonstrated a redistribution of the concanavalin A (Con A) binding sites suggesting movement of molecules inside the membrane (clumping, band formation, and capping) depending on the extent to which the cell periphery extended into pseudopodia.

Third, it is important to determine if adhesion itself involves a lateral redistribution of membrane molecules. Regarding this phenomenon, Ryter and Hellio[5] demonstrated redistribution of wheat germ agglutin receptors in phagocytic cups, and De Petris et al.[6] demonstrated redistribution of Con A receptors, the mobility of which seemed to play a role in cell-cell binding as shown by Rutishauser and Sachs.[7] In another system, Foa et al.[8] evidenced molecule motility during the specific recognition of target cells by cytotoxic T lymphocytes, and Capo et al.[9] obtained indirect evidence for molecule redistribution during Con A-mediated thymocyte agglutination. McCloskey and Poo[10] clearly revealed that antibody molecules, fluorescent antidinitrophenyl IgE (DNP IgE), underwent a marked lateral redistribution, accumulating at the site of contact between rat basophilic adherent leukemia cells armed with anti-DNP IgE and cell-size phospholipid vesicles bearing DNP lipid haptens. They concluded that this redistribution resulted from passive diffusion-mediated trapping, rather than active cellular response, and could contribute to the formation of stable adhesion between membranes.

One of the exciting cell-cell interaction models is the CTL-target cell conjugate resulting from a specific immunological recognition process. The first step is reversible and Mg^{2+} dependent and followed by an irreversible lytic event which begins with the Ca^{2+} dependent programming for lysis.[11-13] Finally, the target cell swells prior to disintegrating.[14] CTL and target cells are rounded when maintained in suspension, but flattened when adherent together in a very close manner as already described.[15] However, the size and structure of the effector-target cell contact, the mode of intercellular communication, and the initial target lesion leading to its death remain controversial.[16] As it was reported that the extent of apposition between the surface of a cell and a particle or another cell to which it adheres is highly dependent on adhesive forces,[17] we focused our interest on studying the relationship between cell-surface roughness and curvature modifications during CTL target cell adhesion.

Electron microscopic examination of cell-cell contact[18,19] or CTL-target interaction has been often performed,[20] but the authors have mainly observed the aspect of microvilli with conventional[20] or scanning microscope[21] using freeze fracture[22] or immunoelectron microscopic techniques.[8] For our part, we try here to study cell-surface roughness on electron micrographs with a method which allows precise quantification of cell deformation in adhesive zones.[23] More precisely, we shall attempt to answer three main questions: (1) during the onset of adhesion is there a progressive appearance of membrane deformations?, (2) is it possible to demonstrate the existence of a mechanical stress exerted by a cell or another cell, as monitored by the appearance of geometrical deformations, and (3) what fraction of the apparent contact region between CTL and targets (as monitored with optical microscopy) is actually involved in molecular contact; this question is of importance if the methods of determining surface affinities that were described in Chapter 7 are applied to rough nucleated cells.

II. MATERIALS AND METHODS

A. Cells

Two types of cell conjugates were used. First, cytotoxic T lymphocytes were generated by mixed lymphocyte culture of spleen cells from C57B1/6 mice (H-2^b) and irradiated stimulating spleen cells (H-2d). Conjugates were created between effector cells and S194 (H-2d, BALB/c myeloma). Second, we used the lymphoid line TG2OUA2,[24] which was derived from a cloned CTL (B6, 1-2, H-2^b). This line was obtained from spleen cells of a female C57B1/6 mouse activated against syngeneic male cells and was kindly supplied by M. Nabholz. This clone showed previously high lytic activity against S194, which was its most sensitive target, but is no more cytolytic. However, its recognition step is always active.[8]

B. Conjugate Formation

Target cells (10^7) and 2 × 10^7 effector cells were suspended in 2 mℓ of pH 7.2 N-2-hydroxyl ethylpiperazine-N'-2-ethane sulfonic acid (Hepes)-buffered RPMI 1640 medium supplemented with 1 mM EGTA (ethyleneglycol-bis β-aminoethyl ether tetraacetic acid; Sigma, St. Louis) and 0.3 mM MgCl$_2$. Ca^{2+} and Mg^{2+} concentrations were thus expected to be close to 0 and 0.5 mM, respectively,[23] which allowed effector-target adhesion but prevented lethal hit delivery.[13] Cells were centrifuged (250 × g, 5 min) in conical plastic tubes, then transferred into an ice-cold water bath, and supernatants were discarded.

C. Electron Microscopy

Conjugated cells were then fixed with glutaraldehyde (4%) and osmium (1%), embedded in epon, cut with a Reichert (Germany) ultramicrotome, stained with uranylacetate and lead citrate, and observed in a Jeol (Japan) 100 C electron microscope. Conjugated cells were systematically photographed with the following cautions: (1) between each ultrathin section a semithin section was done to avoid the observation of two different sections of the same conjugated cells, (2) CTL and TG2OUA2 are easily distinguishable from S194 targets because of their general aspect;[8] therefore, we retained only CTL-target and TG2OUA2-target conjugated cells and (3) all samples were photographed at the same magnification (× 5000) and printed with the same enlargement (× 3).

D. Principle of the Morphometric Analysis of Electron Micrographs

The analysis involves three sequential steps.

1. Digitization of Cell Contours

Cell contours are digitized with a Calcomp 2000 Digitizer (Calcomp, Calif.) connected to a CBM 64 desk computer.

A BASIC program allows an average storage rate of two pairs of coordinates per second. When the digitization process is completed, a second BASIC program is used to discard a fraction of stored points (about 50%) in order that the curvilinear distance between two neighboring points be higher than 1 mm (this is to prevent any error due to an arrest of the stylus during the digitization). Also, the stored contour is drawn on the monitor to check that no gross error occurred. The checked data are stored on a floppy disk for further analysis.

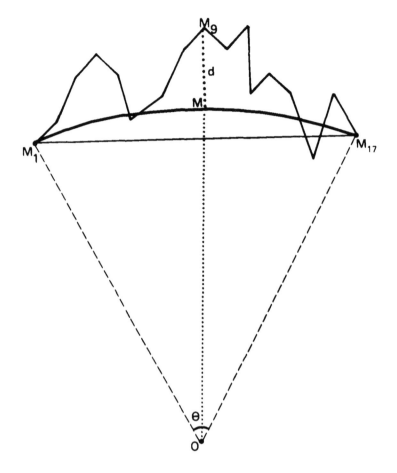

FIGURE 1.   Analysis of individual contours (see complete explanation in Section II).

2. Analysis of Individual Contours

A cell contour is made of N-ordered points. The analysis is performed by considering all possible sets of $2p + 1$ consecutive points (there are N-2p of these sets) where p is an arbitrarily fixed number; 8 was found a convenient value, and the significance of this choice was discussed in a previous report.[23]

The basic principle of the analysis (see Figure 1 for notations) is to construct the circular line $\widehat{M_1M_{2p+1}}$ (here $\widehat{M_1M_{17}}$) enclosing the same area with the straight line M_1M_{17} as the digitized cell contour (i.e., the set of 2p straight lines $M_1M_2 \ldots M_2M_{17}$). Then we define three parameters:

1. The *relative length excess*, e, is the excess of the length of the digitized contour relatively to the length of the circular reference line, namely:

$$e = \sum_{i=1}^{2p} M_iM_{i+1}/\text{length } \widehat{M_1M_{2p+1}} - 1 \tag{1}$$

Clearly, e is expected to be zero if the cell is a smooth sphere. Also, since the area of a given surface enclosed in a box is proportional to the mean length of its intersection with a random plane[26] encountering this box, e is expected to be related to the cell "excess membrane area". However, this is only an approximate view, since the membrane asperities with a characteristic size smaller than the distance between consecutive points or higher than the total contour length are not accounted for by our procedure.

2. The *deviation* d is the distance between the midpoint M_{p+1} (M_9) of the analyzed line and the corresponding point M on the reference circle (M is the intersection between OM_9 and M_1M_{17}, where O is the center of the circle containing M_1M_{17}). The interest of defining this parameter is that the equivalent parameter obtained in measuring the asperities of plane surfaces (i.e., the deviation between each point and the reference plane) plays a key role in theoretical studies done on the adhesion between rough solid surfaces.[42]

3. The *curvature* c is the reciprocal value of the radius R of the circle containing the reference line. If cells were smooth spheres of radius R, the mean curvature of their intersection with a random plane could be

$$<c> = (1/R) \int_0^R dx/(R^2 - x^2)^{1/2} = \pi/(2R)$$

hence, c may be expected to give a figure of the cell size.

Now, when e, d, and c are calculated for each set of 2p + 1 consecutive points of a given cell contour, the characteristic parameters of this contour are defined as: (1) the relative length excess E is the mean value of e, (2) the standard deviation of the deviation D is the standard deviation of d, and (3) the mean curvature C is the mean value of c. Parameters E, D, and C are provided by a BASIC program.

3. Analysis of Paired Contours Representing Interacting Membrane Areas of Bound Cells

The purpose of this analysis is to derive the distribution of intercellular distances in the intercellular contact area. The distance between each point M_i of a contour (defined as contour 1) is calculated as the smallest value of the distances between M_i and every segment of the contour 2. Then as a first approximation, the length of the corresponding fraction of contour 1 is calculated as:

$$1/2(\|\overrightarrow{M_{i-1}M_i}\| + \|\overrightarrow{M_iM_{i+1}}\|)$$

This calculation is repeated for each point M_i of contour 1, and the BASIC program yields as final results:

1. The total length of the fraction of contour 1 that is separated from contour 2 by an apparent distance smaller than k × 100 Å (where k is varied between 1 and 25).

2. The fraction of contour 1 that is separated from contour 2 by a gap of apparent width ranging between (k − 1) × 100 and 100 × k Å, where k is varied between 1 and 25.

3. The number of "contacts", i.e., the number of sets of consecutive points on contour 1 with a distance to 2 smaller than 500 Å. The contact area is defined as the contour portion where the apparent distance between CTL and target membranes is less than 2500 Å.

E. Reproducibility of the Analysis

Errors may be due to an intrinsic limitation of the digitizer or a difficulty to appreciate the exact cell contour. In order to check the reliability of our procedure, several checks were performed:

1. Four parallel lines of 100 mm length and 10 mm distance were drawn on a graph paper and digitized. The mean distance between the first line and the following three lines was then determined. The results were 9.8 mm ($\pm$ 0.15 SD), 19.9 ($\pm$ 0.17 SD), and 39.6 ($\pm$ 0.15 SD), respectively, suggesting a precision of about 0.15 mm for the overall procedure.

2. Using a representative electron micrograph (15,000 $\times$ magnification), a cell contour was digitized five sequential times and analyzed for determination of roughness parameters. The obtained results are shown below:

Experiment	Contour length	No. points	Relative length excess (E)	SD of deviation (D)	Mean curvature (C)
1	4.0	58	0.08	0.047	461
2	4.1	63	0.09	0.047	478
3	4.1	50	0.08	0.054	372
4	4.0	54	0.08	0.044	377
5	4.1	54	0.08	0.054	381
Mean	4.1	56	0.08	0.049	413
SD	0.05	5	0.004	0.0045	51

3. Finally, an intercellular adhesion area was studied and we calculated the distance between each point of contour 1 (there were 58) and five sets of coordinates obtained after fivefold repeat of contour 2 digitization. For each point of contour 1, the SD of the five distances obtained with these sets of coordinates was calculated. The mean value of this SD was 0.14 mm (ranging between 0.08 and 0.22 mm). It is concluded that the random error involved in the determination of the intercellular distances is about 100 Å (i.e., $0.15 \times 10^7/15,000$) when electron micrographs of 15,000 $\times$ magnification are used.

III. RESULTS AND DISCUSSION

First, we tried to determine if, during the onset of adhesion, there is a progressive appearance of membrane deformations. With electron microscopy, we observed the ultrastructure of contacts, both on fresh CTL-target (Figure 2) and cultured CTL-target conjugates (Figure 3). We choose to evidence the different lengths of contact areas involving simultaneously one CTL and one target (Figure 2a, b, c, Figure 3a, b), three CTL and one target (Figure 2d), and one CTL and two targets (Figure 3c and d). As shown on these micrographs, the contact lengths depend both on the total involved area and on the level of the section. Therefore, only mean values can be compared. As defined in Tables 1 and 2, the length of the apparent contact approximately corresponds to the contact area visualized with optical microscopy, and the length of close contact corresponds to the portion where intermembrane distance between the two conjugated cells is less than 500 Å. It results from these tables that a stable adhesive configuration is reached within the first minute as these two parameters do not vary significantly with time, in the two conjugated CTL-target studied models. This finding is consistent with those obtained with other models since extensive cell deformations

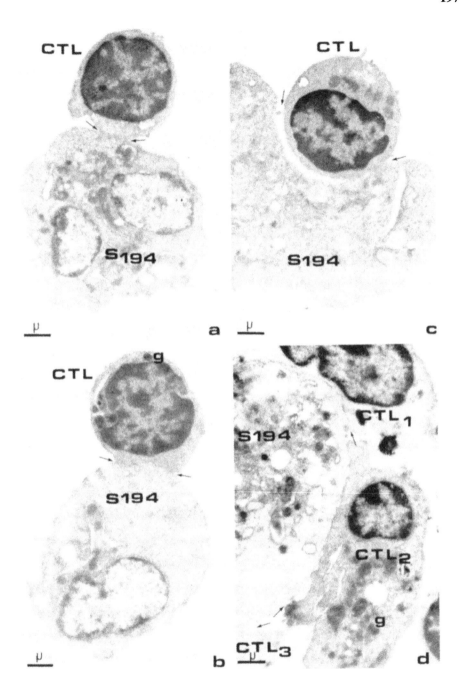

FIGURE 2. Interacting zones of fresh CTL and S194 target conjugated cells (→ ←). The two types of cells are easily recognizable because of their general aspect. The CTL are generally small with a round nucleus containing a dense euchromatin. Depending on the section, the cytoplasm shows a few granules (g). The targets are larger in size and their nuclei exhibit dispersed heterochromatin. The four photographs show various aspects of the interacting areas, the size of which part depends on the section. Times of contact: 5' for a and b, 1' for c and d. (a and b) The interacting zone measures, respectively, 1 and 2.5 μm and the "contact" is very tight; (c) the contact is less intimate, but the interacting zone is extensively developed (>7 μm) and it is possible to observe two close contacts; (d) three CTL are adhering to the same target. The contacts between the CTL and the target are simultaneously extremely tight and extensively long: 4.5 and 6.5 μm. (Magnification a, b, c, d × 7200.)

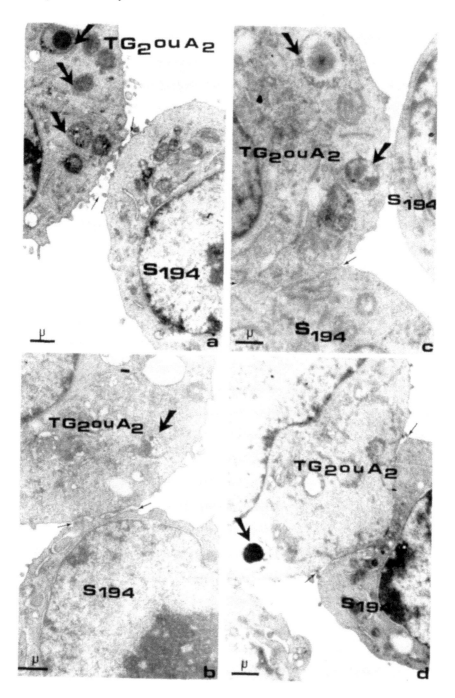

FIGURE 3.   Interacting areas between cultured CTL (TG2OUA2) and S194 targets (→ ←). TG2OUA2 cells are characterized by a cytoplasm filled with numerous heterogeneous granules. Most of them contain a granulous material and a few of them show a dense homogeneous core (⬥). We retain only the conjugated cells formed by TG2OUA2 cells characterized by these granules and targets without granules (see Section II). Time of contact between CTL and target a, b, 1′; c, 15′; d, 5′. (a) The interacting area is badly defined because of the presence of numerous microvilli arising from the TG2OUA2 cell; (b) the contact zone measures about 2.5 μm; (c) the contact zone reaches about 4.5 μm, the gap is tight, and two targets are adhering to the same TG2OUA2 cell; (d) the interacting area is extensively developed (about 7.5 μm). (Magnification a, b, c, and d × 7500.)

Table 1
KINETIC OF CONTACT FORMATION BETWEEN FRESH CTL AND
S194 TARGET CELLS

Time of contact	No. of contact areas		Length of apparent contact (μm) (LA)[a]	Length of close contact (μm) (LC)[b]	% of close contact[c]
1 min					
Expt 1	11	Mean	3.58	2.65	67
		SE	0.44	0.45	6
Expt 2	14	Mean	3.59	2.30	65
		SE	0.47	0.32	5
5 min					
Expt 1	14	Mean	3.18	1.92	59
		SE	0.27	0.29	5
Expt 2	12	Mean	3.95	2.46	63
		SE	0.63	0.40	4
15 min					
Expt 2	7	Mean	4.58	2.35	53
		SE	1.12	0.56	7
30 min					
Expt 2	9	Mean	2.50	1.28	46
		SE	0.22	0.31	9
60 min					
Expt 1	9	Mean	4.14	2.36	51
		SE	1.11	0.83	6
Expt 2	10	Mean	4.37	2.10	47
		SE	0.70	0.45	7

[a] This was defined as the length of contour portion where the distance between CTL and target membranes was less than 2500 Å. This portion was assumed to correspond to the region of intercellular contact as visualized with optical microscopy.

[b] This was defined as the length of contour portion where intermembrane distance was less than 500 Å. This distance is compatible with occurrence of intermolecular binding (Chapters 1 and 4).

[c] This is 100 × LA/LC.

such as phagocytosis may be completed within such a limited amount of time. This was shown by Evans,[27] who used micromanipulation methods to put in close contact granulocytes and zymosan particles under continuous observations.

However, both the lengths of apparent contact and of close contact are greater in fresh CTL-target conjugates than in cultured CTL-target ones. This result is not surprising, as the recognition step and the strength of the stability of the adhesion are related to the expression of T-lymphocyte antigens, the amounts of which may vary in culture.[28]

Second, we searched for the existence of a mechanical stress exerted by a cell on another cell, as monitored by the appearance of geometrical deformations. Considering the macrovilli, ultrastructural observations suggest that the CTL forces the target (Figure 2c and Figure 3c and d), but the opposite situation, i.e., that the target imposes its surface shape on the CTL which tightly fits to it, can also be observed (Figure 2b and d). Studying the microvilli with our method, due to the stability of the adhesive configuration with time, it was possible to pool the results of a great number of determinations as seen in Table 3. It appears that the surface asperities of cytotoxic cells were different from those of the targets. The CTL (fresh or cultured) exhibited rougher surface than myeloma cells, whatever the examined region. An interesting point is that in cultured CTL, the surface asperities were

Table 2
KINETIC OF CONTACT FORMATION BETWEEN CTL LINE TG2OUA2 AND S194 TARGET CELLS

Time of contact		No. of contact areas	Length of apparent contact (μm) (LA)[a]		Length of close contact (μM) (LC)[b]	% of close contact[c]
1 min						
	Expt 1	12	Mean	2.62	0.66	26
			SE	0.42	0.23	5
	Expt 2	6	Mean	4.35	1.76	41
			SE	0.55	0.30	6
5 min						
	Expt 1	12	Mean	2.82	1.39	47
			SE	0.46	0.32	7
	Expt 2	13	Mean	3.20	0.80	26
			SE	0.27	0.15	5
15 min						
	Expt 2	11	Mean	3.98	1.85	44
			SE	0.44	0.43	8
30 min						
	Expt 2	16	Mean	4.24	1.96	47
			SE	0.44	0.43	7
60 min						
	Expt 1	7	Mean	3.35	0.71	16
			SE	0.55	0.35	7
	Expt 2	10	Mean	3.10	1.28	38
			SE	0.51	0.39	9

[a]   This was defined as the length of contour portion where the distance between CTL and target membranes was less than 2500 Å. This portion was assumed to correspond to the region of intercellular contact as visualized with optical microscopy.

[b]   This was defined as the length of contour portion where intermembrane distance was less than 500 Å. This distance is compatible with occurrence of intermolecular binding (Chapters 1 and 4).

[c]   This is $100 \times$ LA/LC.

Table 3
MEMBRANE FOLDING DURING CELL CONTACT

	CTL killer	S194 target	TG2OUA2 killer	S194 target
Free region				
Mean relative length excess (μm)	0.150 ± 0.018	0.0072 ± 0.017	0.188 ± 0.020	0.123 ± 0.016
No. of determinations	72	86	87	81
Bound region				
Mean relative length excess (μm)	0.143 ± 0.011	0.072 ± 0.009	$0.106^a \pm 0.014$	0.098 ± 0.019
No. of determinations	81	87	85	82

[a]   The difference between free and bound regions is highly significant ($t = 3.35$, $p < 0.01$).

different in contact areas and free regions whereas no such difference was found in target cells. This conclusion is consistent with preliminary results described on a similar model.[23] Also, in an electron microscopic study of the binding of glutaraldehyde-treated sheep erythrocytes by rat macrophages, the relative length excess was found to be significantly lower in the contact regions than in the free-membrane areas of the phagocytes.[29]

Hence, smoothing of the cell surface may be an active cell process that could be triggered by the stimulation of suitable cell surface receptors. On the contrary, no obvious stretching of target membranes by cytotoxic T-lymphocytes could be demonstrated, suggesting that actively adhering cells, at least at the level of microvilli, yield to their substrate rather than pulling at it.

Third, we tried to evaluate what portion of the apparent contact region between CTL and target could be involved in molecular contact. Before analyzing quantitatively the interacting area, it was of importance to notice that the whole contact corresponds to the portion visible with the optical microscope. At the ultrastructural level, inside this contact, the intermembrane distances vary. On both CTL target junctions shown in Figure 4, there simultaneously appear close appositions (about 100 Å) and broad regions of contact (between 500 and 1000 Å) as calculated with the used magnification. Moreover, we observed some contact zones where the linear aspect of the membrane is discontinuous, which suggests penetration by the CTL of the target membrane as proposed by Sanderson and Glauert.[30] Such disruption of the target membrane could be the initiating phenomenon leading to the cell death. However, such images could also result from the limits of the fixation or from cutting-incidence modifications in the samples. In our experimental model, the recognition step is not necessarily followed by the target cell destruction.[8] The accuracy of our procedure is probably not sufficient to allow a more precise conclusion on the actual distribution of the intermembrane distances, but is consistent with the view that lipid bilayers were separated by a low-density gap of a few hundred angstroms width. The lanthanum staining of CTL-targets contact[15] known to evidence the polysaccharide-rich surface coating, clearly demonstrated that the contact is completely stained. As the glycocalyx is not revealed with conventional electron microscope staining, the lowest distance of 100 to 200 Å observed mostly between the cells may correspond to the juxtaposition of the two cell coats.

Taking into account these considerations, we decided to evaluate the evolution of close contact during time (Figure 5). It resulted from the two studied models that an average of four portions of close contact (Table 4) corresponding to about 50% of the whole contact length, were observed within the first minute, without significant modification with time. These results are consistent with the finding that in normal culture medium, the nearest cell-substrate approach is never less than 100 Å as measured with the interference reflection microscope,[31] and only edges of the cells come into the 100 Å adhesion with the substrate. As the time for separation is proportional to the area of contact, this phenomenon could be related to the rapidity of deadhesion as proposed by Curtis.[31] Therefore, it would be of high interest to detail on an electron microscopical level the mechanism of target separation that were found to occur after 20 to 40 min.[32]

Finally, we evaluated the limits of the biological model. We took into account the problems of interpretation due to the limits of the biological model which has to be fixed and embedded in epon before observation. An important problem in our approach is the possibility that glutaraldehyde or osmium might alter the cell-surface geometry. Also, the width of the gap separating bilayers in contact areas might be modified. Admittedly, there is no rigorous way of ruling out these possibilities until our results can be obtained with different methods. Indeed, interference reflection microscopy,[31] time lapse cinematography,[32] resonance energy transfer,[33] or total internal reflection fluorescence[34] may yield some information on the distribution of distances between cell surface and flat substrates without any fixation (see also Chapter 9).

However, some earlier reports suggest that fixation may not involve major changes of the parameters we measure. Indeed, electron microscopic study of the distance between erythocyte membranes bound by immunoglobulins[18] or dextran[19] leads to an average of 250

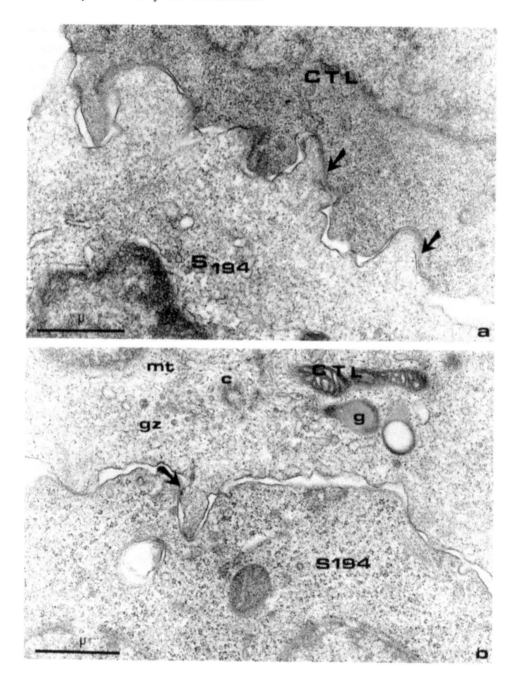

FIGURE 4.   Two aspects of the interacting area between CTL and target. (a) Microvilli from the CTL are deeply anchored in the S194 target. The maximum distance between the two membranes is about 0.15 μm but in places the contact is so tight that it is not possible to clearly distinguish the CTL cytoplasmic membrane from the target cytoplasmic membrane (◆); (b) a deep microvillus (◆) arising from the CTL is visible. The centriole (c), microtubules (mt), Golgi zone (gz), and the granules (g) known to be associated with the cytolytic process are present. (Magnification × 22,800.)

Å, which is compatible with the results obtained with the previously quoted works. Further, glutaraldehyde treatment was reported not to alter the membrane protrusions of leukocytes as monitored by phase contrast microscopy.[35] Osmium tetroxide fixation resulted in a final separation of 100 to 200 Å between cell and substrate.

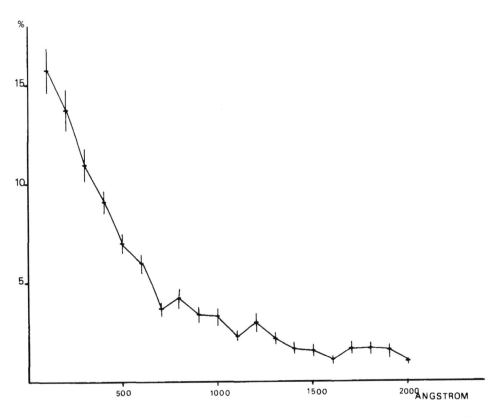

FIGURE 5. Distribution of intermembranar distances in fresh CTL-target conjugates, where 86 regions of inter-action between CTL and target cells were analyzed for determination of the percent of the length of lymphocyte contours separated from the target by a distance of 100, 200, 300 . . . 2500 Å. Mean values are shown. Vertical bar length is 2 × standard error of the mean.

Table 4
NUMBER OF CLOSE CONTACTS DURING TIME

Time of contact (min)	No. of expts	CTL-S194 conjugated cells	No. of expts	TG2OUA2-S194 conjugated cells
1	25	3.75 + 0.48	18	3.40 + 0.47
5	26	4.11 + 0.48	25	3.50 + 0.42
15	7	4.14 + 1.15	11	4.50 + 0.72
30	9	2.10 + 0.43	16	3.70 + 0.59
60	19	4.64 + 0.74	17	2.70 + 0.52

Curtis[31] suggested that osmium tetroxide might reduce the gap between the two surfaces by decreasing their negative charges. Alternatively, osmium tetroxide might produce a complete reversal in electric charges, and it is possible, considering the results obtained with other techniques, that the gap found between fixed cells is an artifact, but one which closely reproduces the situation in life.

IV. CONCLUSION

As the CTL-target recognition is a specific binding process, it was of interest to search for its particularities compared with the general adhesion phenomenon. Many works have been done to describe and try to understand the mechanisms of binding and killing in this model. The authors have particularly focused their interest on the preeminent CTL macrovilli more or less deeply anchored in the target. In the present chapter, we considered the CTL-target binding in the forthcoming view of molecular adhesion. Therefore, we studied the precise roughness of both cell surfaces as revealed with electron microscopy using a quantitative method. We studied precisely the organization and distribution of the microvilli and their behavior during cell binding.

In conclusion we can state that CTL-target binding shares some common features with other adhesion processes such as cell-cell and cell-substrate contacts. First, the CTL-target binding occurred very rapidly (within 1 min). Second, we evidenced that binding involved some close contacts (<500 Å), which corresponded to about 50% of the total interacting area. We therefore clearly demonstrated that only a fraction of the cell membrane that would seem to be bound when examined with light microscopy is really involved in molecular adhesion. These results have to be considered in view of one of the chronologically following events, which is deadhesion of CTL from the target and which easily can be compared to cell locomotion on a substrate. As a consequence, the actual work of adhesion is higher than the value that would be obtained by deformation analysis made on light micrographs, as described in the previous chapter. Some caution is required to apply these methods to rough cells, although these are perfectly valid when applied to smooth erythrocytes or lipid vesicles.[36,37]

In contrast, some features seem to be specific to this original model: CTL-target binding simultaneously involves macrovilli (1 μm in size) and microvilli (0.1 μm in size) on the CTL. In the contacts between cultured CTL and targets, most of the microvilli disappeared and the region of intimate contact displayed a smoothing of the CTL surface. This phenomenon was not detected with other methods. Very close membrane appositions without evidence of membrane fusion were observed, but the accuracy of our method is probably too low to bring a more precise conclusion on the possibilities of existence of an equilibrium intermembranar distance which would be of high interest to test theories of cell adhesion. Of particular interest is the demonstration that the rapid binding leads immediately to a stable configuration which allows a rapid occurrence of the following events: the reorientation of the nucleus, granules, and Golgi zone begins within 6 to 10 min,[38] and this phenomenon is completed within 30 min.[39] Also, the contact area of cytotoxic T-lymphocytes and target cells was found to be spatially related to the microtubule-organizing centers.[40] Thus, it would be suitable to repeat our experiences with fibroblasts that were demonstrated to pull at their substrate by Harris and colleagues,[41] who cultured different cell species on elastic substrates.

ACKNOWLEDGMENT

We thank Dr. M. Nabhloz for providing the TG2OUA2 clone.

REFERENCES

1. **Jones, G. E., Gillett, R., and Partridge, T.,** Rapid modification of the morphology of cell contact sites during the aggregation of limpet haemocytes, *J. Cell Sci.,* 22, 21, 1976.
2. **Willingham, M. C. and Pastan, I.,** Cyclic AMP modulates microvillus formation and agglutinability in transformed and normal mouse fibroblasts, *Proc. Natl. Acad. Sci. U.S.A.,* 72, 1263, 1975.
3. **Ben-Shaul, Y. and Moscona, A. A.,** Scanning electron microscopy of aggregating embryonic neural retina cells, *Exp. Cell Res.,* 95, 191, 1975.
4. **Weller, N.,** Visualization of Concanavalin A binding sites with scanning electron microscopy, *J. Cell Biol.,* 63, 699, 1974.

5. **Ryter, A. and Hellio, R.,** Electron-microscope study of dictyostelium discoideum plasma membrane and its modification during and after phagocytosis, *J. Cell Sci.,* 41, 75, 1980.

6. **De Petris, S., Raff, M. C., and Hallucci, L.,** Ligand-induced redistribution of concanavalin A receptors on normal, trypsinized and transformed fibroblasts, *Nature (London) New Biol.,* 244, 275, 1973.

7. **Rutishauser, U. and Sachs, L.,** Receptor mobility and the mechanism of cell-cell binding induced by concanavalin A, *Proc. Natl. Acad. Sci. U.S.A.,* 71, 2456, 1974.

8. **Foa, C., Bongrand, P., Galindo, J. M., and Golstein, P.,** Unexpected cell surface labeling in conjugates between cytotoxic T lymphocytes and target cells, *J. Histochem. Cytochem.,* 33, 647, 1985.

9. **Capo, C., Garrouste, F., Benoliel, A. M., Bongrand, P., Ryter, A., and Bell, G.,** Concanavalin A-mediated thymocyte agglutination: a model for a quantitative study of cell adhesion, *J. Cell Sci.,* 56, 21, 1982.

10. **McCloskey, M. A. and Poo, M. M.,** Contact-induced redistribution of specific membrane components: local accumulation and development of adhesion, *J. Cell Biol.,* 102, 2185, 1986.

11. **Martz, E.,** Mechanism of specific tumor cell lysis by alloimmune T lymphocytes: resolution and characterization of discrete steps, in *Contemporary Topics in Immunobiology,* Vol. 7, Stutman, O., Ed., Plenum Press, New York, 1977, 301.

12. **Plunt, M., Bulbers, J. E., and Henney, C. S.,** Studies on the mechanism of lymphocyte mediated cytolysis. VII. Two stages in the T cell mediated lytic cycle with distinct cations requirements, *J. Immunol.,* 116, 150, 1976.

13. **Golstein, P. and Smith, E. T.,** Mechanism of T-cell mediated cytolysis: the lethal hit stage, *Contemp. Top. Immunobiol.,* 7, 269, 1977.

14. **Zagury, D., Bernard, J., Thierness, N., Feldman, M., and Berke, G.,** Isolation and characterization of individual functionally reactive cytotoxic T lymphocytes: conjugation killing and recycling at the single cell level, *Eur. J. Immunol.,* 5, 818, 1975.

15. **Sanderson, C. J., Hall, P. J., and Thomas, J. A.,** The mechanism of T cell mediated cytotoxicity. II. Morphological studies of cell death by time lapse microcine matography, *Proc. R. Soc. London Ser. B,* 192, 241, 1976.

16. **Grimm, E., Price, Z., and Bonavida, B.,** Studies on the induction and expression of T cell-mediated immunity, *Cell. Immunol.,* 46, 77, 1979.

17. **Capo, C., Bongrand, P., Benoliel, A. M., Ryter, A., and Depieds, R.,** Particle-macrophage interaction: role of surface charge, *Ann. Immunol.,* 132, 165, 1981.

18. **Easty, G. C. and Mercer, E. H.,** An electron microscope study of model tissues formed by the agglutination of erythrocytes, *Exp. Cell Res.,* 28, 215, 1962.

19. **Jan, K. M. and Chien, S.,** Role of surface electric charge in red blood cell interactions, *J. Gen. Physiol.,* 61, 638, 1973.

20. **Sanderson, C. J. and Glauert, A. M.,** The mechanism of T-cell mediated cytotoxicity. V. Morphological studies by electron microscopy, *Proc. R. Soc. London Ser. B,* 194, 417, 1976.

21. **Liepins, A., Faanes, R. B., Lifter, J., Choi, Y. S., and De Harven, E.,** Ultrastructural changes during T-lymphocyte-mediated cytolysis, *Cell. Immunol.,* 28, 109, 1977.

22. **Nicolas, G. and Zagury, D.,** Etude par cryofracture de la zone de contact entre cellule cytolytique et cellule cible, *Biol. Cell.,* 37, 231, 1980.

23. **Mege, J. L., Capo, C., Benoliel, A. M., Foa, C., Galindo, R., and Bongrand, P.,** Quantification of cell surface roughness; a method for studying cell mechanical and adhesive properties, *J. Theor. Biol.,* 119, 147, 1986.

24. **Nabholz, M., Cianfriglia, M., Acuto, D., Conzelmann, A., Haas, W., Van Boehmer, H., MacDonald, H. R., Pohlit, H., and Johnson, J. P.,** Cytolytically active murine T-cell hybrids, *Nature (London),* 287, 437, 1978.

25. **Bongrand, P., Pierres, M., and Golstein, P.,** T-cell mediated cytolysis on the strength of effector-target cell interaction, *Eur. J. Immunol.,* 13, 424, 1983.

26. **Elias, H., Henning, A., and Schwartz, D. E.,** Stereology: applications to biomedical research, *Physiol. Rev.,* 51, 158, 1971.

27. **Evans, E. A.,** Structural model for passive granulocyte behaviour based on mechanical deformation and recovery after deformation tests, in *White Cell Mechanics: Basic Science and Clinical Aspects,* Meiselman, H. J., Lichtman, M. A., and Celle, P. L., Eds., Alan R. Liss, New York, 1984, 53.

28. **Dialynas, D. P., Loken, M. R., Glasebrook, A. L., and Fitch, F. W.,** Lyt2$^-$/Lyt3$^-$ variants of a cloned cytolytic T cell line lack an antigen receptor functional in cytolysis, *J. Exp. Med.,* 153, 595, 1981.

29. **Mege, J. L., Capo, C., Benoliel, A. M., and Bongrand, P.,** Use of cell contour analysis to evaluate the affinity between macrophages and glutaraldehyde treated erythrocytes, *Biophys. J.,* 52, 177, 1987.

30. **Sanderson, C. J. and Glauert, A. M.,** The mechanism of T-cell mediated cytotoxicity. VI. T-cell projections and their role in target cell killing, *Immunology,* 36, 119, 1979.

31. **Curtis, A. S. G.,** The mechanism of adhesion of cells to glass: a study by interference reflection microscopy, *J. Cell Biol.,* 20, 199, 1964.

32. **Rothstein, T. L., Mage, M., Jones, G., and McHugh, L. L.,** Cytotoxic T lymphocyte sequential killing of immobilized allogeneic tumor target cells measured by time-lapse cinematography, *J. Immunol.,* 121, 1652, 1978.

33. **Stryer, L. and Haugland, R. P.,** Energy transfer: a spectroscopic ruler, *Proc. Natl. Acad. Sci. U.S.A.,* 58, 719, 1967.

34. **Axelrod, D.,** Cell-substrate contacts illuminated by total internal reflection fluorescence, *J. Cell Biol.,* 89, 141, 1981.

35. **Lichtman, M. A., Santillo, P. A., Kearney, E. A., Roberts, G. W., and Weed, R. I.,** The shape and surface morphology of human leukocytes. In vitro effect of temperature, metabolic inhibitors and agents that influence membrane structure, *Blood Cell,* 2, 507, 1976.

36. **Evans, E. A. and Metcalf, M.,** Free energy potential for aggregation of mixed phosphatidylcholine/phosphatidyl serine lipid vesicles in glucose polymer (dextran) solutions, *Biophys. J.,* 45, 715, 1984.

37. **Buxbaum, K., Evans, E. A., and Brooks, D. E.,** Quantitation of surface affinities of red blood cells in dextran solutions and plasma, *Biochemistry,* 21, 3235, 1982.

38. **Somers, S. D., Whisnant, C. C., and Adams, D. O.,** Quantification of the strength of cell-cell adhesion: the capture of tumor cells by activated murine macrophages proceeds through two distinct stages, *J. Immunol.,* 136, 1490, 1986.

39. **Kupfer, A. and Dennert, G.,** Reorientation of the microtubules organizing center and the Golgi apparatus in cloned cytotoxic lymphocytes triggered by binding to lysable target cells, *J. Immunol.,* 133, 2762, 1984.

40. **Geiger, B., Rosen, D., and Berke, G.,** Spatial relationships of microtubule-organizing centers and the contact area of cytotoxic T lymphocytes and target cells, *J. Cell Biol.,* 95, 137, 1982.

41. **Harris, A. K., Stopack, D., and Wild, P.,** Fibroblast traction as a mechanism for collagen morphogenesis, *Nature (London),* 290, 249, 1981.

42. **Greenwood, J. A.and Williamson, J. B. P.,** Contact of nominally flat surfaces, *Proc. R. Soc. London Ser. A,* 295, 300, 1966.

Section III
Results and Models

Chapter 9

THE DATA ON CELL ADHESION

A. S. G. Curtis

TABLE OF CONTENTS

I. INTRODUCTION

The other chapters of this book are concerned with the chemistry of cell adhesion or with experimental methods of investigating cell adhesion. The various theories of the mechanisms of cell adhesion must be consistent with the actual data on cell adhesion, derived from experiment or observation. This chapter is devoted to that topic.

It should be appreciated from the outset that any particular chemical which affects adhesion may be doing so directly because it forms part of the molecular structure of the adhesion system or far more indirectly. Inhibitors of cell adhesion may again be acting directly or indirectly. It should also be realized that almost any molecule at the cell's interface with the environment is potentially either a molecule which will be adhesive or one which will diminish adhesion. The reason for this is that all molecules anchored to the outside of the cell surface will have some sets of interaction with surrounding molecules even though these may be weak and unspecific.

Considerable attention is given in Chapters 5 to 8 to a number of important techniques for investigating cell adhesion, but in practice a very large amount of work has been done with much simpler systems. The virtues and shortcomings of these are described below.

II. THE PHYSICAL NATURE OF THE ADHESIVE CONTACT

In order to make sensible interpretations of chemical data on cell adhesion, we need to understand the physical nature of the adhesive contact. I shall discuss these in terms of the techniques used to investigate the contact.

A. Electron Microscopy

The classification of the structure of the cell-cell (and to a lesser extent the cell-substrate) contact was based from the start,[1,2] interestingly, on the concept of the impermeability of the contact to lateral movement of other molecules through the adhesion. Thus, the zonulae adherens was defined as a structure through which tracers such as hemoglobin, ferritin, or peroxidase would pass, while the zonulae occludens[3] was defined as an occluding or impermeable junction. The desmosome or macula adherens, characterized by very extensive cytoskeletal structures close to the contact, does not fit into this classification very well, though there have been suggestions that it, too, is a permeable structure. The gap junction[4] is, of course, a structure in which the cell surfaces have partially fused together so that there is direct structural continuity between the cells to provide the adhesion.

The high permeability of zonulae adherens and desmosomes has been little investigated, but it does suggest that there is little macromolecular material in the gap between the cells.[5] Recently, however, an elegant study has been made[6] of the permeation of proteins of known Stokes' radius through the zonulae adherens linking the photoreceptor and Muller cells of the eye. The pore radius of these contacts is about 3.5 nm. A biotin-avidin system was used to detect the biotinylated tracers, and since permeation was appreciable with a very small hydrostatic head, it can be concluded that few if any large molecules occlude this contact.

B. Interference Reflection Microscopy

In this technique,[7] the interference between light reflected at the surface of the culture dish and at the surface of the cell gives rise to an image which provides some measure of the closeness of approach of the cell surface to a substrate with which it is in adhesion. The reflection from the cell surface presumably arises at the refractive index and dielectric constant discontinuity which is probably placed at the outer end of the hydrocarbon fatty acid chains. Though the type of image seen can be almost unambiguously interpreted, because of an almost unique solution defining the various refractive indexes in the system it is unfortunately the case that zero-order blocks due to a 10 nm gap between the surfaces are almost the same as those arising from a 1 nm gap. Thus, though it tells us that some parts of the cell, the "focal contacts", are close to a surface and some much farther away, it does not really provide wholly useful information about cell contacts and adhesion despite optimistic claims.[8,9]

C. Total Internal Reflection Fluorescence Microscopy

This technique is based upon the concept of allowing a fluorescent tracer molecule to diffuse into the contact between a cell and the substrate to which it is adhering, and stimulating fluorescence by the evanescent wave which penetrates about 100 nm into a second medium when it strikes the interface at the critical angle.[10,11] If a gap open to the tracer molecule exists, then the fluorescence should, if sufficient fluorescent molecules are present, provide a strong indication of the existence of a gap. Use of this method[12] with labeled dextrans suggests that some parts of a fibroblast's contact with a substrate are too close for penetration of appreciable amounts of the fluorescent molecule.

D. FET Microscopy of Cell Contacts (FS)

Fluorescence energy transfer systems provide a "molecular yardstick" for distances less than 5 nm.[13] Thus, if a fluorochrome in the plasmalemma of a cell in adhesion with a substrate labeled with another fluorochrome is optically coupled, the intensity of the secondary fluorescence can be used to measure the distance between them. Recently, Professor C. Wilkinson and I have started to use this system in the microscopy of cell contacts. The initial results for fibroblasts adhering to glass surfaces derivatized with fluorochromes is that the surface proteins are more than 5 nm distant from the labeled substrate.

E. Types of Contact Formed During Adhesion

The various methods used to study adhesions have for the most part short time scales (see Chapters 5 to 8 and Section III below) on the order of 1 to 10 min. Desmosomes are first recognizable morphologically about 30 min after a cell suspension is aggregated,[14] and zonulae occludens form in a longer time period of about 6 hr.[15] Thus, many methods of studying adhesions will be confined to examination of the conditions determining zonulae adherens formation. Intracytoplasmic changes in the distribution of cytoskeletal elements in relation to adhesions have been detected as early as 1 min after settling of a cell.[16]

III. CRITIQUE OF METHODS OF MEASURING CELL ADHESION

A. Introduction

The majority of methods for measuring cell adhesion require the separation of a tissue into individual cells, inhibition of whatever process is used to obtain this dissociation, and then the formation of adhesions either between cells or between cells and nonliving substrates. When adhesions are formed, the cell usually may extend its area of contact (spreading) on the contact surface over an appreciable time interval. The distinction between spreading and adhesion is discussed in Section III.C.

B. Methods of Dissociating Cells

Though the fact that a particular treatment leads to cell dissociation is a direct source of evidence about the adhesive mechanism in itself, there may also be profound effects on the subsequent readhesion due to the dissociation method used. Thus, it has generally been found that cells such as lymphocytes or leukocytes[17] adhere well at low temperatures while tissue cells adhere well only at temperatures above about 20°C.

Unfortunately, one of the most effective methods of dispersing tissues uses trypsin, and the activity of this enzyme must be inhibited before the cells are used for investigation of adhesion. In practice, inhibition of trypsin is often achieved by adding serum which contains antitrypsin as well as various antiadhesive proteins,[18,19] or the medium used for the adhesion measurement contains trypsin. However, it has been shown, as previously mentioned, that serum contains proteins that diminish cell adhesion,[18,19] and cells which have been trypsinized can be dependent on fibronectin for their adhesion immediately therafter.[20] Such observations call into question the interpretation of many experiments in which serum has come in contact with the cells being studied.

C. Spreading and Adhesion

Most cell types tend to spread shortly after they have formed an adhesion. The reasons for this are not very clear. Some authors[21,22] have argued that the cell surface will spread in an adhesion to a minimum surface free-energy configuration. This argument has the hidden implication that (1) the internal cytoskeleton will not normally prevent this taking place, and, perhaps more relevantly when we are considering adhesion, that (2) there is a quasi-infinite number of molecules available for use in adhesion in the surface so that more and more adhesive bonds are formed. If, however, the number of adhesion molecules in the cell surface is rather low and they are mobile, we would expect to find little spreading when the surface on which the cell was attached had a high density of complementary molecular sites. Thus, based on this second interpretation, a spreading cell would seek attachment sites for its rather sparse number of adhesive molecules on a surface of rather low density of complementary sites. Since we do not know whether either of these interpretations is true, nor the extent to which the cytoskeletal properties of the cell affect spreading, we are in something of a quandary when we use methods for studying cell adhesion which include a component of cell spreading in them. Thus, methods which count the number of cells adhering to a surface in a number of minutes in effect measure the resistance of the adhesion recently formed to distraction by the flowing medium. The more the cell is spread, the more likely it is to resist distraction. The method in which adhesions form from a laminar flow[23] and the Couette method[24] are ones in which the interaction is very rapid (0.01 to 0.1 sec), and spreading is likely to be relatively unimportant. In all cases, the geometry of the actual contact should be investigated because contacts that are "smooth" under light microscopy may well be, in reality, convoluted (see Chapter 8).

D. Use of Antibodies and Fab Methods

Antibodies against cell-surface components usually agglutinate cells, though interestingly some cell-surface molecules have been identified as cell-adhesion molecules on the basis of inhibition by divalent, i.e., "normal" antibodies.[25-27] Obviously, if the divalent antibody agglutinates cells, no conclusion can be drawn about the adhesive role of the species of molecule which reacts with the antibody. With this in mind, Gerisch[28] and Beug et al.[29] proposed that Fab fragments should be used for the detection of adhesion molecules, and their methods have been very widely adopted (see, e.g., Section VII and References 30 and 31). Gerisch pointed out correctly that an important control in these experiments is to show that a separate Fab directed against another cell surface antigen present at fairly or very high density does not inhibit adhesion. Even if such a demonstration is made, it is still possible that the Fab inhibits adhesion sterically, and it is also noteworthy that the nonreacting antigen used by Gerisch and Beug as a control is claimed by them to lie deeper in the plasmalemma than the adhesive molecule. This admission weakens the value of the control.

However, a more substantial criticism can be leveled at this type of experiment. Fab molecules are, apart from their antigen binding site, rather hydrophilic molecules capable of sterically hindering interactions. Would a Fab against a synthetic antigen (itself of no adhesive function) incorporated into the plasmalemma inhibit adhesion?

This question has been tested using two different synthetic antigens: in one case a FITC derivative of "monovalent" Concanavalin A, and in the other case a TRITC derivative of stearoylated dextran.[32] The results show that these Fabs directed against artificial antigens which incorporate in the cell surface inhibit cell adhesion.

IV. CRITIQUE OF INTERPRETATION

A. Activation or Bonding

Numerous molecules have been described as cell-adhesion molecules (CAMs) with the implicit and often explicit statement that they directly bond from one cell to another, e.g., fibronectin, laminin, N-CAM, etc. In general, the evidence that they are involved in adhesion arises from demonstrations that:

1. Cells deficient in the molecule are unable to adhere (e.g., see Reference 33).
2. Cells blocked in production of the molecule are unable to adhere (e.g., see Reference 34).
3. Low molecular weight fragments containing the supposed binding site of the molecule block cell adhesion (e.g., see Reference 35).
4. Addition of the molecule in solution or as an adsorbed monolayer to the system enhances cell adhesion, (e.g., see Reference 36).

The precise interpretation of such results depends on other factors because those results in themselves do not specify whether the adhesion molecule is normally embedded (at least in part) in the cell surface or whether it reacts with recognition molecules in the surface to form a "molecular bridge". Indeed, such results do not of themselves exclude other interpretations — for instance, that the molecule binds to a surface receptor not as part of a molecular bridge but as an activator of some other adhesive process.

Consider, for example, work on fibronectin. It has been reported that cells of a line deficient in fibronectin synthesis would adhere only when exogenous fibronectin was added to the adhesion system.[37,38] Thus, it was concluded that cell adhesion depended rigorously on the provision of exogenous or endogenous fibronectin. Later, it was found that fibroblasts would adhere when fibronectin synthesis or secretion[39] were, presumably, inhibited, provided that the test was carried out under serum-free conditions. Finally,[41] it was shown that cells would adhere in the absence of exogenous or endogenous fibronectin or in the presence of the fibronectin cell-binding domain tetrapeptide RGDS, which sould inhibit binding to fibronectin by competition. Thus, since fibronectin is demonstrated to be only an adjunct to cell adhesion, other suggestions as to its mode of action are needed. One new idea due to my colleague, J. G. Edwards, is that fibronectin may be acting as an activator of adhesion. Recently, other workers[40-43] have interpreted effects on adhesion such as calcium ion effects as activation processes. Phorbol esters are thought to influence adhesion by a similar mechanism.[44] Similarly, the chemotactic peptide f-Met-Leu-Phe activates leukocytes[45] and rapidly raises their adhesiveness.[46] The effects of leukotrienes and prostacyclins[47,48] can be similarly interpreted (see also Table 1).

B. Energy of Adhesion and Mechanism

One of the main areas of deficiency in our understanding of cell adhesion lies in the correlation of mechanisms of adhesion and the expected strengths (energies) of adhesion. Yet measurement of adhesion in quantitative terms is a main theme of this book, and it is important that the expectations of various theories correlate with experimental measurements.

Measurements of cell adhesiveness on systems in which zonulae adherens are known, or at least may be presumed to be forming (that is short-term measurements of attachment), show low values of adhesion on the order of 10^{-9} J/cm^{-2} (see Chapter 7). If these values

Table 1
REPORTS OF PROSTAGLANDINS, PROSTACYCLINS,
LEUKOTRIENES, AND THROMBOXANES AFFECTING CELL
ADHESION

Chemical	Cell type affected[a]	Increase (+) or decrease (−)	Effective molarity	Ref.
Arachidonic acid	PMNs	+	10^{-5}	61,63
Prostaglandins				
E1	Ehrlich A.T. cell	− (?)	10^{-5}	62,64
E1	Platelets, PMNs	−	10^{-9}	65
F2a	Ehrlich A.T. cells PMNs	−	10^{-5}	62,64
Prostacyclin (PGI₂)	PMNs	−	10^{-6}	66,67
LTB₄	PMNs to endothelia	+	10^{-7}	68, 69
TXA₂	PMNs	+	10^{-10}	70
PAF-AC ether	PMNs, platelets	+	10^{-9}	71

Note: PMNs = polymorphonuclear leukocytes; Ehrlich A.T. cells = Ehrlich ascites tumor cells; LTB₄ = leukotriene B₄; TXA₂ = thromboxane A₂.

[a] No distinction is made between cell-cell and cell-nonliving substrate measurements in this table.

are to be explained by a molecular bridging mechanism it is clear that low surface densities of receptor-bonding molecule systems are required. For example, Aplin et al.[49] reported that fibroblasts had a surface density of some 50,000 fibronectin receptors per cell. If these receptors were all occupied with fibronectin molecules bound in turn to another similar cell, the energy of adhesion can be estimated from bond energies. If we assume that the bonds in question have an energy in the range 1×10^{-6} J/mol, then the adhesion should have an energy per cell to cell contact in the region 1×10^{-13} J per contact. Obviously, as remarked in the paper by Aplin et al.,[49] cell adhesion can take place with fewer fibronectin sites occupied. Thus, there is no great discrepancy with measurements, but the values are also compatible with other models of adhesion. Clearly, what is required is measurements of the change in adhesion as some factor or component is altered. However, reports suggesting a far higher density of receptor sites for cell binding per cell are in doubt unless careful correlation with measured energies of adhesion is presented as well. Similarly, figures as low as 18 or 28 molecules per cell given for sponge cell aggregation[50] are improbable if the molecules actually act as adhesive bonds.

Few papers report attempts to correlate changes in absolute energies of adhesion with experimental changes in the adhesion system and to interpret these in terms of an actual mechanism of adhesion. Correlations between changes in medium dielectric constant and adhesion were successfully interpreted in terms of a DLVO explanation.[51] It should be noted that the DLVO explanation of adhesion may run into difficulties should the layer of glycoproteins and proteoglycans on the cell surface be of appreciable depth in concentration; layers thinner than 5 nm are not likely to affect the explanation greatly.

C. What Is Unphysiological Adhesion?

Examination of the literature shows that cell adhesion has been studied on dead, fixed cells[52,53] and in various media that do not even approximate in composition the fluids in which the cells live, or of adhesion to surfaces whose chemical composition is substantially unlike that found on other cells. How relevant are such studies?

The question of the relevance of studies on fixed cells can be simply resolved by inquiring whether the method of fixation may not have altered the adhesive mechanism. Most methods of fixation are inherently likely to alter mechanisms of adhesion if only because they may cross-link proteins, damage macromolecular conformation, and in some cases leave reactive sites on the cell surface.

There has long been a view that cell adhesion to substrata is inherently different from cell-cell adhesion, thus, that information obtained from such experiments is misleading.[54] Similarly, is data obtained from studies of cell adhesion of mammalian cells in serum-free media valid? It is true that cells normally adhere in a protein-rich medium, the interstitial fluid, which, however, is substantially different from serum in its protein composition?[55] Protein adsorption takes place in both cellular and other surfaces when they are placed in protein solutions. If cells are placed in protein-free media, do they adhere to ''bare'' substrates or do they secrete sufficient protein to modify the surface appreciably? Preliminary evidence from this laboratory suggests that in a 30-min period, cells such as BHK cells release less protein than would form a monolayer beneath the cells on a polystyrene substrate. The use of serum in media for cell adhesion studies is further complicated by the fact that several serum proteins are nonadhesive for cells and are preferentially adsorbed by certain substrata, e.g., α-1-antitrypsin by bare polystyrene.[56] The effects of serum and its components on cell-adhesion measurement technology have been explored in a number of papers.[56-58]

V. THE ROLES OF LIPIDS

Lipids form a major component of the plasmalemma and thus alterations in membrane lipids might be expected to influence cell adhesion in several different ways. Changes in the bulk dielectric constants of the plasmalemmal lipid might alter electrodynamic forces between membranes[59] or the fluidity of the membranes, and thus the diffusion or mobility of adhesion molecules in the membrane toward (or away from) sites at which adhesion is required. Rather similar explanations have been advanced both for the effects of changing the fatty acid composition of plasmalemmal phospholipids[59] and of the level of cholesterol[60] in the membrane. These effects require relatively large percentages of alterations in the lipids of the membrane and can perhaps best be described as bulk effects.

Unlike the bulk effects described above, arachidonic acid, prostaglandins, prostacyclins, leukotrienes, and thromboxanes (see Table 1) effect the adhesion of leukocytes, and platelets to marked degrees. Unfortunately, little work has been done on the effects of these substances on other cell types. These substances are active often at nanomolar levels. It is clear that their effects are involved in cell activation,[72] and in particular in interactions with protein kinase C[73] and oxidation processes,[74,75] so that the exact route by which adhesion is affected (or prevented) is unclear. However, it is likely that further investigation of these systems may be very worthwhile.

VI. ROLE OF OXIDATION SYSTEMS

It is only in the past few years that evidence has begun to accumulate which suggests that oxidations of other cells performed by cells or oxidations of nonliving substrates may play a role in adhesion.

Leukocytes and lymphocytes are known to produce superoxide anions on activation[76] which decompose into hydrogen peroxide and hydroxyl radicals. It was shown[77] that polystyrene surfaces which had been hydroxylated by a chloric acid reaction were of greatly increased adhesion for cells both in the presence or absence of serum proteins. Later,[58] it was demonstrated that activated leukocytes might be increasing the adhesion of cells to polystyrene surfaces by the action of superoxide anions on the plastic, and that BHK fibroblasts could effect some oxidation of the same substrate. In this and related work, much evidence was produced that cells had high adhesion to a moderate density of hydroxyl groups on the surface.[78] It should be appreciated that though a thorough demonstration was made of the role of surface hydroxyl groups in adhesion, it has not yet been shown that this is the form of oxidation achieved on the cell surface by cellular oxidations. It is interesting to note that polylysine, usually believed to increase cell adhesion because of its high positive charge (see Section X), has recently been discovered to stimulate cellular oxidations.[79]

A different type of explanation of the effects of oxidation has been that cross-linking dialdehydes are generated at the cell surface.[80] Berke[81] showed that such an explanation was unlikely because the reducing agent sodium borohydride did not inhibit the system (cytotoxic lymphocyte-target cell interaction) in which such a reaction was claimed. In this connection, it is interesting to note that arachidonate treatment of leukocytes leads to the release of malonyldialdehyde, and correlates well with the extent of aggregation, but that various inhibitors of aggregation had no effect or even stimulated malonyldialdehyde production.[82] This makes a dialdehyde bridging explanation unlikely.

Thus, it seems worthwhile to accord further investigation into cell-surface oxidations and their effects on adhesion.

VII. THE ROLES OF CELL ADHESION MOLECULES (CAMs)

Though etymologists might argue that the term used in the title of this section and its acronym be applicable to any molecule involved in cell-cell adhesion, usage, at least to date, has confined the term to molecules usually identified by the Fab technique (see Section III.D), and believed to be directly involved in bonding cell-cell. The first molecules to be identified by this technique were described by Gerisch[28] in the slime mold *Dictyostelium discoideum*. Somewhat later, Edelman and colleagues[83,87] described N-CAM, L-CAM and NG-CAM molecules associated with nerve, liver, and glial cells, respectively. Table 2 lists a range of molecules which have been described as CAMs.

Though a considerable amount of work has been carried out describing the distributions of these molecules in various tissues, very little work has been done to actually demonstrate how such molecules effect adhesion. It appears that many workers have assumed that these molecules, unlike the receptors for the various nectins (see Section VIII), interact homotypically. The only attempt to test this was carried out by incorporating the molecules in liposomes,[99,100] and then showing that they increased adhesion.

One test which has been fairly widely used in the characterization of CAMs is to check whether the isolated CAM blocks the inhibition of adhesion by a Fab. Such a result, though perfectly consistent with the Fab blocking of cell adhesion, goes no further in demonstrating that the CAM actually acts directly in cell adhesion than the blocking of adhesion by a Fab species, and the criticisms advanced in Section III.D are still unanswered. Perhaps the most rewarding and critical area of future investigation would be to check the structure of the cell contact in relation to CAMs.

Little correlation has been made between molecules identified by the Fab method and those detected by other methods, though an interesting correlation of neuronal adhesion molecules identified by a variety of methods has been published.[101] Evidence has been produced[102] that a heparin-binding domain in N-CAM is involved in cell-substrate adhesion.

VIII. ROLE OF FIBRONECTINS AND SIMILAR NECTINS

Though it would be perfectly reasonable on etymological grounds to include the various nectins (in particular, fibronectin, vitronectin, chondronectin, and laminin) among the CAMs, I choose to distinguish them partly on grounds of usage, partly on the fact that they are believed to be linking molecules bound to their appropriate receptor (unlike CAMs which

Table 2
CELL ADHESION MOLECULES (CAMs): A LISTING

Name	Cell type(s)	Mol wt (kdalton)	Comments	Ref.
Contact site A	*Dictyostelium*	80	Asulfated glycoprotein	28,29
Contact site B				
N-CAM	Chick nerve cells, etc.	142	Polysialylated protein	83,85
L-CAM	Liver, etc.	124	Acidic glycoprotein	86
NG-CAM	Chick neuron-glia cell	200, 135, & 80		87
LFA-1	Leukocytes	150—177	α-chain variable β-chain	88
p150,95	Lymphocytes	& 95	constant	89
Mol	Monocytes			
Leu-CAM	Leukocytes	90	Not identical to preceding	90
C-AIM	Hepatocyte	120	For collagen adhesion	91
P-AIM	Hepatocyte	105	For plastic (polystyrene?) adhesion	
CAM-105	Hepatocyte	105		92
Uvomorulin	Mouse embryo	123		93
	Mouse liver	102		
	Mouse gut	92		
Cadherin				94
L1	Mouse brain	140		
	Granule cells	200		95,96
BSP-2				
D2				
PH-20	Sperm			97
?	Sponge			98

are thought to be homophilic molecules acting as receptors for themselves), and partly because, on the whole, they have been identified by methods substantially different from those used to detect the CAMs described in Section VII.

The various nectins are listed in Table 3. The methods used in their detection have been

1. Isolation of molecules from serum, intercellular material, and cell surfaces which promote adhesion when added to systems of adhering cells
2. Demonstration that cells naturally deficient in production of the molecule in question have reduced adhesion
3. Demonstration that fragments of the molecule may act as inhibitors of cell adhesion

For fibronectin, all of the above criteria have been satisfied, but work is less complete for chondronectin, laminin, vitronectin, etc. There is extensive literature on the sites of occurrence and on the possible developmental correlations of these molecules which will not be reviewed here.

It is thought that these molecules are bound by surface receptors so that the nectins act as intercellular bridging molecules.[113,114] In the case of fibronectin, good evidence has been produced for the existence of a receptor on a variety of cell types.[114-116] The fibronectin receptor appears to have rather low affinity for the molecule, with a K_d of 0.8 μM and about 100,000 sites per cell. A considerable amount of work has been carried out, very adequately reviewed by Hynes,[104] on the general biochemistry of fibronectin and on the various sequences within it and their various binding properties, e.g., the cell-binding sequence differs from a collagen and these in turn from a heparin-binding domain. The structure of the cell-

Table 3
LISTING OF VARIOUS NECTINS INVOLVED IN CELL ADHESION

Name	Mol wt (kdalton)	Remarks	Associated with cell types	Ref.
Fibronectin	250 (Dimer of this)		F	103,104
Laminin	2×200 (A) + 1×400 (B)		E, N	105
Chondronectin	180		C	106
Vitronectin	70	At present known only to affect cell-glass or cell-plastic adhesion		107,108
Cytotactin			NG	31
Thrombospondin	160 (Trimer)		P	109—111
von Willebrand protein			P	112

Note: F = fibroblasts, M = myoblasts, N = Nerve cells, NG = neurons to glia. P = platelets, C = chondrocytes, H = hepatocytes, E = epithelial cells.

binding domain is discussed further on. A cell-surface receptor for laminin has been isolated as a 70 kdalton complex with a high affinity ($K_d = 2 \times 10^{-9}$ M) for laminin,[103] and a vitronectin receptor[108] has been identified. Fibronectin has been shown to bind to gangliosides.[117]

Though claims have been made at incautious moments that one or another of these nectins is solely responsible for the adhesion of a given cell type, it is clear that at least two, if not many, mechanisms of cell adhesion are open to many, if not all, cell types.[118,119]

The nature of the contact structure in which fibronectin or other nectins may be involved has been investigated by stripping cells off surfaces and using IRM methods,[120] immunological, and other methods.[121] These techniques often show fibronectin concentrated in "focal adhesions" and related to cytoskeletal accumulations and orientations in the nearby cytoplasm. However, some caution should be applied to such interpretations for a number of reasons, e.g., (1) the process of stripping cells off a substrate may well rearrange various species of molecule, (2) the stripping appears to result in fragments of plasmalemma plus subjacent cytoplasm being left on the substrate so that immunochemical localization is rather uncertain, and (3) IRM microscopy should not give unchanged images of the stripped portions because of the whole change of the optical structure of the object, but in practice it appeared to do so,[120] which suggests that the "focal adhesion" structures may have had considerable optical and real depth. The comments made in Section II are relevant.

The region of the fibronectin involved in cell attachment has been identified[35] as a tetrapeptide-sequence RGDS both by the fact that the free tetrapeptide competes with fibronectin in cell-attachment assays and by the fact that this peptide or larger sequences containing the RGDS sequence when cross-linked to adsorbed IgG on polystyrene promoted cell adhesion. The paper[35] does not give exact details on the surface density of adsorbed peptide required for cell adhesion, but it is clearly high. The tetrapeptide has an affinity constant for BHK cells of 6×10^{-4} M. This short sequence is, of course, found in many other proteins, e.g., fibrinogen, collagen, thrombin, vitronectin,[108] and von Willebrand's factor,[122] and may be a widespread recognition signal for cell adhesion. Auffray and Novotny[123] have speculated that because a similar-sequence RFDS and the inverted-sequence SDGR occur in MHC class 1 and 2 molecules, such molecules may play a role in adhesion.

Though there is much evidence that fibronectin, laminin, and chondronectin play a role in development and maintenance of adult tissue structure,[124] no such physiological role has yet emerged for vitronectin. It can be shown that the support of cell attachment and spreading shown by 10% serum in culture on glass or plastic is due not to serum fibronectin but to vitronectin.[91,107,125] Incidentally, vitronectin is perhaps better termed as a cell spreading factor.[107]

As remarked in Section IV.A, it is possible that these molecules act as activators of adhesion rather than as bonding agents. It is clear that serum fibronectin can play a role in opsonization,[126] but this in itself may be an activation phenomenon. The possiblity that nectins may activate cells has been considered,[127] and it was concluded that if this happens, then the activation is rather unlike that produced by reagents such as phorbol myristoyl acetate. If the activation interpretation is correct, the receptors for fibronectin and other nectins then become signal receptor molecules.

IX. ROLE OF CARBOHYDRATE

The CAMs and nectins described in the preceding sections are all glycoproteins, which is not surprising since they are cell surface or secreted molecules. In the case of fibronectin, and probably also of vitronectin and von Willebrand protein, the presence and activity of the RDGS sequence suggests that it is the protein chain rather than the glycosidic units which plays a major role in the adhesive function of this type of molecule.

Nevertheless, a variety of techniques have been used to test whether the glycosidic component of a glycoprotein has a major functional role in a process such as adhesion. Chief among these are

1. Inhibition of glycosylation of the molecules during synthesis, e.g., with tunicamycin, to test whether the aglyco form of the protein is fully functional. Fibronectin, e.g., was shown to retain activity when unglycosylated.[128]
2. Removal of glycosidic groups with various enzymes. This technique is very useful if terminal sugars are being removed, but the variety of glycosidases and range of glycosidic structures makes such work very laborious when applied to anything but short chains of sugars. In at least one instance, removal of sugars enhanced the activity of a CAM[129] in which treatment of N-CAM with an endoneuraminidase increased the activity of the molecule. An α-mannosidase[130] was found to reduce the adhesion of *Dictyostelium* cells.
3. Use of sugars and oligosaccharides in solution to compete with the binding sites for sugars which cells may possess. This type of experiment has been, on the whole, enormously unrewarding. Use of sugars at concentrations around 0.1 M has in many cases partially inhibited cell adhesion, but the general physiological effects of such concentrations of sugars are unclear. In one case, slight inhibition of aggregation was found at low molarities of about $10^{-5} M$, but there was no concentration dependence.[131] More impressively,[132] mannose-6-phosphate and fructose-1-phosphate were effective in inhibiting the adhesion of lymphocytes to air-dried or fixed high endothelial venule cells at 0.01 M. It should be appreciated that the approaches outlined in the first and second tests above are primarily applicable to the study of the function of cell-surface molecules such as CAMs or to nectins, whereas this technique is a less clearly targeted method which might also apply to evidence about the substrate to which cells attach. Thus, a fourth method of experiment has been used for examining substrates suitable for cell adhesion and glycosylation.
4. Preparation and use of synthetic glycoside surface for testing cell attachment. Weigel and co-workers[133,134] have synthesized a range of polyacrylamide surfaces containing specific sugars at chosen surface densities. They found that some specificity in attachment was shown to different types of sugars and that the surface density of sugars

was very critical. It has been pointed out that their data may be susceptible to interpretation simply in terms of a suitable density of surface hydroxyl groups.[39] Weigel
found that an asialoorosomucoid inhibited the attachment of hepatocytes to a galactoside surface. Concentrations of nearly 10 μg/mℓ, i.e., 5 pg per cell, were required
for this. It is interesting to note that Weigel[134] notes that the threshold nature of the
process suggests that some type of response process may be happening. He produces
circumstantial evidence that the cells adhere by an asialoglycoprotein patch on their
surfaces.

Related to this type of experiment are those in which adsorbed layers of lectins[135] or other
proteins involved with carbohydrate recognition, such as glycosidases,[136] have been used in
studies of cell attachment and spreading. Since a very wide range of adsorbed proteins
support spreading and attachment,[18,137] it is unclear whether such experiments provide any
evidence in support of a role for cellular glycosidic compounds in adhesion.

Edwards[113] has reviewed the evidence that cells are bound by proteins which bind cell-
surface oligosaccharides. He concluded that the concept was very poorly established, though
some changes in N-linked glycosylations of cell surface glycoproteins were related to changes
in adhesion. It has been suggested[138] that cell-surface glycoproteins may take part in low
specificity interactions leading to differential adhesion phenomena, in the manner that mixtures of dextrans and polyethylene glycols demix.

X. ROLE OF SMALL MOLECULES

There is a vast literature in this area. To reduce it to manageable proportions, I shall
discuss it in terms of types of interpretation that have have or may be applied to such data.

A. Small Ions

Though it was at one time customary to interpret ionic phenomena in cell adhesion as
explicable in terms of simple electrostatic interactions at the cell surface,[17] and although
this type of explanation is applicable to many types of adhesion,[139] it is now clear that some
cell adhesions have ionic requirements which accord better with other explanations, e.g.,
that calcium activation is taking place. Calcium dependent and independent mechanisms of
adhesion have been described for the same cell type,[140,141] and though the experimental
conditions used do not preclude a requirement for very low levels of calcium it is hard to
apply electrostatic explanations to such phenomena. Similar criticisms can be applied to
demonstrations that manganous ions are potent in promoting adhesion[142,143] at $1 \times 10^{-4} M$.

B. Other Small Molecules

I have already mentioned the possible roles of lipids and their derivatives (Section V),
reagents involved in oxidation systems (Section VI), and incidentally the effects of various
metabolic inhibitors which may affect protein synthesis, or secretion at the surface, or protein
glycosylation (Sections VII, VIII, and IX). These sections embrace, at least potentially, a
large variety of chemicals. Similarly, many small molecules are at least potential activators
of changes in cell adhesion, as mentioned.

The remaining chemicals that may affect cell adhesion are perhaps best considered in
terms of their possible modes of action:

1. Chemical or physical adsorption of a reagent to the plasmalemmal surface may take
    place to effect massive changes to the cell surface, possibly affecting adhesion.[144,145]
2. Reagents may have effects on lipid fluidity, and thus on plasmalemmal structure and
    association and orientation of cell adhesion molecules; detergents may act in such a
    manner.[146]

3. The contact area or the general deformability of the cell may be changed by such reagents. Microtubule and microfilament poisons may affect adhesion by such means.[147]

XI. ADHESION OF LIPOSOMES TO CELLS

I have chosen to separate discussion of this topic from general discussion of cell adhesion between cells or of cells to other nonliving substrates because such systems are now being used to test whether possible CAMs are actual adhesion molecules.[100] Pagano and Takeichi[148] pointed out that liposomes might interact with cells to form stable associations, or might be endocytosed, form fusions, or transfer lipid between themselves and the adjacent cell surface.

Unfortunately, little practical or theoretical work has been done on the mechanisms of adhesion of liposomes bearing proteins. An interesting analysis has been carried out[149] of the N-CAM bearing liposomes used by Rutishauser et al.[99] The results are inconclusive, and fail to demonstrate that these liposomes adhere by a mechanism like that shown by cells. Yet much of the evidence that N-CAM is an adhesive molecule comes from the increased adhesion of liposomes bearing the molecule. Clearly, further work is needed. It is interesting to note that liposomes can exchange lipids with each other by collisional events,[150] and this possiblity makes any interpretation of the interaction of liposomes with cells even more difficult. The field of liposome adhesion has been reviewed in detail.[151]

XII. ADHESION MOLECULES IN SPECIALIZED STRUCTURES

The proteins discussed in Sections VII and VIII have not, in many cases, been associated with any particular type of cell-contact structure, though, of course, fibronectin has been proposed as being located in "focal contacts".[120,121] Nevertheless, it is appropriate to consider adhesion molecules in location in desmosomes, zonulae occludens, and other specialized contacts.

Volk and Geiger[152] have described a 135 kdalton protein from the adherens-type junctions of chick cardiac muscle (in this case, the term adherens junction means a desmosomal structure). A series of desmosomal glycoproteins have been reported for macula adherens.[153-155] These are desmoglein I and II. These results have in part been disputed by Garrod and his co-workers.[156,157] Using the technique of Fab inhibition of adhesion, they have suggested that desmocollin I, 115 kdaltons, and desmocollin II, 100 kdaltons, components of desmoglein II, are directly involved in cell-cell adhesion. The majority of CAMs have not yet been reported to be associated with specialized binding structures of cells, but uvomorulin[93] has been described as being concentrated in the belt desmosomes of small intestinal epithelia,[158] and Arc-1 as being found in junctional complexes in the same cell type.[159] An interesting review of the area has been published by Obrink[160] (see also References 161 to 163). However, as with other CAMs, no demonstration has been made that these molecules are directly involved in forming molecular contacts between surfaces.

REFERENCES

1. **Farquhar, M. G. and Palade, G.**, Junctional complexes in various epithelia, *J. Cell Biol.*, 17, 375, 1963.
2. **Brightman, M. W. and Palay, S. I.**, The fine structure of the ependyma in the brain of the rat, *J. Cell Biol.*, 19, 415, 1963.
3. **Brightman, M. W.**, The distribution within the brain of ferritin injected into the cerebrospinal fluid compartments. II. Parenchymal distributions, *Am. J. Anat.*, 117, 193, 1965.
4. **Pitts, J. D. and Finbow, M.**, The gap junction, *J. Cell Sci.*, 4(Suppl.), 239, 1986.

5. **Curtis, A. S. G.,** *The Cell Surface,* Academic Press, New York, 1967, 83.

6. **Bunt-Milam, A. H., Saari, J. C., Klack, I. B., and Garwin, G. G.,** Zonulae adherentes pore size in the external limiting membrane of the rabbit retina, *Invest. Ophthalmol. Vis. Sci.,* 26, 1377, 1985.

7. **Curtis, A. S. G.,** The adhesion of cells to glass: a study by interference reflection microscopy, *J. Cell Biol.,* 19, 199, 1964.

8. **Gingell, D.,** The interpretation of interference-reflection images of spread cells: significant contributions from thin peripheral cytoplasm, *J. Cell Sci.,* 49, 237, 1981.

9. **Bereiter-Hahn, J., Fox, C. H., and Theorell, B.,** Quantitative reflection contrast microscopy of living cells, *J. Cell Biol.,* 82, 767, 1979.

10. **Axelrod, D.,** cell-substrate contacts illuminated by total internal reflection fluorescence, *J. Cell Biol.,* 89, 141, 1981.

11. **Weisz, R. M., Balakrishnan, K., Smith, B. A., and McConnell, H. M.,** Stimulation of fluorescence in a small contact region between rat basophil leukemia cells and planar lipid membrane targets by coherent evanescent radiation, *J. Biol. Chem.,* 257, 6440, 1981.

12. **Gingell, D., Todd, L., and Bailey, J.,** Topography of cell-glass apposition revealed by total internal reflection fluorescence of volume markers, *J. Cell Biol.,* 100, 1334, 1985.

13. **Stryer, L.,** Fluorescence energy transfer as a spectroscopic ruler, *Ann. Rev. Biochem.,* 47, 819, 1978.

14. **Skerrow, C. J.,** Desmosomal proteins, in *Biology of the Integument,* Vol. 2, *Vertebrates,* Bereiter-Hahn, J., Matoltsy, A. G., and Richards, K. S., Eds., Springer-Verlag, Berlin, 1985, chap. 38.

15. **Talmon, A. and Ben-Shaul, Y.,** Tight junctions of dissociated and reaggregated embryonic lung cells, *Cell Differ.,* 8, 437, 1979.

16. **Heaysman, J. E. M. and Pegrum, S. M.,** Early contacts between fibroblasts: an ultrastructural study, *Exp. Cell Res.,* 78, 71, 1973.

17. **Curtis, A. S. G.,** Cell adhesion, *Prog. Biophys. Mol. Biol.,* 27, 315, 1973.

18. **Curtis, A. S. G. and Forrester, J. V.,** The competitive effects of serum proteins on cell adhesion, *J. Cell Sci.,* 71, 17, 1984.

19. **Forrester, J. V., Lackie, J. M., and Brown, A. F.,** Neutrophil behaviour in the presence of protease inhibitors, *J. Cell Sci.,* 59, 213, 1983.

20. **Curtis, A. S. G. and MacMurray, H.,** *J. Cell Sci.,* in press.

21. **van Oss, C. J., Gillman, C. F., and Neumann, A. W.,** *Phagocytic Engulfment and Cell Adhesiveness as Cellular Surface Phenomena,* Marcel Dekker, New York, 1975.

22. **Steinberg, M. S. and Poole, T. J.,** Strategies for specifying form and pattern: adhesion-guided multicellular assembly, *Philos. Trans. R. Soc. London,* 295, 451, 1981.

23. **Lackie, J. M. and Forrester, J. V.,** Neutrophil adhesion: studies using a flow chamber assay, *Agents Actions,* 16, 1, 1985.

24. **Curtis, A. S. G. and Greaves, M. F.,** The inhibition of cell aggregation by a pure serum protein, *J. Embryol. Exp. Morphol.,* 13, 309, 1965.

25. **Martz, E.,** Lytic granules, adhesion molecules, and other recent insights, *Immunol. Today,* 5, 254, 1984.

26. **Swanborg-Eden, C., Marild, S., and Korhonen, T. K.,** Adhesion inhibition by antibodies, *Scand. J. Infect. Dis.,* 33(Suppl.), 72, 1982.

27. **Grady, S. R., Nielsen, L. D., and McGuire, E.,** Organ and class specificity of cell adhesion blocking antisera, *Exp. Cell Res.,* 142, 169, 1982.

28. **Gerisch, G.,** Univalent antibody fragments as tools for the analysis of cell interactions in *Dictyostelium, Curr. Top. Dev. Biol.,* 14, 243, 1980.

29. **Beug, H., Katz, F. E., Stein, A., and Gerisch, G.,** Quantitation of membrane sites in aggregating *Dictyostelium discoideum* by use of tritiated univalent antibody, *Proc. Natl. Acad. Sci. U.S.A.,* 70, 3150, 1973.

30. **Neumeier, R., Josic, D., and Reutter, W.,** Integral membrane antigens involved in cell-substratum adhesion of hepatocytes and hepatoma cells, *Exp. Cell Res.,* 151, 567, 1984.

31. **Grumet, M., Hofman, S., Crossin, K. L., and Edelman, G. M.,** Cytotactin, an extracellular matrix protein of neural and non-neural tissues that mediates glia-neuron interaction, *Proc. Natl. Acad. Sci. U.S.A.,* 82, 8075, 1985.

32. **Curtis, A. S. G.,** manuscript in preparation.

33. **Pouyssegur, J. M. and Pastan, I.,** Mutants of Balb/c fibroblasts defective in adhesiveness to substratum: evidence for alteration in cell surface proteins, *Proc. Natl. Acad. Sci. U.S.A.,* 73, 544, 1976.

34. **Moscona, M. H. and Moscona, A. A.,** Inhibition of adhesiveness and aggregation of dissociated cells by inhibitors of protein and RNA synthesis, *Science,* 142, 3595, 1963.

35. **Pierschbacher, M. D. and Ruoslahti, E.,** Cell attachment activity of fibronectin can be duplicated by small fragments of the molecule, *Nature (London),* 309, 30, 1984.

36. **Grinnell, F. and Feld, M. K.,** Initial adhesion of human fibroblasts in serum-free medium: possible role of secreted fibronectin, *Cell,* 17, 117, 1979.

37. **Yamada, K. M., Yamada, S. S., and Pastan, I.,** Cell surface protein partially restores morphology, adhesiveness, and contact inhibition of movement to transformed fibroblasts, *Proc. Natl. Acad. Sci. U.S.A.,* 73, 1217, 1976.

38. **Pena, S. D. J. and Hughes, R. C.,** Fibronectin-plasma membrane interactions in the adhesion and spreading of hamster fibroblasts, *Nature (London),* 276, 80, 1978.

39. **Curtis, A. S. G., Forrester, J. V., McInnes, C., and Lawrie, F.,** Adhesion of cells to polystyrene surfaces, *J. Cell Biol.,* 97, 1500, 1983.

40. **Sherhan, K., Broekman, M. J., Korchak, H. M., Smolen, J. E., Marcus, A. J., and Weissmann, G.,** Changes in phosphatidylinositol and phosphatidic acid in stimulated human neutrophils. Relationship to calcium mobilisation, aggregation and superoxide radical generation, *Biochim. Biophys. Acta,* 762, 420, 1983.

41. **Argov, S., Hebdon, M., Cuatrecasas, P., and Koren, H. S.,** Phorbol ester-incuded lymphocyte adherence: selective action on NK cells, *J. Immunol.,* 134, 2215, 1985.

42. **Dunham, P., Anderson, C., Rich, A. M., and Weissman, G.,** Stimulus-response coupling in sponge cell aggregation: evidence for calcium as an intracellular messenger, *Proc. Natl. Acad. Sci. U.S.A.,* 80, 4756, 1983.

43. **Kruskal, B., Shak, S., and Maxfield, F. R.,** Spreading of human neutrophils is immediately preceded by a large increase in cytoplasmic free calcium, *Proc. Natl. Acad. Sci. U.S.A.,* 83, 2919, 1986.

44. **Pattaroyo, M. and Jondal, M.,** Phorbol ester-induced adhesion (binding) among human mononuclear leukocytes requires extracellular Mg and is sensitive to protein kinase C, lipoxygenase and ATPase inhibitors, *Immunobiology,* 170, 305, 1985.

45. **van Epps, D. E., Bender, J. G., Steinkamp, J. A., and Chenoweth, D. E.,** Modulation of neutrophil-reduced pyridine nucleotide content following stimulation with phorbol myristate acetate and chemotactic factors, *J. Leukocyte Biol.,* 38, 587, 1985.

46. **O'Flaherty, J. T., Kreutzer, D. L., and Ward, P. A.,** Neutrophil aggregation and swelling induced by chemotactic agents, *J. Immunol.,* 119, 232, 1977.

47. **Hoover, R. L., Karnovsky, M. J., Austen, K. F., Corey, E. J., and Lewis, R. A.,** Leukotriene B_4 (action on endothelium mediates augmented neutrophil-endothelial adhesion, *Proc. Natl. Acad. Sci. U.S.A.,* 81, 2191, 1984.

48. **Buchanan, M. R., Vazquez, M. J., and Gimbrone, M. A.,** Arachidonic acid metabolism and the adhesion of human polymorphonuclear leukocytes to cultured endothelial cells, *Blood,* 82, 889, 1983.

49. **Aplin, J. D. and Hughes, R. C.,** Cell adhesion on model substrata: threshold effects and receptor modulation, *J. Cell Sci.,* 50, 89, 1981.

50. **Burger, M. M., Burkart, W., Weinbaum, G., and Jumblatt, J.,** Cell-cell recognition: molecular aspects. Recognition and its relation to morphogenetic processes in general, in *Cell-Cell Recognition,* Curtis, A. S. G., Ed., Society for Experimental Biology, Cambridge, 1978, 1.

51. **Jones, G. E.,** Intercellular adhesion: modification by dielectric properties of the medium, *J. Membr. Biol.,* 16, 297, 1974.

52. **Oppenheimer, S. B. and Meyer, J. B.,** Isolation of species-specific and stage-specific adhesion promoting component by disaggregation of intact sea urchin embryo cells, *Exp. Cell Res.,* 137, 472, 1982.

53. **Woodruff, J. J. and Rasmussen, R. A.,** In vitro adherence of lymphocytes to unfixed and fixed high endothelial cells of lymph nodes, *J. Immunol.,* 123, 2369, 1979.

54. **Berwick, L. and Coman, D. R.,** Some chemical factors in cell adhesion and stickiness, *Cancer Res.,* 22, 982, 1962.

55. **Carter, R. D., Joyner, W. L., and Renkin, E. M.,** Effects of histamine and some other substances on molecular selectivity of the capillary wall to plasma proteins and dextran, *Microvasc. Res.,* 7, 31, 1974.

56. **Grinnell, F. and Feld, M. K.,** Fibronectin adsorption on hydrophilic and hydrophobic surfaces detected by antibody binding and analysed during cell adhesion in serum-containing medium, *J. Biol. Chem.,* 257, 4888, 1982.

57. **Grinnell, F.,** Cellular adhesiveness and extracellular substrata, *Int. Rev. Cytol.,* 53, 65, 1978.

58. **Curtis, A. S. G., Forrester, J. V., and Clark, P.,** Substrate hydroxylation and cell adhesion, *J. Cell Sci.,* in press.

59. **Curtis, A. S. G., Chandler, C., and Picton, N.,** Cell surface lipids and adhesion. III. The effects on cell adhesion of changes in plasmalemmal lipids, *J. Cell Sci.,* 18, 375, 1975.

60. **Ohki, S. and Leonardis, K. S.,** A possible role of cholesterol in cell adhesion, *Biochemistry,* 23, 5578, 1984.

61. **O'Flaherty, J. T., Showell, H. J., Becker, E. L., and Ward, P. A.,** Neutrophil aggregation and degranulation. Effect of arachidonic acid, *Am. J. Pathol.,* 95, 433, 1970.

62. **O'Flaherty, J. T., Kruetzer, D. L., and Ward, P. A.,** Effect of prostaglandin E_1 and F_{2a} on neutrophil aggregation, *Prostaglandins,* 17, 201, 1970.

63. **Cohen, H. J., Chovaniec, M. E., Takahashi, K., and Whitin, J. C.,** Activation of human granulocytes by arachidonic acid: its use and limitations for investigating granulocyte functions, *Blood,* 67, 1103, 1986.

64. **Weiss, L.,** Studies on cellular adhesion in tissue culture. XIIA. Some effects of prostaglandins and cyclic nucleotides, *Exp. Cell Res.,* 81, 57, 1973.

65. **Bruno, J. J., Taylor, L. A., and Droller, M. J.,** Effect of prostaglandin E_2 on human platelet adenyl cyclase and aggregation, *Nature (London),* 251, 721, 1974.

66. **Chuang, H. Y. K., Mohammad, S. F., and Mason, R. G.,** Prostacyclin (PGI_2) inhibits the enhancement of granulocyte adhesion to cuprophane induced by immunoglobulin G, *Thromb. Res.,* 19, 1, 1980.

67. **Jones, G. and Hurley, D.,** The effect of prostacyclin on the adhesion of leucocyte to injured vascular endothelium, *J. Pathol.,* 142, 51, 1984.

68. **Fricke, D., Damerau, B., and Vogt, W.,** Adhesion of guinea pig polymorphonuclear leukocytes to autologous aortic strips: influence of chemotactic factors and of pharmacological agents which affect arachidonic acid metabolism, *Int. Arch. Allergy Appl. Immunol.,* 78, 429, 1985.

69. **Palmblad, J., Malmsten, C. L., Uden, A.-M., Radmark, O., Engstedt, L., and Samuelsson, B.,** Leukotriene B_4 is a potent and stereospecific stimulator of neutrophil chemotaxis and adherence, *Blood,* 58, 658, 1981.

70. **Spagnuolo, P. J., Ellner, J., Hassid, A., and Dunn, M. J.,** Thromboxane A2 mediates augmented polymorphonuclear leukocyte adhesiveness, *J. Clin. Invest.,* 66, 406, 1980.

71. **Ford-Hutchinson, A. W.,** Neutrophil aggregating properties of PAF-acether and leukotriene B_4, *Int. J. Immunopharmacol.,* 5, 17, 1983.

72. **Nishizuka, Y.,** The role of protein kinase C in cell surface signal transduction and tumour promotion, *Nature (London),* 308, 693, 1984.

73. **Mege, J. L., Capo, C., Benoliel, A. M., and Bongrand, P.,** Self-limitation of the oxidative burst of rat polymorphonuclear leukocytes, *J. Leukocyte Biol.,* 39, 599, 1986.

74. **Cox, J. A., Jeng, A. Y., Blumberg, P. M., and Tauber, A. I.,** Activation of the human neutrophil nicotinamide adenine dinucleotide phosphate (NADPH)-oxidase by protein kinase C, *J. Clin. Invest.,* 76, 1932, 1985.

75. **Badwey, J. A., Curnutte, J. T., and Karnovsky, M. L.,** Cis-polyunsaturated fatty acids induce high levels of superoxide production in human neutrophils, *J. Biol. Chem.,* 256, 12640, 1982.

76. **Wroggman, K., Weidemann, M. J., Peskar, B. A., Staudinger, H., Rietschel, E. T., and Fischer, H.,** Chemiluminescence and immune activation. I. Early activation of rat thymocytes can be monitored by chemiluminescence and measurements, *Eur. J. Immunol.,* 8, 749, 1978.

77. **Curtis, A. S. G., Forrester, J. V., and Clark, P.,** Substrate hydroxylation and cell adhesion, *J. Cell Sci.,* 86, 9, 1986.

78. **Lydon, M. J., Minett, T. W., and Tighe, B. J.,** Cellular interactions with synthetic polymer surfaces in culture, *Biomaterials,* in press.

79. **Ohno, K., Kanoh, T., and Uchino, H.,** Poly(L-lysine) enhances the release of reactive oxygen metabolites from human neutrophils, *Acta Haematol. Jpn.,* 48, 969, 1985.

80. **Bradley, T. P. and Bonavida, B.,** Mechanism of cell-mediated cytotoxicity at the single cell level. III. Evidence that cytotoxic lymphocytes lyse both antigen-specific and nonspecific targets pretreated with lectins or periodate, *J. Immunol.,* 127, 208, 1980.

81. **Berke, G.,** Cytotoxic T-lymphocytes. How do they function?, *Immunol. Rev.,* 72, 5, 1983.

82. **Panus, P. C., Eddy, L. J., and Longenecker, G. L.,** Measurement of malonyldialdehyde production during sodium arachidonate-induced polymorphonuclear leukocyte aggregation, *J. Pharmacol. Methods,* 13, 1779, 1985.

83. **Edelman, G. M.,** Cell adhesion molecules, *Science,* 219, 450, 1983.

84. **Rutishauser, U.,** Developmental biology of a neural cell adhesion molecule, *Nature (London),* 310, 549, 1984.

85. **Thiery, J.-P., Brackenbury, R., Rutishauser, U., and Edelman, G. M.,** Adhesion among neural cells of the chick embryo. II. Purification and characterisation of a cell adhesion molecule from neural retina, *J. Biol. Chem.,* 252, 6841, 1977.

86. **Gallin, W. J., Edelman, G. M., and Cunningham, B. A.,** Characterisation of L-CAM, a major cell adhesion molecule from embryonic liver cells, *Proc. Natl. Acad. Sci. U.S.A.,* 80, 1038, 1983.

87. **Grumet, M., Hoffman, S., Chuong, C.-M., and Edelman, G. M.,** Polypeptide components and binding functions of neuron-glia adhesion molecules, *Proc. Natl. Acad. Sci. U.S.A.,* 81, 7989, 1984.

88. **Gallin, J.,** Leukocyte adherence-related glycoproteins, LFA-1, MOl, and p150,95: a new group of monoclonal antibodies, a new disease, and a possible opportunity to understand the molecular basis of leukocyte adherence, *J. Infect. Dis.,* 152, 661, 1985.

89. **Todd, R. F. and Arnaout, M. A.,** Monoclonal antibodies that identify Mol and LFA-1: two human leukocyte membrane glycoproteins: a review, in *Leukocyte Typing,* Vol. 2, Reinherz, E. I., Haynes, B. F. Nadler, L. M., and Burnstein, I. D., Springer-Verlag, New York, 1985.

90. **Pattaroyo, M., Beatty, P. G., Serhan, C. N., and Gahmberg, C. G.,** Identification of a cell-surface glycoprotein mediating adhesion in human granulocytes, *Scand. J. Immunol.,* 22, 619, 1985.

91. **Neumeier, R. and Reutter, W.,** Hepatocyte adhesion on plastic, *Exp. Cell Res.,* 160, 287, 1985.
92. **Vestweber, D., Ocklind, C., Gossler, A., Odin, P., Obrink, B., and Kemler, R.,** Comparison of two cell-adhesion molecules, uvomorulin and cell-CAM 105, *Exp. Cell Res.,* 157, 451, 1985.
93. **Hayafil, F., Morello, D., Babinet, C., and Jacob, F.,** A cell surface glycoprotein involved in the compaction of embryonal carcinoma cells and cleavage stage embryos, *Cell,* 21, 927, 1981.
94. **Yoshida-Noro, C., Suzuki, N., and Takeichi, M.,** Molecular nature calcium-dependent cell-cell adhesion system in mouse teratocarcinoma and embryonic cells studied with a monoclonal antibody, *Dev. Biol.,* 101, 19, 1984.
95. **Faissner, A., Kruse, J., Goridis, C., Bock, E., and Schachner, M.,** The neural cell adhesion molecule L1 is distinct from the N-CAM related group of surface antigens BSP-2 and D2, *EMBO J.,* 3, 733, 1984.
96. **Hirn, M., Pierre, M., Deagostini-Bazin, H., Hirsch, M., and Goridis, C.,** Monoclonal antibody against cell surface glycoprotein of neurons, *Brain Res.,* 214, 433, 1981.
97. **Primakoff, P., Hyatt, H., and Myles, D.,** A role for the migrating sperm surface antigen PH-20 in guinea pig sperm binding to the egg zona pellucida, *J. Cell Biol.,* 101, 2339, 1985.
98. **Gramzow, M., Bachmann, M., Uhlenbruck, G., Dorn, A., and Muller, W. E. G.,** Identification and further characterisation of the specific cell binding fragment from sponge aggregation factor, *J. Cell Biol.,* 102, 1344, 1986.
99. **Rutishauser, U., Hoffman, S., and Edelman, G. M.,** Binding properties of a cell adhesion molecule from neural tissue, *Proc. Natl. Acad. Sci. U.S.A.,* 79, 685, 1982.
100. **Hoffman, S. and Edelman, G. M.,** Kinetics of homophilic binding by embryonic and adult forms of the neural cell adhesion molecule, *Proc. Natl. Acad. Sci. U.S.A.,* 80, 5762, 1983.
101. **Garrod, D. R. and Nicol, A.,** Cell behaviour and molecular mechanisms of cell-cell adhesion, *Biol. Rev.,* 56, 199, 1981.
102. **Cole, G. J. and Glaser, L. A.,** Heparin-binding domain from N-CAM is involved in neural cell-substratum adhesion, *J. Cell Biol.,* 102, 403, 1986.
103. **Yamada, K. M.,** Cell surface interactions with extracellular materials, *Ann. Rev. Biochem.,* 52, 761, 1983.
104. **Hynes, R.,** Molecular biology of fibronectin, *Ann. Rev. Cell Biol.,* 1, 67, 1986.
105. **Terranova, V. P., Rohrbach, D. H., and Martin, G. R.,** Role of laminin in the attachment of PAM 212 (epithelial) cells to basement membrane collagen, *Cell,* 22, 719, 1980.
106. **Hewitt, A. T., Warner, H. H., Silver, M. H., Dessau, W., Wilkes, C. M., and Martin, G. R.,** The isolation and partial characterisation of chondronectin: an attachment factor for chondrocytes, *J. Biol. Chem.,* 257, 2330, 1982.
107. **Barnes, D. W., Mousetis, L., Amos, B., and Silnutzer, J.,** Gas 11-bead affinity chromatography of cell attachment and spreading-promoting factors of human serum, *Anal. Biochem.,* 137, 196, 1984.
108. **Pytela, R., Pierschbacher, M. D., and Ruoslahti, E. A.,** 125/115 kDa cell surface receptor specific for vitronectin interacts with the arginine-glycine-aspartic acid adhesion sequence derived from fibronectin, *Proc. Natl. Acad. Sci. U.S.A.,* 82, 5766, 1985.
109. **Dixit, V. M., Haverstick, D. M., O'Rourke, K. M., Hennessy, S. W., Grant, G. A., Santoro, S. A., and Frazier, W. A.,** A monoclonal antibody against human thrombospondin inhibits platelet aggregation, *Proc. Natl. Acad. Sci. U.S.A.,* 82, 3472, 1985.
110. **Ginsberg, M. H., Wolf, R., Marquerie, G., Coller, B., McEver, R., and Plow, E. F.,** Thrombospondin binding to thrombin-stimulated platelets. Evidence for a common adhesive protein binding mechanism, *Clin. Res.,* 32, 308a, 1983.
111. **Jaffe, E. A., Leung, L. L. K., Nachman, R. L., Levin, R. L., and Mosher, D. F.,** Thrombospondin is the endogenous lectin of human platelets, *Nature (London),* 295, 246, 1982.
112. **De Marco, L., Girolami, A., Zimmerman, T. S., and Ruggieri, Z. A.,** Interaction of purified type IIB von Willebrand factor with the platelet membrane glycoprotein Ib induces fibrinogen binding to the glycoprotein IIb/IIIa complex and initiates aggregation, *Proc. Natl. Acad. Sci. U.S.A.,* 82, 7424, 1985.
113. **Edwards, J. G.,** The biochemistry of cell-adhesion, *Prog. Surf. Sci.,* 13, 125, 1983.
114. **Chapman, A. E.,** Characterization of a 140 kD cell surface glycoprotein involved in cell adhesion, *J. Cell Biochem.,* 25, 109, 1984.
115. **Akiyama, S. K. and Yamada, K. M.,** The interaction of plasma fibronectin with fibroblastic cells in suspension, *J. Biol. Chem.,* 260, 4492, 1985.
116. **Akiyama, S. K., Yamada, S. S., and Yamada, K. M.,** Characterization of a 140 kD avian cell surface antigen as a fibronectin-binding molecule, *J. Cell Biol.,* 102, 442, 1986.
117. **Matyas, G. R., Evers, D. C., Radinsky, R., and Morre, D. J.,** Fibronectin binding to gangliosides and rat liver plasma membranes, *Exp. Cell Res.,* 162, 296, 1986.
118. **Rubin, K., Johansson, A., Hook, M., and Obrink, B.,** Substrate adhesion of rat hepatocytes. On the role of fibronectin in cell spreading, *Exp. Cell Res.,* 135, 127, 1981.
119. **Grinnell, F.,** Studies on the mechanism of cell attachment to a substratum: evidence for three biochemically distinct processes, *Arch. Biochem. Biophys.,* 160, 304, 1974.

120. **Rees, D. A., Badley, R. A., and Woods, A.,** Relationships between actomyosin stress fibres and some cell surface receptors in fibroblast adhesion, in *Cell Adhesion and Motility*, Curtis, A. S. G. and Pitts, J., Eds., Cambridge University Press, London, 1980, 389.

121. **Culp, L. A., Ansbacher, R., and Domen, C.,** Adhesion sites of neural tumor cells: biochemical composition, *Biochemistry*, 19, 5899, 1980.

122. **Haverstick, D. M., Cowan, J. F., Yamada, K. M., and Santoro, S. A.,** Inhibition of platelet adhesion to fibronectin, fibrinogen, and von Willebrand factor substrates by a synthetic tetrapeptide derived from the cell-binding domain of fibronectin, *Blood*, 66, 946, 1985.

123. **Auffray, C. and Novotny, J.,** Speculations on sequence homologies between the fibronectin cell-attachment site, major histocompatibility antigens, and a putative AIDS virus polypeptide, *Human Immunol.*, 15, 381, 1986.

124. **Ruoslahti, E., Hayman, E. G., and Pierschbacher, M. D.,** Extracellular matrices and cell adhesion, *Arteriosclerosis*, 5, 581, 1985.

125. **Silnitzer, J. and Barnes, D. W.,** Effects of fibronectin-related peptides on cell spreading *in vitro*, *Cell. Dev. Biol.*, 21, 73, 1985.

126. **Takashima, A. and Grinnell, F.,** Human keratinocyte adhesion and phagocytosis promoted by fibronectin, *J. Invest. Dermatol.*, 83, 352, 1984.

127. **Marino, J. A., Pensky, J., Culp, L. A., and Spagnuolo, P. J.,** Fibronectin mediates chemotactic-factor stimulated neutrophil substrate, *J. Lab. Clin. Med.*, 105, 725, 1985.

128. **Olden, K., Pratt, R. M., and Yamada, K. M.,** Role of carbohydrate in biological function of the adhesive glycoprotein fibronectin, *Proc. Natl. Acad. Sci. U.S.A.*, 76, 3343, 1979.

129. **Rutishauser, U., Watanabe, M., Silver, J., Troy, F. A., and Vimr, E. R.,** Specific alteration of NCAM-mediated cell adhesion by an endo-neuraminidase, *J. Cell Biol.*, 101, 1842, 1985.

130. **Hirano, T., Yamada, H., and Miyazaki, T.,** Direct implication of surface mannosyl residues in cell adhesion of *Dictyostelium discoideum*, *J. Biochem.*, 908, 199, 1985.

131. **Asao, M. I. and Oppenheimer, S. B.,** Inhibition of cell aggregation by specific carbohydrates, *Exp. Cell Res.*, 120, 101, 1979.

132. **Stoolman, L. M., Tenforde, T. S., and Rosen, S. D.,** Phosphomannosyl receptors may participate in the adhesive interaction between lymphocytes and high endothelial venules, *J. Cell. Biol.*, 99, 1535, 1984.

133. **Weigel, P. H., Schnaar, R. L., Kuhlenschmidt, M. S., Schnell, E., Lee, R. T., Lee, Y. C., and Roseman, S.,** Adhesion of hepatocytes to immobilized sugars. A threshold phenomenon, *J. Biol. Chem.*, 254, 10830, 1979.

134. **Weigel, P. H.,** Rat hepatocytes bind to synthetic galactoside surfaces via a patch of asialoglycoprotein, *J. Cell Biol.*, 87, 855, 1980.

135. **Grinnell, F. and Hays, D. G.,** Induction of cell spreading by substratum-adsorbed ligans directed against the cell surface, *Exp. Cell Res.*, 116, 275, 1978.

136. **Carter, W. G., Rauvala, H., and Hakamori, S. I.,** Studies on cell adhesion and recognition. II. The kinetics of cell adhesion and cell spreading on surfaces coated with carbohydrate-reactive proteins (glycosidases and lectins) and fibronectin, *J. Cell Biol.*, 88, 138, 1981.

137. **Hayman, E. G., Engvall, E., Ahearn, E., Barnes, D., Pierschbacher, M., and Ruoslahti, E.,** Cell attachment on replicas of SDA polyacrylamide gels reveals 2 adhesive plasma proteins. *J. Cell Biol.*, 95, 20, 1982.

138. **Edwards, P. A.,** Differential cell adhesion may result from nonspecific interactions between cell surface glycoproteins, *Nature (London)*, 271, 248, 1978.

139. **Rutter, P. R.,** The physical chemistry of the adhesion of bacteria and other cells, in *Cell Adhesion and Motility*, Curtis, A. S. G. and Pitts, J., Eds., Cambridge University Press, London, 1980, 103.

140. **Takeichi, M., Ozaki, H. S., Tokunaga, K., and Okada, T. S.,** Experimental manipulation of cell surface to affect cellular recognition mechanisms, *Dev. Biol.*, 70, 195, 1979.

141. **Takeichi, M., Atsumi, T., Yoshida, C., Uno, K., and Okada, T. S.,** Selective adhesion of embryonal carcinoma cells and differentiated cells by Ca^{2+}-dependent sites, *Dev. Biol.*, 87, 340, 1981.

142. **Rabinovitch, M. and DiStefano, M. J.,** Manganese stimulates adhesion and spreading of mouse sarcoma I ascites cells, *J. Cell Biol.*, 59, 165, 1973.

143. **Grinnell, F.,** Manganese-dependent cell-substratum adhesion, *J. Cell. Sci.*, 65, 61, 1984.

144. **Burns, R. G.,** Interaction of microorganisms, their substrates and their products with soil surfaces, in *Adhesion of Microorganisms to Surfaces*, Academic Press, New York, 1979, 109.

145. **Brooks, D. E., Greig, R. S., and Janzen, J.,** Mechanisms of erythrocyte aggregation, in *Erythrocyte Mechanics and Blood Flow*, Alan R. Liss, New York, 1980, 119.

146. **Curtis, A. S. G., Campbell, J., and Shaw, F. M.,** Cell surface lipids and adhesion. I. The effects of lysophosphatidyl compounds, phospholipase A_2 and aggregation-inhibiting protein, *J. Cell Sci.*, 18, 347, 1975.

147. **Ozaki, H., Okada, T. S., and Yasuda, K.,** Does colchicine affect aggregation of normal and transformed BHK cells differently?, *Dev. Growth Differ.*, 20, 55, 1978.

148. **Pagano, R. E. and Takeichi, M.,** Adhesion of phospholipid vesicles to Chinese hamster fibroblasts, *J. Cell Biol.,* 74, 531, 1977.

149. **Bell, G. I. and Torney, D. C.,** On the adhesion of vesicles by cell adhesion molecules, *Biophys. J.,* 48, 939, 1985.

150. **Mutsch, B., Gains, N., and Hauser, H.,** Interaction of intestinal brush border membrane vesicles with small unilamellar phospholipid vesicles. Exchange of lipids between membranes is mediated by collisional contact, *Biochemistry,* 25, 2134, 1986.

151. **Nir, S., Bentz, J., Wilschutz, J., and Duzgunes, N.,** Aggregation and fusion of phospholipid vesicles, *Prog. Surf. Sci.,* 13, 1, 1983.

152. **Volk, T. and Geiger, B.,** A 135-kd membrane protein of intercellular adherens junctions, *EMBO J.,* 3, 2249, 1984.

153. **Giudice, G. J., Cohen, S. M., Patel, N. H., and Steinberg, M. S.,** Immunological comparison of desmosomal components from several bovine tissues, *J. Cell Biochem.,* 26, 35, 1984.

154. **Gorbsky, G. and Steinberg, M. S.,** Isolation of the intercellular glycoproteins of desmosomes, *J. Cell Biol.,* 90, 243, 1981.

155. **Gorbsky, G., Cohen, S. M., Shida, H., Giudice, G. J., and Steinberg, M. S.,** Isolation of the non-glycosylated protein of desmosomes and immuno-localisation of a third plaque protein, desmoplakin III, *Proc. Natl. Acad. Sci. U.S.A.,* 82, 810, 1985.

156. **Cowin, P., Mattey, D., and Garrod, D.,** Identification of desmosomal surface components (desmocollins) and inhibition of desmosome formation by specific Fab, *J. Cell Sci.,* 70, 41, 1984.

157. **Garrod, D. R.,** Desmosomes, cell adhesion molecules and the adhesive properties of cells in tissues, *J. Cell Sci.,* 4(Suppl.), 221, 1986.

158. **Boller, K., Vestweber, D., and Kemmler, R.,** Cell-adhesion molecule uvomorulin is localized in the intermediate junction of adult intestinal epithelial cells, *J. Cell Biol.,* 100, 327, 1985.

159. **Imhof, B. A., Vollmers, P. H., Goodman, S. L., and Birchmeier, W.,** Cell-cell interactions and polarity of epithelial cells: specific perturbation using a monoclonal antibody, *Cell,* 35, 667, 1983.

160. **Obrink, B.,** Epithelial cell adhesion molecules, *Exp. Cell Res.,* 163, 1, 1986.

161. **Steinberg, M. S., Shida, H., Giudice, G. J., Shida, M., Patel, N. H., and Blaschuk, O. W.,** On the molecular organization, diversity and functions of desmosomal proteins, in *Junctional Complexes of Epithelial Cells,* Ciba Foundation Symp. 125, Bock, G. and Clark, S., Eds., John Wiley & Sons, New York, 1987, 3.

162. **Franke, W. W., Cowin, P., Schmelz, M., and Kapprell, H.-P.,** The desmosomal plaque and the cytoskeleton, in *Junctional Complexes of Epithelial Cells,* Ciba Foundation Symp. 125, Bock, G. and Clark, S., Eds., John Wiley & Sons, New York, 1987, 24.

163. **Mattey, D. L., Suhrbier, A., Parrish, E., and Garrod, D. R.,** Recognition, calcium and the control of desmosome formation, in *Junctional Complexes of Epithelial Cells,* Ciba Foundation Symp. 125, Bock, G. and Clark, S., Eds., John Wiley & Sons, New York, 1987, 49.

Chapter 10

MODELS OF CELL ADHESION INVOLVING SPECIFIC BINDING*

George I. Bell

TABLE OF CONTENTS

*   The U.S. government (the author's employer) retains a nonexclusive, royalty-free license to publish or reproduce this chapter, or to allow others to do so, for U.S. government purposes. This chapter was written under the auspices of the U.S. Department of Energy.

I. INTRODUCTION

A. Meaning of *Specific* Binding

All cells of multicellular animals have on their surfaces various complex molecules such as glycoproteins. As such a cell approaches another cell or a substrate, these macromolecules will interact with their counterparts on the other cell or with the substrate as described in earlier chapters. In general, such interaction will not involve details of the exact amino acid or sugar composition of the macromolecule; they are thus said to be nonspecific. In contrast, some of the attractive interactions may be highly specific and depend on the detailed sequence and structure of two or more reacting partners. Prototypes of such interactions are those of enzyme and substrate, antibody and antigen, or lectin and oligosaccharide. These specific interactions do not involve the formation of covalent bonds between the binding partners; rather, a variety of hydrogen bonds, electrostatic and van der Waals interactions are responsible for the affinity of binding.

We shall frequently refer to the molecules on a cell surface that are responsible for specific binding as *receptors*.

It is likely that binding interactions of cells in multicellular organisms are generally specific. Specific interactions play a major role during development and in determining the organization of tissues. In some cases, the precise molecules responsible for the interactions are known, such as antibodies and antigens or neural cell adhesion molecules (NCAMs). Sometimes the interacting molecules give rise to specialized structures between the cells, such as desmosomes, tight junctions, and gap junctions linking epithelial cells. In other cases, the precise molecules mediating adhesion have not yet been identified, but F_{ab} fragments of monoclonal antibodies as antiserums against minority surface components will block adhesion, presumably because they interfere with specific interaction. Finally, in many instances cells have been found to adhere preferentially to cells of the same species or tissue, suggesting that binding is occurring between like receptors on like cells, perhaps directly or perhaps mediated by soluble or other intercellular components. All of these lines of evidence point more or less strongly to the role of specific binding in cell adhesion.

B. Comparison of Cell-Cell and Cell-Substrate Adhesion

The macromolecules associated with a cell and outside of its plasma membrane are collectively called the cell *glycocalyx* or cell coat, and may be revealed by a variety of stains, such as ruthenium red. When two cells adhere to each other there will be many interactions between the molecules of their glycocalyxes, some involving specific binding and others nonspecific repulsion (see Chapter 2). However, even adjacent cells in tissues are often separated by an *extracellular matrix*,[1] a complex network of macromolecules including the fibrous protein collagen, various polysaccharide or proteoglycan molecules (forming a highly hydrated gel), and various glycoproteins of which some, such as fibronectin and laminin, play important roles in cell adhesion to the matrix. Since matrix molecules are synthesized by the adjacent cells, it is somewhat arbitrary to say where the cell glycocalyx ends and the extracellular matrix begins. Nevertheless, it is clearly useful to distinguish in vivo cell-cell and cell-substrate interactions.

When a highly mobile cell such as a lymphocyte or macrophage, or a cell which spends a good portion of the time not in contact with other cells such as blood cells, adheres to another cell, this is probably best viewed as cell-cell adhesion. On the other hand, the fibroblasts resident in connective tissue are clearly better thought of as interacting with the extracellular matrix. Some cells adhere both to a matrix and, more directly, to other cells. A good example is found in a sheet of epithelial cells which adhere directly to one another in various specialized junctions but also to an underlying basal lamina which is a specialized form of the extracellular matrix. The basal lamina also surrounds muscle cells and is important in stabilizing synapses between nerve and muscle cells.

In vitro, one can, of course, also study the adhesion of cells to artificial surfaces including those to which ligands have been attached and to which in turn cell receptors can bind.

C. Competition between Specific Binding and Nonspecific Repulsion

If cells were to stick tightly and indiscriminately to one another, there would be little opportunity for specific binding interactions. Thus, specificity of binding implies that the nonspecific interactions between cells or of cells to biological substrates are generally repulsive. We have seen in Chapters 1 and 2 that such nonspecific repulsion is expected because the cell glycocalyx and extracellular matrix are composed of hydrophilic and negatively charged polymers. Hence, there is competition between specific binding and nonspecific repulsion. This means that the adhesiveness of a cell may be modified either by changing the properties of its receptors, their number, charge, mobility, distribution, etc., or by changing the properties of the glycocalyx, e.g., the number or kind of proteoglycan molecules associated with the cell. To some extent, the cell environment, especially pH and ion concentrations, will affect properties of both the receptor and the glycocalyx.

Of particular importance are the relative extensions of the cell receptors and glycocalyx from the cell plasma membrane. If, for example, receptors are deeply buried in the glycocalyx, they are unlikely to be able to bind to another cell, and modification of the glycocalyx by compression, lateral redistribution, or shedding might be required to permit adhesion.

D. Importance of Receptor Mobility

Lateral mobility of receptors in the plane of the cell membrane is an important variable in cell adhesion. Freedom of a receptor to translate in the plane of the membrane and to rotate about an axis perpendicular to the membrane is important in enabling the receptor to locate and bind to another molecule on an adjacent cell or matrix. As predicted by the fluid mosaic model of the cell membrane[2] and confirmed by numerous experiments, receptors are often free to diffuse over the whole cell membrane. Such receptors will tend to accumulate in any areas of cell adhesion, partly because it is only there that they can stick, but also because the nonspecific repulsion will tend to minimize areas of adhesion and confine the bound receptors therein (see Section V).

Translational diffusion coefficients of integral membrane glycoproteins range from $\sim 10^{-9}$ cm^2/sec to $\leq 10^{-12}$ cm^2/sec with the precise value reflecting the extent of receptor coupling to the cell cytoskeleton. For example, the motion of membrane proteins on red cells is strongly hindered by the cytoplasmic spectrin network adjacent to the cell membrane.

E. Qualitative Effects of Specific Adhesion

1. Primary Effects

The most obvious effect of binding by specific receptors is that the resulting adhesion will reflect the specificity of the receptor recognition. Depending on the number and mobility of the ligands or other receptors to which the receptors in question bind, there may be a marked clustering of receptors in regions of adhesion (see Section V). When the binding is between mobile receptors on two adjacent cells, a high concentration of bound receptors in a small area of cell-cell contact is to be expected. Examples of such self-organizing structures caused by specific cell-cell adhesion may include the specialized junctions between epithelial cells (desmosomes, tight and gap junctions), synaptic junctions, and the localized contacts between T lymphocytes and their targets. An important point is that marked clustering of receptors in regions of cell-cell contact is a simple thermodynamic consequence of the competition between specific binding and nonspecific repulsion; it does not necessarily imply that adjacent receptors on the same cell attract each other.

When a cell binds to a substrate such as the basal lamina, on which the ligand molecules are relatively immobile, the cell will tend to spread until its receptors can be mostly bound by ligands on the contact area. Such a cell will be polarized, in the sense that its membrane adjacent to the basal lamina will be enriched in receptors and any associated molecules, while the rest of its membrane will be depleted of such molecules. Such contact-induced polarization appears to be one element in the general polarization of epithelial cells attached to basal lamina. In a simpler system, it was shown that the apical surfaces of macrophages become depleted of receptors for antibody, when placed on an antibody-coated surface.[3]

2. Secondary Effects

A variety of secondary effects may arise from the clustering of cell surface receptors. In many systems, clustering of receptors through cross-linking by multivalent ligands leads to pronounced effects on cell behavior. For example, cross-linking of $F_{c\epsilon}$ receptors on basophils or mast cells by multivalent antibody-antigen complexes leads to a Ca^{2+} influx and degranulation of the cell.[4] Similarly, binding of a mast cell to a lectin-coated particle can lead to receptor clustering and a localized degranulation.[5] As another example, cross-linking or clustering of F_c receptors on macrophages has been reported to create cation-selective channels.[6]

When a cell spreads upon a surface (or on another cell), the region between the cell and the surface may become a specialized microenvironment, not only because of the high concentration of receptors in the adjacent membrane but also because ion pumps in the membrane may alter the concentration of ions in the small volume between cell and surface.[7] The following argument may be relevant to pH changes in such a microenvironment. Consider what happens when a piece of plasma membrane, originally in a coated pit, becomes interiorized to become an endosome. Proton pumps in the endosome membrane rapidly reduce the pH inside the endosome, thereby uncoupling receptors from ligands, etc.[8] Such pumps are presumably present and active in the plasma membrane prior to interiorization. The distance between a cell and a surface to which it is attached may be smaller than the diameter of an endocytic vesicle (~ 0.1 μm). Thus, proton pumps should reduce the extracellular pH in this specialized microenvironment to a noticeable extent. Such acidification has been reported in the space between osteoclasts and bone,[9] and may be a factor in macrophage phagocytosis.[7] There are, of course, other ion pumps in cell membranes, notably the Na^+-K^+ ATPase. These, together with possible ion channels opened by receptor clustering, offer a rich repertoire for producing specialized microenvironments exterior to the cell. An essential feature for establishing such a microenvironment must be that ion flow parallel to the membrane should be slow enough to permit maintenance of a significant gradient in that direction. Some numerical considerations are given in the Appendix.

Just as pH changes are used in endosomes to promote dissociation of receptors from their ligands, so also pH changes exterior to the cell may enable it to detach from other cells or substrates. Whether this is an important process, cell detachment must involve some active processes such as ion transport, enzymatic cleavage, or tension exerted by cytoskeletal elements, since otherwise cells once attached would tend to remain in place (see Section III).

Clustering of cell-surface receptors appears to play a role in the organization of the cytoskeleton, especially actin and intermediate filaments. First of all, there is the capping phenomenon, whereby receptors cross-linked by various antibodies or lectins can be moved, presumably by the cellular actin network, to one pole of the cell.[10] In related observations, the mobility of cell surface molecules was observed to be retarded following the local binding of lectin-coated particles to a lymphocyte surface; this was attributed to an overall anchorage modulation of surface proteins caused by a mobilization of the cytoskeleton triggered by binding.[11] Cross-linking of IgE receptors on basophilic leukemia cells leads rapidly to their immobilization,[12,13] an event that appears in turn to be associated with their release of serotonin, histamine, etc. Specialized junctions between cells are frequently associated with the termination of cytoskeletal filaments, e.g., desmosomes appear anchored to keratin filaments. Finally, we note that structure of filaments in the extracellular matrix is frequently correlated with structure of the cytoskeleton. For example, fibronectin fibers in the extracellular matrix are correlated with actin stress fibers.[14] While each of these phenomena could be interpreted in alternative ways, a unifying explanation would involve clustered receptors serving as a nucleation center for cytoskeletal organization.[10]

F. Similarities to Vesicle Adhesion and Aggregation

Insights concerning cell adhesion can sometimes be gained by studying the adhesion of vesicles that contain, on their surfaces, molecules of significance for cell adhesion. This has the advantage of simplifying the system, and if purified components are used in reconstituted vesicles, of focusing attention on the properties of specific cell-adhesion molecules. For example, in a series of experiments, the aggregation of lipid vesicles containing N-CAMs was studied as a function of N-CAM concentration.[15] By analyzing the results, it was possible to conclude that the N-CAM molecules formed trimers on the surface of a vesicle and that adhesion involved trimer-trimer interactions in a rate-limiting step.[16] Care must, of course, be taken in extrapolating from results with vesicles to expectations regarding cells. For example, if components simulating the repulsive glycocalyx are not incorporated into the vesicles, it will be difficult to draw any conclusions concerning the competition between nonspecific repulsion and specific binding.

Of course, there are a variety of in vivo scenarios that require the binding of subcellular particles or vesicles to cell membranes at an early stage. As examples, we mention the binding of viruses to cell-surface receptors prior to internalization, the binding of platelets to endothelial cells or to basement membranes, and the binding of vesicles to plasma membranes as a prelude to exocytosis. All of these events and many more are likely to involve the participation of *specific* macromolecules on the binding partners.

II. SOME MOLECULES INVOLVED IN CELL ADHESION

The purpose of this section is to simply summarize properties of some of the molecules that are important in cell adhesion and to provide references for further information. Emphasis is placed on some molecules important for in vitro experiments as well as in vivo (antibodies and lectins), receptors for cell-cell adhesion, and molecules mediating adhesion to the extracellular matrix.

A. Antibodies and Related Molecules

Antibodies are undoubtedly the best-characterized molecules that play a role in cell adhesion. The typical antibody molecule is a glycoprotein composed of four different polypeptide chains, two identical heavy chains, and two identical light chains. The amino acid sequences of many antibodies have been completely determined,[17] the three-dimensional structures of several have been determined by X-ray diffraction,[18] and the elaborate recombinational events that are responsible for the generation of antibody diversity have been elucidated at the DNA

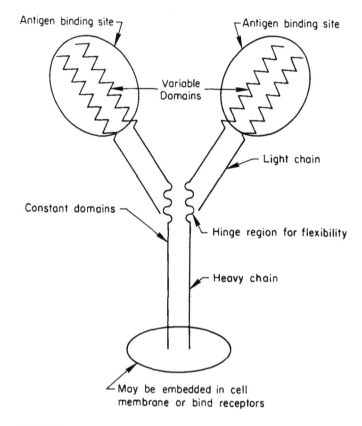

FIGURE 1.   Schematic drawing of an antibody molecule. Each light chain has about 220 amino acids and each heavy chain has about 440. Papain will cleave in the hinge region to produce two identical antigen binding fragments (F_{ab} fragments) and one F_c fragment.

level.[19] There are various antibody or immunoglobulin (Ig) classes such as IgG, IgM, IgA, IgE, and IgD; these differ in their heavy chains, binding to receptors, and function. The amino terminal ends of the light and heavy chains are extremely variable and determine the antigen-binding properties of the molecule. Because of this variability, a mammal has a potential repertoire of at least 10^8 different antibody specificities.

A typical antibody is shown in Figure 1. It is evidently a "Y" shaped molecule. A typical B-lymphocyte has about 10^5 antibody molecules, having identical antigen-binding properties, anchored in its membrane by their stems and serving as receptors for the generation and regulation of immune responses. In this configuration, the antibodies can mediate the adhesion of B-lymphocytes to any cells or surfaces which have complementary antigens exposed.

Antibodies are also soluble glycoproteins, secreted in large number by activated B cells and constituting about 20% of plasma protein by weight. Since each molecule has two (or more for polymeric immunoglobulins such as secreted IgM or IgA) binding sites, they can cross-link and agglutinate molecules, particles, or cells that express appropriate antigens. Moreover, soluble antibodies can also bind by their Fc stems to receptors on other cells of the immune system, notably macrophages and basophils, thereby endowing these cells with diverse recognition specificities that can be used in binding to and phagocytosing pathogens and cells.

Antibodies have desirable properties for cell-adhesion studies in that one can obtain purified preparations of uniform specificity (monoclonal antibodies) against almost any antigen. Moreover, they can be attached to surfaces or to other cells such as macrophages in controlled amounts.

A variety of other cell-surface molecules have some similarities to the immunoglobulins. In particular, the antigen receptor on T-lymphocytes is composed of two polypeptides, α and β, each having a variable N-terminal region for antigen recognition plus a constant region anchored in the membrane.[20] This receptor is not known to have any secreted counterpart, and its function is the recognition of antigens on other cells (even in the presence of soluble antigen). Thus, the T-cell receptor would seem a splendid example of a specific molecule mediating cell-cell recognition and adhesion. Unfortunately, the interaction between a T-cell and its target involves a variety of other molecules as well. For one thing, the T-cell receptor seems to recognize not only antigen but also (simultaneously) determinants of histocompatibility antigens. Moreover, associated with the receptor are other molecules such as the T-3 complex[21] that are involved in the adhesion, at least in the sense that F_{ab} fragments directed to such molecules can block adhesion. Thus, T-cell recognition and adhesion, though of great importance for immune responses, is still somewhat mysterious.

Also somewhat similar to immunoglobulins are the major histocompatibility antigens.[22,23] Although highly polymorphic, these antigens inherit their variability in Mendelian fashion rather than generating it by shuffling of DNA pieces. Nevertheless, they have a domain structure rather similar to that of the immunoglobulins and T-cell receptor,[20] and are clearly implicated in the cell-cell interactions that T-cells are designed to experience.

Of great importance for functioning of the immune system, but probably unrelated to immunoglobulins, are receptors on the surfaces of lymphocytes that mediate their adhesion to high endothelial cells of the postcapillary venules. Approximately 25% of the lymphocytes entering such a venule in the blood will attach to the endothelium, and crawl between endothelial cells to emigrate into a lymph node. A lymphocyte receptor implicated in this adhesion is a branched-chain glycoprotein carrying the small polypeptide, ubiquitin, on one or more of its branches.[24] Ubiquitins (~8450 daltons) are widely found in cells, and their encoding genes show remarkable evolutionary conservation. One may thus speculate that these unusual molecules play a more general role in cell adhesion. It may be noted that different kinds of lymphocytes show preferences for adhesion to gut-associated high endothelium as opposed to that in peripheral nodes. Thus, lymphocyte-endothelium recognition shows a range of specificities that probably reflects diversity in the lymphocyte receptors and/or that which they recognize on the endothelial cells.

B. Lectins

Lectins are soluble glycoproteins with two or more binding sites that recognize a specific sequence of sugar residues.[25] They are commonly isolated from plants and play important roles in the adhesion of plant cells. For example, a plant lectin probably mediates adhesion of the nitrogen-fixing bacterium *Rhizobium* to legume root cells.

Lectins are favorable experimental cell-adhesion molecules because they are readily available in pure form and promote adhesion between a large variety of cells. Moreover, if used at modest concentration, interesting differences show up in the extent to which cells can be aggregated. For example, if was found in the early 1970s that lectins such as concanavalin A (derived from jack beans) more readily agglutinate tumor cells as compared to normal cells.[26] A disadvantage in the use of lectins for cell-adhesion studies is the lack of specificity in the cell receptors to which they bind; any surface macromolecule with an appropriate exposed oligosaccharide is fair game for lectin binding. In addition, lectins will cross-link molecules on the same cell as well as form bonds between cells.

Lectins have also been described that mediate cell adhesion in multicellular animals.[25]

Another interesting class of molecules that has been found on cell surfaces is that of the glycosyl transferases, which catalyze the addition of sugar to specific oligosaccharides.[27] In the absence of free sugar, such enzymes could bind to complementary oligosaccharides that may be present on, for example, the glycoproteins of adjacent cells, and thus serve as specific cell-adhesion molecules. Whatever the receptors involved, it is clear that cells can recognize and bind to specific sugars when they are presented on a substrate.[28]

C. Neural Cell Adhesion and Related Molecules[29,30]

N-CAMs mediate adhesion between neurons in a homophilic manner, i.e., N-CAMs on one cell bind to similar molecules on other cells. These molecules appear to be important determinants in neural development and perhaps quite generally in embryonic development; the number of molecules per unit area and their adhesiveness vary in a systematic pattern during embryonic development. They are expressed on neurons and muscle cells. N-CAMs are glycoproteins, and a number of polypeptides ranging from 120 to 180 kdaltons have been identified from different tissues but with similar binding properties.[30] All N-CAMs are remarkable in containing a large carbohydrate component consisting of polysialic acid, i.e., a polymer of the negatively charged sugar sialic acid. The sialic acid contributes between 10 and 30% by weight to the glycoprotein, with the larger amounts being somewhat associated with embryonic tissues and the lower amounts with adult tissue. It is not surprising that the more highly charged N-CAMs are less efficient at aggregating vesicles than their less-charged counterparts.[15] Analysis of such experiments as a function of the N-CAM surface density on the vesicles suggests that N-CAMs are predominantly associated as trimers before initiating adhesion.[16]

Other molecules believed to be important in cell adhesion during embryonic development are L-CAM or liver cell adhesion molecule and Ng-CAM mediating neuron-glia cell adhesion. These appear to be less well characterized than N-CAM.

D. Receptors for Fibronectin[14] and Laminin

Fibronectin is a major glycoprotein component of the extracellular matrix. The molecule exists in the matrix as a dimer, each subunit of about 250 kdaltons. It contains various domains that bind to collagen and fibrin, two other glycoproteins that are important components of extracellular matrices. Cells such as fibroblasts can in turn bind to fibronectin utilizing among others a receptor complex of three glycoproteins of about 140 kdaltons.[31] This receptor recognizes a specific tripeptide (arginine-glycine-aspartate) on the fibronectin molecule; adhesion is blocked by the soluble tripeptide. Indeed, there is considerable evidence[32] that this tripeptide is recognized by other receptors, e.g., on platelets or monocytes, and that it is present and active on other matrix components including vitronectin and laminin.[33] Receptors for fibronectin are transmembrane proteins, and it is likely that they are coupled to the actin filament network within the cell so that the matrix structure can influence that of the cytoskeleton and vice versa.

Fibronectin-rich pathways guide and enhance the migration of many kinds of cells during embryonic development. During embryonic development, an interesting correlation has been reported between loss of extracellular fibronectin and appearance of cell-surface N-CAM; the cells walk on fibronectin and are bound together by N-CAM.[29] Initial interest in fibronectin was generated by the observation that tumor cells are deficient in the production or organization of fibronectin. As a result, they suffer from cytoskeletal disorganization which together with lack of association with the matrix may contribute to their invasive character.

Laminin is a major glycoprotein of the basal lamina. Laminin receptors are found on cells that normally interact with basal lamina[34] as well as on cells, including metastasizing tumor cells[35,36] and the pathogenic bacterium *Staphylococcus aureus*.[37] Indeed, there is a good correlation between the metastatic potential of a tumor cell and the abundance of laminin receptors on its surface.

We may expect that receptors for elements of the extracellular matrix will become rapidly better characterized, and that many additional ones will be discovered to explain specificity in cell adhesion. For example, tumor cells preferentially metastasize to various organs, suggesting that they can recognize organ-specific characteristics of the endothelium or basal lamina lining the capillaries. Additionally, during embryonic development, neural and other cells often perform extravagant migrations as if directed by specific pathways of yet uncharacterized cell-adhesion molecules.

E. Specific Cell Adhesion in Simple Systems

Even social unicellular organisms make use of specific cell-cell adhesion molecules during the aggregation phase of their life cycle.[1] For example, the myxobacteria execute coordinated motion requiring cell-cell recognition and adhesion. More complicated is the life cycle of the slime mold amoebae *Dictyostelium discoideum* which, while food is abundant, move about as free living amoebae, but during times of scarcity aggregate into migrating, multicellular slugs. Species-specific cell-adhesion molecules have been implicated in the organization, and in addition the cells secrete a lectin, discoidin 1, that serves as a pathway to guide the migration. The interaction between these cells and discoidin 1 has interesting parallels with that of fibroblasts and fibronectin. In particular, the arg-gly-asp sequence on discoidin 1 is recognized by a receptor on the amoeba.[38] In addition, the lectin can recognize *N*-acetylgalactosamine on the cells.

Cell adhesion has also been extensively studied between sponge cells. The species-specific aggregation of dissociated sponge cells requires the presence of Ca^{2+} and a large extracellular aggregation factor, thought to be a complex of glycoprotein and/or proteoglycan molecules.[39]

III. PROPERTIES OF SPECIFIC BONDS

A. Maximum Rate of Bond Formation[40]

When two cells are adjacent to each other, a great deal must be known about the geometry of the configuration before one can estimate the rate of bond formation between complementary receptors on the two cells. In particular, one would like to know the distance between the cell membranes (or distribution of distances), the extension of the receptors from the membrane, and their mobility in all directions. Such information is seldom available.

However, one can estimate an upper limit to the rate of bond formation by assuming that the rate-limiting step is diffusion of the receptors in the planes of the cell membranes. Since diffusion of glycoproteins in membranes is slow compared to that of similar molecules in solution, this assumption may not be too bad, provided that the membrane separation is such that the receptor binding sites can readily find each other. Obviously, if receptors are deeply embedded in the cell's glycocalyx, there will be serious problems in initiating any adhesion; mechanical forces or environmental changes (pH or ionic) might be required to compress or partially remove the glycocalyx prior to adhesion.

The diffusion-limited reaction rate, d_+, for two complementary receptors, 1 and 2, is given in terms of their diffusion coefficients, $D(1)$ and $D(2)$ for translation in the membrane:

$$d_+ = 2\pi[D(1) + D(2)] \tag{1}$$

Thus, if n_1 and n_2 represent the number of free receptors per unit area on the two cells, while n_b is the number of bonds per unit area, the maximum rate of bond formation per unit area $(dn_b/dt)_{max}$, will be given by:

$$\left.\frac{dn_b}{dt}\right|_{max} = 2\pi[D(1) + D(2)]\, n_1 n_2 \tag{2}$$

For numerical values, we may be able to take measured values of D for isolated cells — a representative value being $D \sim 10^{-10}$ cm²/sec — although it is possible that the diffusion of receptors in the contact area may be somewhat hindered relative to that on the free cell. As an example, consider cells such as lymphocytes having $\sim 10^5$ receptors on an area of 200 μm² such that $n_1 = n_2 = 500$ μm⁻². With the above value of D, $(dn_b/dt)_{max} \cong 3 \times 10^4$ μm⁻²/sec.

We are not aware of experimental data demonstrating such rates of bond formation, though it is clear that when cells adhere to a surface such as the endothelium from a flowing fluid, they must form bonds rather rapidly. However, Equation 2 has been used for estimating the reaction rate for interaction between two species on the same membrane,[41,42] and there is evidence that reaction rates on the disc membranes of photoreceptors are near the diffusion limit.

Comparison of reaction rates in two or three dimensions is made in Reference 40, where it is further argued that if one knows a reaction rate in three dimensions, e.g., between an antibody and antigen in solution, then one may be able to estimate reaction rates when the partners can translate only in two dimensions. That is, corrections to the diffusion limited rate constants in two dimensions are given, based on rate constants in three dimensions. However, it is still assumed that the reacting partners are appropriately spaced and mobile on the two membranes. An equilibrium constant, K_m, for reactants on membranes can also be related to that (K_s) for reactants in solution by the relation $K_m \cong K_s/r_{12}$ where r_{12} is roughly the thickness of a layer parallel to the membrane within which the receptors are located. If, for example, $K_s = 10^6$ ℓ/mol $= 1.7 \times 10^{-3}$ μm³ and $r_{12} = 5$ nm, then $K_m \cong 0.3$ μm². Thus, if the receptor concentrations on the cells are $\geq 10/\mu$m², even for this moderate equilibrium constant, most receptors are likely to be bound. At equilibrium, the *free* receptor concentration is expected to be about equal within the contact area and on the rest of the cell, since otherwise there would be a net flow of receptors into or out of the contact area. Since bound receptors can exist only in the contact area and they can easily predominate over the free receptors therein, we are thus led to expect marked concentration of receptors in the contact area. In Section V we shall see that a theory of the competition between nonspecific repulsion and specific bonding will enable us to predict the contact area and extent of receptor redistribution.

B. The Force Required to Break Bonds[40]

The bonds between cells are properly viewed as reversible. Any particular bond will be breaking and often reforming perhaps every second or every hour, depending on the reverse rate constant for the binding reaction. However, once a number of bonds have formed, it is unlikely that they will all be broken at once, and it will in general be necessary to stress the bonds, i.e., to exert force on them, in order to detach a cell from a substrate. (Alternatively we might modify the environment, e.g., reduce Ca^{2+} or pH, so as to destroy bonding.) When a force is imposed on a cell, it will in general be difficult to predict the stress distribution on all the bonds since the cell is a complex rheological object (see Chapter 5). However, for purposes of orientation we may assume that a number (N_0) of bonds are all subject to the same stress and inquire what force is required to break them.

The order of magnitude of the required force to rapidly break a bond is given by the free energy of the bond divided by the range over which the bonding force acts. If we measure the energy E_0 in kcal/mol and the range r_0 in nanometers, then the force f_0 is numerically:

$$f_0 = 7 \times 10^{-7} \, E_0/r_0 \text{ dyn/bond} \qquad (3)$$

Thus, for a representative antigen-antibody bond, $E_0 \cong 8.5$ kcal/mol and $r_0 = 0.5$ nm, we have $f_0 \cong 1.2 \times 10^{-5}$ dyn per bond. (Note that if parameters for a covalent bond are inserted, $E_0 \cong 70$, $r_0 \cong 0.14$, $f_0 \cong 3 \times 10^{-4}$ dyn. Hence, the weak antigen-antibody bond will break for a much smaller force than will a covalent bond.)

The force f_O will rapidly break a bond, but a somewhat smaller force per bond can destroy bonding on a longer time scale. A theory for this has been developed using results from the kinetic theory of the strength of solids,[43] according to which the lifetime of a bond is written

$$\tau = \tau_O \exp[(E_O - \gamma f)/kT] \tag{4}$$

where τ_O is the reciprocal of a natural frequency of oscillation of atoms in solids ($\sim 10^{13}$ sec), E_O is the bond energy, f is the applied force per bond, and γ is an empirical parameter. We have interpreted this to mean that the reverse rate constant k_-, for bond breakage in the presence of a force, is related to that in its absence, k_-°, by

$$k_- = k_-^\circ \exp(F/kT\, r_O\, N_b) \tag{5}$$

where F is the total force exerted on N_b bonds so that F/N_b is the force per bond.[40] The result is that for weak forces ($F/N_b \ll f_O$) the result of the force is simply to reduce the equilibrium constant and thus somewhat reduce N_b. For a larger force, N_b will be further reduced and the force per bond increased, and there will be critical force F_c which is just sufficient to detach the cell. For bonding parameters given earlier, I have estimated the critical force to be about one third of f_O or 4.0×10^{-6} dyn per bond.

There are other ways in which a stressed receptor-receptor bond or receptor-ligand bond might fail. For one thing, the force might cause the receptor to partially unfold or denature, thereby destroying the binding site.[44] The force required to do this depends on the detailed structure of the receptor and its binding site, and cannot be reliably estimated in general. It could be comparable to the energy of folding of a protein receptor divided by its average dimension.

Alternatively, the receptor might pull out of the membrane. If the receptor is anchored by a short hydrophobic α-helix tail, one can estimate the force from Equation 3 with E_O free energy difference between having the hydrophobic helix in membrane and water and r_O the membrane thickness. Various estimates for E_O have been made, perhaps best in Reference 45, where the value $E_O \cong 15$ kcal/mol was found. Using this together with $r_O \cong 4$ nm, we have $f_O \cong 2.6 \times 10^{-6}$ dyn or less than our earlier estimate of the force to break the bond. However, if the receptor has an appreciable cytoplasmic tail, it may be harder to pull the tail into the membrane than the hydrophobic helix out. For glycophorin, I estimated that the tail contribution would roughly triple the energy or force required. Hence, the force required to extract a receptor from the membrane may be expected to be comparable to that required to break the bond and to depend substantially on the cytoplasmic tail. Receptors with substantial tails may extract a plug of lipid if they are pulled through the membrane.[40]

It is of interest to compare the above forces with those to which a cell may be subject in hydrodynamic flow. For example, the force required to hold a cell in laminar flow at velocity v is given by Stokes' law:

$$F = 6\pi\eta r v \tag{6}$$

where η is the fluid viscosity ($\sim 10^{-2}$ g/cm sec for water), r the cell radius, and v the fluid velocity. If $r = 4\ \mu$m, $F \cong 7.5 \times 10^{-5}$ v dyn with v in centimeters per second. Hence, it will take ~ 20 v of our representative, uniformly stressed bonds to tether a cell against the fluid flow.

Quantitative experimental comparisons with these estimates are lacking, partly because of difficulty in measuring the absolute number of bonds involved and partly because of problems in calculating the expected stress distribution in bonds (see Chapter 4 and Reference 46).

IV. EFFECTS OF CELL MECHANICAL PROPERTIES AND SHAPE

Some cells, such as red blood cells, have quite smooth surfaces, whereas others, such as lymphocytes, macrophages, neurons, and cells of the intestinal epithelium, have highly irregular surfaces with various protrusions such as microvilli, microspikes, ruffles, etc. These specialized protrusions are in turn organized by special cytoskeletal structures including organized bundles of actin filaments. In any case when a cell makes contact with something to which it may adhere, it may do so with a representative portion of its membrane, as for the red cell, or with a protrusion that may be highly specialized for purposes of adhesion, as for microspikes on the growth cone of a neuron. In the latter case, there are some complications in making a general theory. For one thing, the tips of processes are likely to be specialized so that glycocalyx thickness, receptor abundances, or diffusion constants measured over the cell membrane may not apply to these tips. Moreover, the mechanical properties of such structures, and indeed their existence, may depend on cell activity. It would appear that more experimental data are needed before one can prepare a theoretical framework for treating adhesion to such special cell structures. Of course, if a cell is forced onto a surface, e.g., by centrifugation, there can be a large increase in force per unit area if the cell makes contact only at the tips of a few microprocesses.

As a cell adheres to a flat surface, perhaps that of another cell, the adhesion will tend to make the cell surface flatter. If the cell surface is fairly smooth (or we are considering adhesion of a vesicle), then the resulting deformation of the cell membrane will consume energy which can be calculated using the theory of elastic membranes as described in Chapter 4. In particular, Evans has analyzed the stresses in the membrane and the resulting deformation near the edge of a flat contact area, and for two alternative assumptions regarding the cell bonds. In the first case,[47] the bonds were taken to be of finite length but continuously distributed over the cell contact area, while in the second case[48] they are spaced a definite distance apart, as on a lattice. In the latter case, it was noted that there may be a large difference in that membrane tension which is just sufficient to prevent further cell spreading, and that which is required to decrease the contact area (i.e., to break bonds), while in the former case there is none.

When a rough cell adheres to a smooth surface, the cell surface will be expected to become somewhat smoother, and such results were observed in a quantitative study of the adhesion of macrophages to glutaraldehyde-treated eythrocytes.[49] Since a rough cell has a lot of excess membrane, it is not so clear what can be said of the energy of membrane deformation when such a cell adheres to a flat surface. However, when macrophages or neutrophiles are aspirated into suitable micropipettes, they show a smooth dependence of membrane tension on apparent area,[49,50] which was used in interpreting the smoothing effect.

When the contact area is small, as will often be the case when there is substantial nonspecific repulsion, or when the cell is flaccid, then the energy of mechanical deformation will be expected to be small compared to the adhesion energy. The contact area will then be determined by the balance between specific binding and nonspecific repulsion as described in the following section. For larger contact areas, favored by weak, nonspecific repulsion and/or immobile receptors or binding sites, the mechanical deformation will be more important and will eventually limit the size of the contact area. At that point, any decrease in free energy by expanding the contact area will be balanced by the gain due to increasing deformation energy. As a useful approximation, we will assume in the next section that the deformation energy is negligible for small contact areas, but imposes a definite upper limit to the contact area.

V. COMPETITION BETWEEN NONSPECIFIC REPULSION AND SPECIFIC BONDING

A. General Effects

When competition is present, there will be an adhesion threshold which may be expressed in the number of receptors per unit membrane area or for fixed receptor numbers in terms of their binding affinity, extension from the cell surface, or environmental parameters. Thus, an adhesion threshold[28] may be taken as evidence for competition with nonspecific repulsion.

Two kinds of adhesion thresholds may be distinguished: a kinetic threshold and an equilibrium threshold. In the kinetic case we consider what happens when two cells, say, first make contact. Those receptors initially in the contact area are assumed able to move about locally and may be able to establish some bonds, but unless the cells become stuck together or are held together by other forces, there will be little time available for a general accumulation of receptors in the contact area. Thus, in evaluating the kinetic threshold, we must consider whether the receptors *initially* in the contact area are sufficient to establish adhesion (see following Section B.4).

If the cells are kept in contact for a long enough time so that the receptors can diffuse into the contact area and reach an equilibrium concentration therein, a different and less stringent threshold for equilibrium binding may be defined, namely, when the total number of receptors and other conditions are just such that adhesion is possible in the equilibrium state.

To what extent is it possible for cells that fail to meet the kinetic threshold criteria to nevertheless come to adhere to each other in an equilibrium configuration? Of course, if the cells are held together, that will favor the nucleation of adhesion centers where by chance or design the receptors are particularly numerous or the glycocalyx is particularly thin. Presumably, the adhesion will always start from such centers. Thus, through local as well as general modulation of their surface properties, cells may be able to reach adhesion thresholds at precisely defined places and times during development.[29] Once adhesion has begun locally, presumably a contact area can enlarge due to the diffusion of receptors thereto, and perhaps aided by cell activity. When living cells are being considered, one may doubt whether the theoretical equilibrium configuration is ever really reached because of the cell activity. Nevertheless, by understanding the equilibrium configuration, we will have a basis for judging how much of what is seen is driven by purely thermodynamic considerations and how much requires active processes.

Another general effect of competition between specific bonding and nonspecific repulsion is the marked accumulation of receptors to be expected in a contact area, and the resulting polarization of the cell surface and possibly the cytoskeleton, discussed in Section I.E. Accumulation of receptors in contact areas has been observed for a number of experimental systems.[51-54] However, some amount of receptor accumulation is expected whether or not there is any nonspecific repulsion (see Section III.A), and the data on redistribution are not sufficiently quantitative to test the theoretical models. Additional evidence comes from the clustering of the connexns or dyads comprising a gap junction. Analysis indicates[55] that the forces between dyads are repulsive, and it was suggested that they are compressed in a more or less close-packed array by the nonspecific repulsion of the cell membranes or glycocalyxes outside of the gap junction.

B. Equilibrium Theory

1. The Free Energy and Constraints

In this section we will develop a general thermodynamic approach for predicting equilibrium-binding configurations and then apply it to several cases in which various assumptions are made regarding mobility of the receptors and the glycocalyx. In order to focus on the

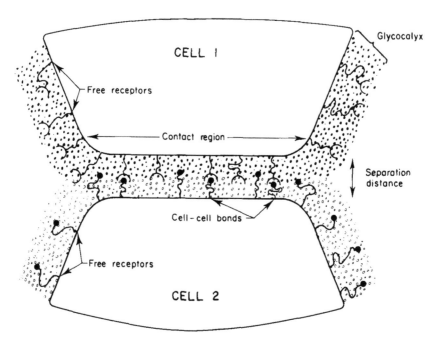

FIGURE 2.   Model for the interaction of cell-surface molecules in the region of cell contact. In the contact area, nonspecific repulsion is produced by interaction of the hydrophilic polymers comprising the glycocalyx. The repulsive forces are counterbalanced by the formation of bonds between mobile surface receptors.

ideas and results, we shall keep the development as simple as possible. In particular, we will use simple expressions for the chemical potentials of the molecules involved. Also, the mechanical energy of membrane deformation will be neglected except that a constraint will be placed upon the maximum allowable contact area.

Many of the limitations of a purely thermodynamic theory of cell adhesion have been mentioned earlier. For one thing, there may be kinetic barriers to overcome before equilibrium can be approached. Additionally, we have noted a number of secondary effects of adhesion (Section I.E.2) that in turn will perturb any equilibrium state. Finally, active cells do not necessarily tend to equilibrium states. Nevertheless, we suggest that thermodynamic models provide us "with a powerful method for integrating various notions about the complementary receptors on the two surfaces (e.g., lateral mobility, heterogeneity, total number per cell), notions about the cell-cell bridges that mediate adhesion (e.g., spring constant, length, binding constant), notions about the repulsive forces between cells (e.g., compressibility, thickness, and lateral mobility of the glycocalyx), and notions about the purely geometrical parameters of adhesion (e.g., maximum contact area, total surface area of the two cells, heterogeneity vs. uniformity of contact distance)".[56]

Our first task is to define a Gibbs free energy for an isolated system consisting of two cells in contact over an area A or alternatively of a cell in contact with a portion of a substrate. For the moment, we will adopt the nomenclature of cell-cell adhesion. The geometry is then as depicted in Figure 2.

Receptors on either cell can exist in two states, namely, bound or free, and if free they may be within the contact area or outside of it. Let N_b denote the number of bonds, N_i (i = 1,2), the number of free receptors on either cell, of which N_{ic} are in the contact area and N_{io} outside the contact area. The total number of receptors N_{it} on either cell is evidently:

$$N_{it} = N_b + N_{ic} + N_{io} \qquad i = 1, 2 \qquad (7)$$

The corresponding number of receptors per unit area will be denoted by n_{it}, n_b, n_{ic}, and n_{io}, respectively, so that $n_{it} = N_{it}/A_i$, $n_b = N_b/A$, $n_{ic} = N_{ic}/A$, and $n_{io} = N_{io}/(A_i - A)$.

For the present, let us ignore any lateral movement of the glycocalyx (see Section 5). The state of the system is then defined by the contact area A, the cell-cell separation distance S, the number of bonds N_b, and the partition of free receptors in and out of the contact area. Let N_j and n_j denote generic receptor numbers and densities (j = b, 1c, 1o, 2c, 2o). Then we assume that the Gibbs free energy for the system can be written

$$G(A, S, N_b, \ldots) = \sum_j N_j \, \mu_j(n_j) + A\Gamma(S) \tag{8}$$

where μ_j is the chemical potential of the j species, assumed to be in ideal solution form:[57]

$$\mu_j = \mu_j^\circ + kT \, \ell n(n_j) \tag{9}$$

In these equations, μ_j° is the chemical potential at some standard receptor density such as 1 μm^{-2}, and $\Gamma(S)$ is the energy of nonspecific repulsion per unit area[56] (see Chapter 2). These equations involve a number of assumptions. First of all, the ideal solution form of the chemical potential is expected to be reasonable for receptors which do not interact on the cell membrane, hence at low receptor densities. More generally, one could introduce a receptor activity[57] in place of the surface density or use a lattice model in which each lattice site could be occupied by only one receptor (or repeller) molecule.[58] Another point is that the free energy properly should have terms $\sim \int_o^N \mu(n')dN'$ and not $N\mu(n)$, but this makes only a small difference. In addition, we have allowed only a single cell-cell separation distance (S), ignored mobility of the glycocalyx, and omitted any contribution to the energy from membrane deformation. With these caveats in mind, we continue.

The standard chemical potentials are all constants except for μ_b° which is a function of the separation distance. Expanding μ_b° in a Taylor series about its value at the natural bond length L,

$$\mu_b^\circ(S) = \mu_b^\circ(L) + \frac{1}{2} \kappa(S - L)^2 + \ldots \tag{10}$$

where κ is a spring constant for stretching of the bonds. We have suggested[56] that a reasonable approximation to the energy of nonspecific repulsion is

$$\Gamma(S) \cong \frac{\gamma}{S} \exp(-S/\tau) \tag{11}$$

See Chapter 2 and Reference 59.

In addition, there are constraints on the independent variables that define the state of the system. Clearly, the separation distance must be nonnegative:

$$0 \leqq S \tag{12a}$$

In addition, the number of bonds cannot exceed the total number of receptors on either cell, or:

$$0 \leqq N_b \leqq \min(N_{1t}, N_{2t}) \tag{12b}$$

Finally, instead of including membrane-deformation energy in the free energy, we impose a constraint on the allowable contact area:

$$0 \leqq A \leqq A_{max} \leqq \min (A_1, A_2) \tag{12c}$$

so that if the deformation energy is negligible, at least the contact area cannot exceed that of the smaller cell.

We are now almost prepared to seek the equilibrium state by minimizing the free energy (Equations 8 to 11) with respect to the state variables, N_b, A, and S, taking the constraints (Equation 12) into account as well as the conservation of receptors (Equation 7). Several interesting cases can be considered which determine the partition of receptors between the contact and other areas. In the first case, we assume the receptors can move freely over each cell, thereby obtaining equilibrium conditions for globally mobile receptors on both cells. In the second case, we assume the receptors are mobile on one cell but immobile on the other, or that the latter represents a substrate with fixed sites to which a cell is adhering. In the third case, we assume that neither receptor is free to move into the contact area, although they have sufficient local mobility to find each other and bond. Equilibrium conditions for this case define the early stages of adhesion, and may be used to analyze kinetic thresholds.

2. The Equilibrium State for Mobile Receptors on Both Cells[56]

When receptors are free to diffuse into and out of the contact area, we expect that at equilibrium the number of receptors per unit area will be the same in the contact area and outside of it, or $n_{ic} = n_{io}$ ($i = 1,2$). (Actually, this result can also be derived setting $N_{io} = N_{it} - N_b - N_{ic}$ and by minimizing the free energy with respect to N_{ic}.) To minimize G with respect to N_b, A, and S, we set the partial derivations of G with respect to these three variables to zero. From $\partial G/\partial N_b = 0$, we find:

$$n_b = n_1 n_2 \, K(S) \tag{13}$$

where $K(S)$ is an equilibrium constant,

$$K(S) = \exp[(\mu_1^0 + \mu_2^0 - \mu_b(S) + kT)/kT]$$

$$= K(L) \exp\left[-\frac{1}{2} \kappa (S - L)^2/kT \right] \tag{14}$$

Equation 13 resembles a typical mass-action equilibrium result with the equilibrium constant a function of the separation S. From $\partial G/\partial A = 0$, we find:

$$n_b = \Gamma(S)/kT = \frac{\gamma}{kTS} \exp(-S/\tau) \tag{15}$$

indicating that at equilibrium the contact area is determined by a balance between the pressure of bonds $n_b kT$, tending to expand the contact area, and that of nonspecific repulsion $\Gamma(S)$, tending to compress it. Also, n_b is independent of the properties of individual bonds, e.g., the binding constant, and is thus a colligative property[54] of the system.

Finally, on setting $\partial G/\partial S = 0$, we find:

$$n_b = \frac{\dfrac{d}{ds} \Gamma(S)}{\dfrac{d}{ds} \mu_b^0(S)} = \frac{\gamma(S + t) \exp(-S/\tau)}{\kappa \tau S^2 (S - L)} \tag{16}$$

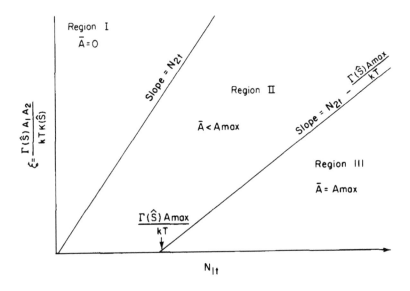

FIGURE 3.   Phase diagram for cell-cell adhesion with mobile receptors. For parameters
in region I, adhesion is not thermodynamically possible. In region II, adhesion is possible,
but the area of contact is $< A_{max}$; Equations 18 to 20 apply. In region III, the contact area
is its maximum value, A_{max}.[56] (With permission.)

Equations 13, 15, and 16 can be solved for the values $\hat{N}_b$, $\hat{A}$, and $\hat{S}$ that minimize G.
One must then check to see if any constraints, notably that on A, are violated for this
solution, and if so seek a minimum for which $A = 0$ or $A = A_{max}$. The results depend on
a nondimensional parameter, $\xi(S)$, that measures the ratio of repulsion to binding:

$$\xi(S) = A_1 A_2 \Gamma(S)/kT \, K(S) \tag{17}$$

It is then found that

$$\hat{S} = \frac{1}{2}\left[L + \frac{kT}{\kappa\tau} + \sqrt{\left(L + \frac{kT}{\kappa\tau}\right)^2 + \frac{4k\tau}{\kappa}} \right] \tag{18}$$

$$\hat{N} = \frac{1}{2}\left[N_{1t} + N_{2t} - \sqrt{(N_{1t} - N_{2t})^2 + 4\xi(\hat{S})} \right] \tag{19}$$

$$\hat{A} = \frac{kT}{\Gamma(\hat{S})}\hat{N}_b \tag{20}$$

Analysis of Equations 18 to 20, together with the constraint Equation 12c on A, shows
that there are two thresholds (or phase transitions), one for adhesion to take place at all,
and the second for the contact area to reach A_{max}. This result may be depicted in a phase
diagram for adhesion as shown in Figure 3. The threshold for adhesion occurs when $\xi(\hat{S})$
$= N_{1t} N_{2t}$; as can be seen from Equations 19 and 20, this value gives $\hat{N}_b = \hat{A} = 0$, while
slightly smaller values of ξ give positive values of $\hat{N}_b$ and $\hat{A}$. This result can also be understood
by noting, from Equation 13, that $n_b \leqq n_{1t}n_{2t}K(\hat{S})$ (since n_i cannot exceed n_{it}); on using this
inequality in Equation 15 we require $\Gamma(\hat{S})/kT \leqq n_{1t} \, n_{2t}K(\hat{S})$ for adhesion, which is the same
criterion. Note that the unconstrained values $\hat{S}$, $\hat{N}_b$, and $\hat{A}$ apply only in region II of the
phase diagram shown in Figure 3. The state variables giving minimum free energy with
constraints are denoted $\overline{S}$, $\overline{N}_b$, and $\overline{A}$.

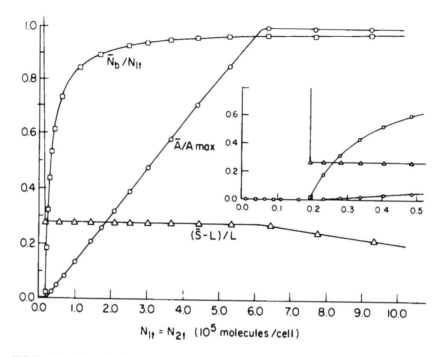

FIGURE 4. Dimensionless contact area (A/A_{max}), number of bonds (N_b/N_{1t}), and cell separation distance ($S/L-1$) as a function of the number of receptors $N_{1t} = N_{2t}$. Other parameters as listed in Table 1.[56] (With permission.)

A number of calculations are shown in Reference 56 of the expected bonding behavior as one or another model parameter is changed. Among the easiest to change experimentally is the number of receptors per cell, and in Figure 4 are shown the predicted contact area, number of bonds, and separation distance at a function of N_{1t} ($= N_{2t}$). As expected, there is a threshold in the number of receptors per cell at which adhesion is possible, and a further threshold at which the contact area is A_{max}. If A_{max} is determined by the energy of deformation of the membrane, then the latter threshold is only approximate; alternatively, if it is determined by the size of the smaller cell, it will represent a true threshold for engulfment.

Capo et al.[53] measured the contact area of thymocytes agglutinated by various surface concentrations of the lectin Con A. Model parameter (A_1, A_2, γ, τ, K_L, L, and k) were all estimated for this system, as summarized in Table 1, and the predicted contact area vs. number of Con A molecules is shown in Figure 5, for several values of $\Gamma(\hat{S})$. It is seen that the appropriate values of $\Gamma(\hat{S}) \cong 0.2$ erg/cm² are in the range estimated earlier.[59]

Thus, these experiments provide some support for this model of cell bonding. It should be noted that results in Figure 5 are insensitive to all model parameters save $\Gamma(\hat{S})$. Indeed, for this system the contact area is approximately

$$\bar{A} \cong \frac{kT\, N_{1t}}{\Gamma(\hat{S})} - \frac{A_1 A_2}{K(\hat{S})\, N_{2t}} \qquad (21)$$

where n_{1t} is the number of Con A molecules per cell and N_{2t} the number of "Con A receptors" per cell. The second term is only a small correction. Hence, according to this model, the measured contact area gives a rather direct way of obtaining the nonspecific repulsive energy.

Surface energies such as shown in Figure 5 imply, according to Equation 15, high concentrations of receptors in the contact area. For example, $\Gamma(\hat{S}) = 0.2$ erg/cm² gives, from Equation 15, $n_b = 0.05$/nm⁻². For bulky surface molecules such as Con A or various surface

Table 1
ESTIMATED PARAMETER VALUES[56]

Parameter	Symbol	Best estimate	Range
Surface area of cell	A_1, A_2	2×10^{-6} cm^2	10^{-7}—10^{-4}
Maximum contact area	A_{max}	10^{-6} cm^2	0 min (A_1, A_2)
Number of receptors on cell	N_{1t}, N_{2t}	10^5	10^3—10^7
Compressibility coefficient of glycocalyx	γ	10^{-6} dyn	0—10^{-4}
Thickness of glycocalyx	τ	10^{-6} cm	0.5—2.0 $\times$ 10^{-6}
Binding constant for formation of unstrained bridges	K_L	10^{-8} cm^2	10^{-11}—10^{-5}
Unstrained length of cell-cell bridge	L	2×10^{-6} cm	1—3 $\times 10^{-6}$
Force constant for stretching of bridges	K	0.1 dyn/cm	10^{-2}—10^3

receptors, such a surface density may be near or exceed close packing, indicating that the interaction between receptors should not be ignored (see Section 5). However, the conclusion that high receptor densities are to be expected in contact areas is robust.

3. The Equilibrium State for Cell Binding to Substrate[60]

In this section we wish to consider the binding of a cell having mobile receptors to a substrate having immobile binding sites, or alternatively to a cell having fixed receptor sites. Thus, on the cell as before, $n_{1c} = n_{10}$. However, on the substrate, denoted by the subscript 2, there is no redistribution of sites and $n_{2t} = n_b + n_{2c}$ in the contact area and $n_{2t} = n_{20}$ outside the contact area. The constraints are somewhat different for N_b and A. The number of bonds must fall in the range

$$0 \leqq N_b \leqq \min (N_{1t}, An_{2t}) \tag{22a}$$

since the number of bonds cannot exceed the total number of binding sites in the contact area. In addition, the maximum contact area cannot exceed half the cell area A_1, so that

$$0 \leqq A \leqq A_{max} \leqq \frac{1}{2} A_1 \tag{22b}$$

The same expressions for the chemical potentials may be used even for immobile sites.

The unconstrained minimum free energy may as before be found by setting the partial derivatives of G with respect to N_b, A, and S to zero. Differentiation with respect to N_b leads to Equation 15 as before, with the caution that on the cell $n_{1f} = (N_{1t} - N_b)/A_1$, while on the substrate $n_{2f} = n_{2t} - N_b/A$. Similarly, differentiation with respect to S gives Equation 16, as before. However, differentiation with respect to A leads to:

$$n_b = n_{2t}\left[1 - \exp - \frac{\Gamma(S)}{n_{2t} \, kT}\right] \tag{23}$$

Only if the exponent is small, i.e., $\Gamma(S) \ll n_{2t} \, kT$, is this equivalent to Equation 15. It is no longer possible to solve analytically for the unconstrained variables $\hat{N}_b$, $\hat{S}$, and $\hat{A}$ at the free-energy minimum, but numerical solutions are readily obtained. In particular, an analytic solution for $\hat{S}$ cannot be found. Once a numerical value is known, it can be used in Equations 13 and 23 to solve for $\hat{N}_b$, obtaining:

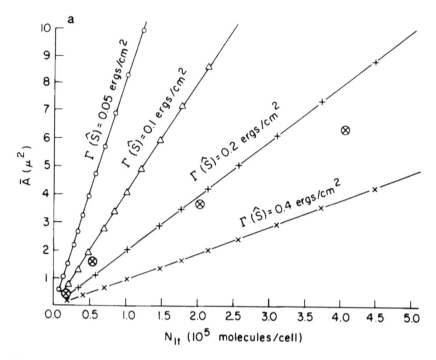

FIGURE 5. Contact area as a function of Con A molecules per cell. Data points $\otimes$ are from Reference 53. Theoretical curves show predictions of Equation 20 for several values of $\Gamma(S)$; other parameters in Table 1.[56] (With permission.)

$$N_b = N_{1t} - \zeta(\hat{S}) \tag{24}$$

where $\zeta(\hat{S})$ is a nondimensional parameter measuring the ratio of repulsion to binding:

$$\zeta(\hat{S}) = A_1[\exp(\Gamma(\hat{S})/kT\, n_{2t}) - 1]/K(\hat{S}) \tag{25}$$

Knowing $\hat{N}_b$, $\hat{A}$ may be found from Equation 23. From Equation 24 it can be seen that the threshold for adhesion occurs when $N_{1t} = \zeta(\hat{S})$; for small values of the exponent in Equation 25, this is the same criterion as in the previous section.

The constraint on N_b (Equation 22a) leads to an interesting variation of the contact area with density of surface sites shown in Figure 6. As the density of surface sites is increased, we first come to a threshold for adhesion. For the parameter in Figure 6 ($N_{1t} = 2 \times 10^5$, $A_{max} = 100\ \mu^2$ and other values as in Table 1), as the density of sites is further increased the cell spreads to its maximum contact area because it has an excess of receptors over the number of sites available in A_{max} ($N_{1t} > n_{2t}\, A$). As the site density is further increased, the cell is predicted to round up since it can saturate all of its receptors with a contact area smaller than A_{max}. This behavior should be readily testable experimentally by measuring the contact area vs. site density. It may be noted that even when the contact area cannot reach its maximum value, a decrease of contact area is to be expected once one is well past the threshold for adhesion. This can be seen by reference to the phase diagram in Figure 7. Figure 6 had $N_{1t} = 2 \times 10^5$; therefore, the vertical line generated as n_{2t} increase included all three phases of the diagram. For a smaller value of N_{1t}, such as 10^5, $\overline{A}$ never reaches A_{max} as n_{2t} is increased, but it does go through a maximum and then decreases.

4. The Equilibrium State for Globally Immobile Receptors, a Kinetic Threshold[61]

In this case we consider that the receptors on each cell are immobile except for a limited

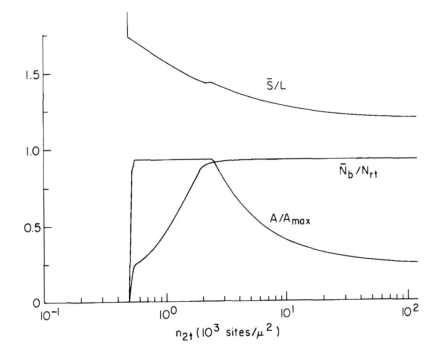

FIGURE 6.   Variation of configuration variables as the density of adhesive sites on the substrate n_{2t} is varied; other variables as in Table 1 except $N_{1t} = 2 \times 10^5$. Note that as n_{2t} is increased beyond a critical value, $\bar{A}$ begins to decrease, i.e., the cell begins to ''round up''.[60] (With permission.)

amount of local mobility so that they can find their complementary partners. It will be argued in Section C that this situation may describe the kinetic adhesion threshold in which receptors will not have the time to diffuse over large distances and establish bonding. Moreover, if the cells cannot establish adhesion in the absence of receptor redistribution, it may be difficult for them to nucleate and propagate adhesive contacts.

For immobile receptors, $n_{ic} = n_{it} - n_b$ for $i = 1,2$. For reasons that will become clear, it is now appropriate to consider n_b, S, and A as the variables of the system, and the constraint on n_b is

$$0 \leqq n_b \leqq \min (n_{1t}, n_{2t}) \tag{26}$$

In terms of these variables, the free energy function of Equation 8 is

$$
\begin{aligned}
G(n_b, S\ A) = \ & n_{1t}(A_1 - A)[\mu_1^0 + kT\ \ell n(n_{1t})] \\
& + (n_{1t} - n_b)\ A[\mu_1^0 + kT\ \ell n(n_{1t} - n_b)] \\
& + n_{2t}\ (A_2 - A)\ [\mu_2^0 + kT\ \ell n(n_{2t})] \\
& + (n_{2t} - n_b)\ A\ [\mu_2^0 + kT\ \ell n(n_{2t} - n_b)] \\
& + n_b\ A[\mu_b^0(S) + kT\ \ell n\ n_b] + A\Gamma(S) \tag{27}
\end{aligned}
$$

where the successive terms represent contributions from free receptors outside the contact area on cell 1, free receptors in the contact area on cell 1, the same two terms for cell 2, the bound receptors and the nonspecific repulsion.

On setting $\partial G/\partial n_b = 0$ we find as in Equation 13:

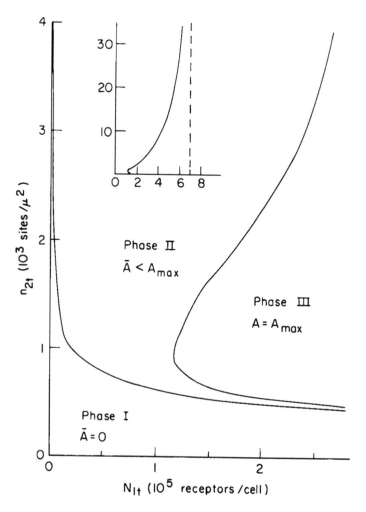

FIGURE 7.   Phase diagram of cell-substrate adhesion in the plane formed by control parameters n_{2t} and N_{1t}, showing three regions. The insert shows the phase II-phase III separation line for large values of the variables.[60] (With permission.)

$$n_b = (n_{1t} - n_b)(n_{2t} - n_b) K(S) \qquad (28)$$

with the equilibrium constant $K(S)$ given by Equation 14.

Similarly, on setting $\delta G/\delta S = 0$ we have Equation 16, here written as

$$n_b = - \frac{d\Gamma}{dS} \Big/ \frac{d\mu_b^0(S)}{dS} \qquad (29)$$

Note that Equations 28 and 29 can be solved for $\hat{S}$ and $\hat{n}_b$. (This was not possible for mobile receptors, because then $n_{1t} = (N_{1t} - N_b)/A$, an expression that involves the unknown A as well.) We are no longer able to set $\partial G/\partial A = 0$ for

$$\frac{\partial G}{\partial A} = n_b[\mu_b^0(S) - \mu_1^0 - \mu_2^0] + \Gamma(S)$$

$$+ kT[-n_{1t} \ln(n_{1t}) + (n_{1t} - n_b) \ln(n_{1t} - n_b) - n_{2t} \ln(n_{2t})$$

$$+ (n_{2t} - n_b) \ln(n_{2t} - n_b) + n_b \ln(n_b)] \qquad (30)$$

Thus $\partial G/\partial A$ does not involve the state variable A which is sought. The values $\hat{S}$ and $\hat{n}_b$ determined from Equations 28 and 29 can be inserted in Equation 30. If the resulting value of $\partial G/\partial A > 0$, this means that $\hat{A} = 0$ and there is no adhesion, while for $\partial G/\partial A < 0$, $\hat{A} = A_{max}$. Thus, the condition

$$\frac{\partial G}{\partial A}(\hat{S}, \hat{n}_b) = 0 \tag{31}$$

is the criterion for the adhesion threshold.

Suppose that a solution $\hat{S}$ has been found to Equations 28 and 29. The corresponding value of $\hat{n}_b$ is then from Equation 28:

$$\hat{n}_b = \frac{n_{1t} + n_{2t} + K^{-1}(\hat{S}) - \sqrt{(n_{1t} + n_{2t} + K^{-1}(\hat{S}))^2 - 4n_1 n_2}}{2} \tag{32}$$

Equation 30 may be simplified by using Equation 28 with K(S) defined in Equation 14 to find:

$$\frac{\partial G}{\partial A} = \Gamma(\hat{S}) + kT\left[n_{1t}\,\ell n\left(1 - \frac{\hat{n}_b}{n_{1t}}\right) + n_{2t}\,\ell n\left(1 - \frac{\hat{n}_b}{n_{2t}}\right) + \hat{n}_b\right] \tag{33}$$

Note that if $n_b \ll n_{1t}$ and $n_b \ll n_{2t}$, we may expand the logarithms to obtain

$$\frac{\partial G}{\partial A} \cong \Gamma(\hat{S}) - \hat{n}_b\,kT \tag{34}$$

Thus, the criterion for the adhesion threshold is that $\partial G/\partial A$ as given by Equation 33 be zero with $\hat{n}_b$ given by Equation 32. For small values of n_b, such that Equation 34 applies, the threshold criterion is as given for mobile receptors in Section 2, except that $\hat{n}_b$ is now given by Equation 32 rather than by $n_1 n_2 K(\hat{S})$. Only for very small values of $K \ll (n_{1t} + n_{2t})^{-1}$ are the two expressions equivalent. For larger values of K, the present criterion is more stringent.

It is of interest to consider how many more receptors one needs for adhesion when the receptors are immobile as compared to mobile. Suppose the cells are identical so that $n_{1t} = n_{2t}$ and, neglecting possible differences in $K(\hat{S})$ for the two cases, we introduce the dimensionless variables $x = K\,n_{1t}$, $y = K\,n_b$. Then Equation 32 gives

$$y = \frac{2x + 1 - \sqrt{4x + 1}}{2} \tag{35}$$

and the adhesion threshold, from Equation 33, is

$$\frac{K(\hat{S})\,\Gamma(\hat{S})}{kT} = -2x\,\ell n\left(1 - \frac{y}{x}\right) - y \tag{36}$$

If $x \ll 1$, then from Equation 35, $y \cong x^2$ and the right side of Equation 36 equals x^2, as is the case of mobile receptors. However, for large values of x, from Equation 35, $y \cong x - \sqrt{x}$ and the right side of Equation 36 is $x(\ell nx - 1)$. If the required value of x at the adhesion threshold for mobile receptors is denoted x_m, so that

$$x_m^2 = \frac{K(\hat{S}) \, \Gamma(\hat{S})}{kT}$$

while the required value for immobile receptors is x_f, then for $x \gg 1$:

$$x_m^2 = x_f(\ell n \, x_f - 1) \tag{37}$$

If, for example, we let $\Gamma(\hat{S}) = 0.1$ erg/cm^2 and $K(\hat{S}) = 10^{-8}$cm^2, then $x_m \cong 160$ while $x_f \cong 3500$ so that we would need around 20-fold more immobile receptors than mobile to achieve adhesion.

We have noted that at the adhesion threshold there is a transition from $A = 0$ to $A = A_{max}$, thus a first-order phase transition. Since intermediate values of the contact area are not permitted in this model, it is likely that the mechanical properties of the membrane will be more important for determining the contact area than in previous cases. That is, as one begins to exceed the adhesion threshold, the free energy available for membrane deformation will be small.

5. Other Effects

In the models that we have described, it was assumed that the cell glycocalyx is immobile and enters only in determining the energy, $\Gamma(S)$, of nonspecific repulsion. However, if the elements of the glycocalyx are largely glycoprotein molecules that are free to move in the plane of the cell membrane, there will be a tendency for the glycocalyx molecules, which we call *repellers*, to leave the contact area just as for the receptors to accumulate therein. This possibility can be analyzed by including terms in the free energy that represent the repeller molecules in and out of the contact area. First of all, there will be terms for the chemical potentials of such molecules. Second, the term $\Gamma(S)$ will be proportional to the repeller density on each cell in the contact area. Finally, as repellers leave the contact area, the resulting crowding of glycocalyx molecules on the rest of the cell may also contribute to the free energy.

A model has been worked out[62] for a cell with mobile repellers and receptors adhering to a surface with fixed sites. Chemical potentials were included for repellers, and $\Gamma(S)$ was assumed proportional to the repeller density in the contact area. The state variables of the system now include the repeller density in the contact area, and four simultaneous equations are obtained on setting to zero the variations of G with respect to this new variable and the three previous ones, N_b, A, and S. Numerical results were obtained and displayed. It was found that there is a substantial loss of repellers from the contact area and consequent lowering of the equilibrium adhesion threshold, provided that the energy of interaction per repeller molecule is $\geq kT$. This is because the contribution to the free energy from the entropy of unmixing of repeller molecules is on the order of kT per molecule (see Section C). If the surface energy $\Gamma(S) \geq 0.1$ erg/cm^2 and the repeller density is 10^{12}/cm^2 (or one repeller per 100 nm^2), then the energy per repeller is $\geq 10^{13}$ erg per molecule or ≥ 2.5 kT, so that substantial redistributions can be expected for mobile repellers. When this energy exceeds 10 kT, only a negligible fraction of repellers remain in the contact area.[62]

The phenomenon of "rounding up" was again predicted (see Section 3) for a wide range of model parameters. However, a new phenomenon was also predicted, namely, the existence of multiple equilibrium states with different values of the separation distance and other state variables, but very similar free energies. One state has a relatively large value of $\hat{S}$ and very little receptor redistribution; the other has a smaller $\hat{S}$ and substantial redistribution. While these dual steady states were found only for a limited range of model parameters, they do point to the richer spectrum of solutions that may arise in more complicated models.

It was noted in Section 2 that predicted bond densities may be unrealistically large in the contact area. One way to avoid this problem is to let the receptor and repeller molecules occupy lattice sites, with no more than one molecule permitted at a site. The free energy for such a lattice model has been worked out.[58] As expected, the entropic terms are now more complicated, but it can be shown that they reduce to the dilute solution form when most lattice sites are unoccupied. The receptor density has, of course, a definite upper limit, namely, when all lattice sites are occupied by receptor molecules. Solutions to the equations obtained on minimizing the free energy with respect to the state variables have not been investigated in any generality.

When soluble ligands such as antibodies or lectins are used to agglutinate cells, they will be able to both cross-link receptors on the same cell as well as to form specific bonds between cells. In the absence of nonspecific repulsion, this competition for ligand binding sites has been analyzed.[63] When considering the agglutination of thymocytes by the lectin Con A,[53] it was simply assumed that the number of sites available for intercellular bonding was equal to the number of Con A molecules bound to the cells. Since each Con A molecule has four binding sites, one or more can bind to each cell.

The *shape* of the contact area has not yet been considered. In many circumstances, it will be determined by the roughness of the cell surfaces and/or by the experimental procedures used to bring about the contact. Entropic considerations would favor many small contact areas over one large one, but a greater penalty would be paid in membrane deformation energy. At the edge of each contact area there is a region where bonds are being stressed and there is substantial membrane curvature.[47,48] Thus, an energy penalty is to be expected for the boundary of each contact region.

In the foregoing models we have considered adhesion between two cells or between a single cell and a surface. Under many circumstances, a cell will have multiple neighbors to which it adheres. Might this lead to qualitatively different conclusions? In particular, might adhesion between the first two cells so deplete their free surfaces of mobile receptors that they will be unable to adhere to otherwise identical cells? Preliminary analysis suggests that this is not so. Near the adhesion threshold, A is small and few receptors are removed from the rest of the cell, and it remains able to bind identical cells.[58]

C. Kinetic Effects on Cell Adhesion

When a cell makes contact with a surface or with another cell, there are a number of important considerations in determining whether adhesion results, in addition to those of equilibrium models. Of initial importance are (1) the force bringing the cells together or cell onto the surface, (2) the surface roughness of the cell, which together with the force will determine the force per unit area, and (3) the lateral extension of glycocalyx and receptors which will determine whether or not glycocalyx compression will be required before the receptors can find each other. In addition, there is the problem of whether glycocalyx and receptors are nonuniformly distributed on the cell surface, perhaps favoring establishing adhesion at certain cell protrusions. We ignore this latter problem for lack of information.

Will a force pushing two cells together cause expulsion of repeller molecules from the contact area? If the repeller molecules are mobile, then if the work done by the force in compression of the glycocalyx is greater than, or on the order of kT per repeller molecule, substantial redistribution of the repeller will be expected. In considering this mechanically induced phase separation we may ignore receptor molecules altogether and consider two objects having mobile repeller molecules brought into contact over an area A by a force F. Let us first consider that the force can bring the cells to a separation distance S_0 in the absence of any repeller redistribution. If repellers move out of the contact area, thereby increasing the free energy due to entropy of mixing, the repulsive interaction will decrease, permitting the cells to come closer together so that further work is done. For a fixed contact area, this will go on until the rate of free energy increase equals the rate at which the force does work.

For each cell let there be N_r repellers in the contact area A and N_{rt} on the whole cell. We may then write the free energy:

$$G(N_r, S) = 2N_r[\mu_r^0 + kT \ell n(N_r/A)]$$

$$+ 2(N_{rt} - N_r) \left[\mu_r^0 + kT \ell n \left(\frac{N_{rt} - N_r}{A_1 - A} \right) \right] + \gamma(S) N_r^2/A + F S \quad (38)$$

where

$$\gamma(S) = \Gamma(S)(A_1/N_{rt})^2 \quad (39)$$

so that the third term in Equation 38 is simply $\Gamma(S)A$ if the repellers are uniformily distributed on the cell surface.

We first seek the value of N_r that will minimize G for fixed S and then allow S to vary. On setting $\partial G/\partial N_r = 0$, we find

$$kT \ell n \left(\frac{N_r}{A} \frac{A_1 - A}{N_{rt} - N_r} \right) + \gamma(S) N_r/A = 0 \quad (40)$$

which for fixed A may be solved for N_r. If we let $N_r = N_0(1 - \Delta)$, where N_0 is the number of repellers initially in the contact area and thus Δ is the fraction that leaves, then for A $<< A_1$, Equation 40 may be written

$$\frac{\ell n(1 - \Delta)}{1 - \Delta} + \frac{\gamma(S)N_0}{kT A} = 0 \quad (41)$$

The quantity $\gamma(S) N_0/A$ is just the repulsive energy per repeller molecule, on either cell, so that Equation 41 tells us that if the energy per repeller is greater than or on the order of kT, we expect significant repeller movement from the contact area.

On setting $\partial G/\partial S = 0$, we have

$$F = - \frac{N_0^2(1 - \Delta)^2}{A} \frac{d\gamma}{dS} \quad (42)$$

Equations 41 and 42 must be solved for equilibrium values of Δ and $\hat{S}$. For small forces, Δ is small, and using our canonical form for $\Gamma(S)$ in Equation 11, we find:

$$\hat{\Delta} \cong \frac{(\hat{S} + \tau)F}{N_0 kT} \quad (43)$$

Thus, approximately, Δ is the work done by the force per repeller molecule divided by kT.

Suppose that we are centrifuging cells together with a force of G times that of gravity. For a cell of mass 10^{-9} and density 1.05 g/cm², $F = 5 \times 10^{-8}$ G dyn. If $\hat{S} + \tau = 30$ nm and we let $N_0 = xA$, then $\Delta \cong 4G/xA$. If, for example, A = 1 μm² while x = $10^4/$ μm², then $\Delta \cong 4 \times 10^{-4}$ G and 10% of the repellers will leave the contact area at G = 250. If contact could be maintained at the tip of a microvillus with A = 0.01 μm² while x = $10^4/$μm², then $\Delta \cong 0.04$ G. In general, it is likely that rather large centrifugation or other forces would be required in order to bring about a significant phase separation in the contact area. Such forces, in turn, would favor large contact areas. Thus, mechanical redistribution of repellers is expected to be important only when cells are strongly forced together.

Redistribution of repellers would take some time, which could be estimated from the receptor mobility, size of contact area, and force.

We return to the problem of initiating cell adhesion. Evidently, the first barrier that may have to be overcome is bringing the cells close enough together, with or without repeller redistribution, so that their receptors can begin to interact. Suppose that this has been done and the cells are at a separation distance such that bonds can begin to form and perhaps stabilize the contact so that it can be maintained in the absence of an applied force. The time constant for bond formation is roughly $(2\pi[D(1) + D(2)]n)^{-1}$ in the diffusion limit (from Equation 2) which is less than or on the order of seconds for many receptor systems. Thus, for favorable values of S, many bonds may form in such times, drawing the cells closer together and nucleating centers of adhesion. Bonds formed in short times must be those from locally available receptors, i.e., those in or near the initial contact area.

The time for receptors to diffuse to a contact area in numbers sufficient to equal those already there is on the order of A/D, for which A in square microns and $D = 10^{-10}$ cm²/ sec is 100 A. Thus, for a contact area of several square microns, as might be expected for smooth cells, it takes minutes for substantial numbers of receptors to diffuse to the contact area. If the force is not maintained for such long periods it would seem that our criterion for an adhesion threshold with immobile receptors (Section B.4) would determine whether stable adhesion would result. If so, additional receptors could accumulate and bond in the contact area as analyzed for the case of mobile receptors (Section B.2). If adhesion is not predicted to result for immobile receptors, there remains a question as to whether and how rapidly centers of adhesion may nucleate on areas much smaller than A; these could be stabilized by a local inflow of receptors on much smaller time scales.

For rough cells, the situation is even less clear since local contact areas may be comparable with the area associated with a few receptors so that times for bond formation and for receptor flow into the contact area could be comparable. Thus, the adhesion criteria for mobile receptors may be more nearly applicable in this case, but fluctuations may be important in view of the small number of receptors involved. However, as is apparent from this discussion, a theory for the kinetics of adhesion, in the presence of nonspecific repulsion is not in a satisfactory state.

Once an adhesive contact has formed, how rapidly will it spread and approach an equilibrium area? For immobile receptors on a cell with significant membrane tension, Evans[48] has argued that substantial spreading will not take place, as observed for lectin-linked red cells. For a continuum of molecular cross-bridges, spreading is predicated to be "rapid" and complete.[47] The more realistic case of cells having a finite number of somewhat mobile receptors has been considered.

An additional time scale for cell spreading may be that required for cell activation. For professional phagocytes such as macrophages or neutrophils, the time for activation is often a few minutes. The cell then actively seeks to spread over the adhering object, probably stimulated by some of the secondary effects noted in Section I.E.2. However, the object will be engulfed only if its whole surface is covered with sites to which the cell receptors can bind.[64] The time scale for active spreading of fibroblasts on a substrate is hours. Once cell activity has begun, substantial departures from equilibrium conditions may be expected.

VI. EFFECTS OF CELL ACTIVITY

We have just discussed the fact that cell activity is likely to enlarge the contact area and perhaps lead to adhesion over an even larger area than that which would be present at equilibrium. In any case, if the cell is actively moving its membrane in the vicinity of a sticky cell or surface, it will tend to increase the area over which is becomes stuck. In Reference 56, this was called the tar-baby effect, in reference to a Southern folk tale describing the plight of one who wishes to become unstuck from a sticky object, namely, the tar baby.

Once a cell has become stuck to another cell or to a surface, it appears that either the cell or the experimenter must do something active in order to break the adhesive contact. A third possibilty is that environmental changes, such as pulsating blood flow or muscular contraction, might so stress the bonds as to break the adhesion. Although we can estimate the force required to break individual bonds (see Section III.B), it is more complicated to predict the distribution of bond stresses produced by a force upon a cell. When cells such as fibroblasts crawl upon an adhesive substrate, they routinely break off trailing tails that remain stuck to the surface; thus, amputation is one way to be rid of an unwanted adhesive contact.

One can imagine any number of less drastic ways in which a cell could terminate adhesion, of which the modulation of its microenvironment by ion pumps is one of the most interesting, since it represents a mechanism that is used intracellularly to terminate binding in endocytic vesicles. Enzymatic cleavage of receptors or even their cytoplasmic tails is another possible mechanism.

Many functional cell adhesions must be broken, e.g., as cells move about during development or wound repair. Lymphocytes must not permanently adhere to endothelial cells, to antigen-presenting cells, or to their target cells, but transient and specific adhesion is vital in all those interactions. We suggest that the study of how cells terminate adhesions will prove rewarding.

VII. CONCLUSIONS

In earlier sections, we discussed a number of conclusions from theoretical models of specific cell adhesion. Among the most robust are the following.

Receptors mediating cell adhesion will, if mobile, become concentrated in regions of cell adhesion; specialized self-organizing structures may thereby come about as well as a polarization of the cell membrane.

Specificity of cell adhesion implies competition between nonspecific repulsion and specific bonding. In the presence of nonspecific repulsion, there will be a threshold for adhesion, whether stated in the number of receptors per cell, binding sites on the substrate, or other model parameters. Predictions can be made concerning the threshold and the equilibrium state of binding.

APPENDIX

Formation of A Specialized Microenvironment

Consider a cell in contact with a surface over a disc of radius R, such that the distance between the cell membrane and surface is d.

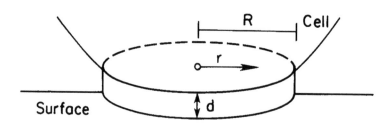

Suppose that the cell is pumping out protons at the rate s/μm²/sec. For d << R, this may be taken to be a volume source of S = s/d protons/μm³/sec. If n(r) is the proton concentration as a function of distance from the axis of the disc, then the steady state diffusion equation is

$$D \frac{1}{r} \frac{d}{dr} \left(r \frac{dn}{dr} \right) = -S \tag{A1}$$

where D is the proton diffusion constant in the disc. The solution is

$$n(r) = N(R) + \frac{S}{4D} (R^2 - r^2) \tag{A2}$$

where n(R) is the general extracellular proton concentration. The maximum increase in proton concentration Δn_{max} is thus for r = 0,

$$\Delta n_{max} = \frac{s}{4dD} R^2 \tag{A3}$$

We may safely assume d ≧ 0.01 μm (= 100 Å) and R ≦ 10 μm, while we set D = εD₀ with D₀ = 10⁻⁴cm²/sec = 10⁴ μm²/sec, the diffusion constant of protons in water and ε < 1. With these figures,

$$\Delta n_{max} \lesssim \frac{S}{4\epsilon} \mu m^{-3} \tag{A4}$$

To obtain a rough estimate of the proton pump rate, s, suppose that the proton pumps in a spherical endosome of radius r = 0.05 μm reduce the pH from 7 to 5 in 1 min and ignore other diffusion or transport across the membrane. If the concentration change is Δc in time t, we find

$$s = \frac{\Delta c \cdot r}{3t} \tag{A5}$$

Thus, for Δc ≅ 10⁻⁵ 6 × 10²⁰/cm³ = 6 × 10³/μm³, s = 1.7/μm²/sec.

If we assume this value of s applies to the plasma membrane and insert it into Equation A4, we see that ε ≦ 7 × 10⁻⁵ to give Δn_{max} = 6 × 10³/μm³ corresponding to pH = 5 or ε ≦ 7 × 10⁻⁴ for pH = 6. Thus, the provision of a specialized extracellular pH requires a marked reduction of proton mobility parallel to the membrane. Of course, a larger value of s would also lead to a proportionally larger value of Δn_{max}. For example, the oxyntic cells of the stomach have larger proton pumping rates by ~10⁷.

A recent report* indicates pronounced lateral proton conduction along a lipid-water interface and a steep pH gradient transverse to the interface. The relevance of these observations to more complex cell surfaces remains to be seen.

ACKNOWLEDGMENT

The author thanks the Aspen Center for Physics where this chapter was conceived and written.

* Prats, M., Teissié, J., and Tocanne, J.-F., Lateral proton conduction at lipid-water interfaces and its implications for the chemiosmotic-coupling hypothesis, *Nature (London)*, 322, 756, 1986.

REFERENCES

1. **Alberts, B., Bray, D., Lewis, J., Raff, M., Roberts, K., and Watson, J. D.,** *Molecular Biology of the Cell,* Garland Press, New York, 1983, 12.
2. **Singer, S. J. and Nicolson, G. L.,** The fluid mosaic model of the structure of cell membranes, *Science,* 175, 720, 1972.
3. **Michl, J., Pieczonka, M. M., Unkeless, J. C., Bell, G. I., and Silverstein, S. C.,** F_c receptor modulation in mononuclear phagocytes maintained on immobilized immune complexes occurs by diffusion of the receptor molecule, *J. Exp. Med.,* 157, 2121, 1983.
4. **Mazurek, N., Schindler, H., Schurholtz, T., and Pecht, I.,** The cromolyn binding protein constitutes the Ca^{++} channel of basophils opening upon immunological stimulus, *Proc. Natl. Acad. Sci. U.S.A.,* 81, 6841, 1984.
5. **Lawson, D., Fewtrell, C., and Raff, M. C.,** Localized mast cell degranulation induced by concanavalin A-sepharose beads, *J. Cell Biol.,* 79, 394, 1978.
6. **Young, J. D.-E., Unkeless, J. C., Young, T. M., Mauro, A., and Cohn, Z. A.,** Role for mouse macrophage IgGFc receptor as ligand-dependent ion channel, *Nature (London),* 306, 186, 1983.
7. **Wright, S. D. and Silverstein, S. C.,** Phagocytosing macrophages exclude proteins from zones of contact with opsonized targets, *Nature (London),* 309, 359, 1984.
8. **Stone, D. K., Xie, X.-S., and Racker, E.,** An ATP-driven proton pump in clathrin-coated vesicles, *J. Biol. Chem.,* 258, 4059, 1983.
9. **Baron, R., Neff, L., Louvard, D., and Courtoy, P. J.,** Cell-mediated extracellular acidification and bone resorption: evidence for a low pH in resorbing lacunae and localization of a 100 kD lysosomal membrane protein at the osteoclast ruffled border, *J. Cell Biol.,* 101, 2210, 1985.
10. **Bourguiguon, L. T. W. and Singer, S. J.,** Transmembrane interactions and the mechanism of capping of surface receptors by their specific ligands, *Proc. Natl. Acad. Sci. U.S.A.,* 74, 5031, 1977.
11. **Edelman, G. M.,** Surface modulation in cell recognition and cell growth, *Science,* 192, 218, 1976.
12. **Pfeiffer, J. R., Seagrave, J. C., Davis, B. H., Deanin, G. G., and Oliver, J. M.,** Membrane and cytoskeletal changes associated with IgE-mediated serotonin release from rat basophilic leukemia cells, *J. Cell Biol.,* 101, 2145, 1985.
13. **Menon, A. K., Holowka, D., Webb, W. W., and Baird, B.,** Cross-linking of receptor-bound IgE to aggregates larger than dimers leads to rapid immobilization, *J. Cell Biol.,* 102, 541, 1986.
14. **Hynes, R. O.,** Fibronectins, *Sci. Am.,* 254(6), 42, 1986.
15. **Hoffman, S. and Edelman, G. M.,** Kinetics of homophilic binding by embryonic and adult forms of neural cell adhesion molecules, *Proc. Natl. Acad. Sci. U.S.A.,* 80, 5762, 1983.
16. **Bell, G. I. and Torney, D.,** On the adhesion of vesicles by cell adhesion molecules, *Biophys. J.,* 48, 939, 1985.
17. **Kabat, E. A., Wu, T. T., Bilofsky, H., Reid-Miller, M., and Perry, H.,** *Sequences of Proteins of Immunological Interest,* Vol. 1, U.S. Department of Health and Human Services, Washington, D.C., 1983.
18. **Amit, A. G., Mariuzza, R. A., Phillips, S. E. V., and Poljak, R. J.,** Three-dimensional structure of an antigen-antibody complex at 2.8 Å resolution, *Science,* 233, 747, 1986.
19. **Leder, P.,** The genetics of antibody diversity, *Sci. Am.,* 246(5), 102, 1982.
20. **Marrack, P. and Kappler, J.,** The T-cell and its receptor, *Sci. Am.,* 254(2), 36, 1986.
21. **Ohashi, P., Mak, T. W., Van den Elsen, P., Yanagi, Y., Yoshikai, Y., Calman, A. F., Terhorst, C., Stobo, J. D., and Weiss, A.,** Reconstitution of an active T3/T-cell antigen receptor by DNA transfer, *Nature (London),* 316, 606, 1985.
22. **Nathenson, S. G., Uehara, H., Euenstein, B. M., Kindt, T.-J., and Colligan, J. E.,** Primary structural analysis of the transplantation antigens of the murine H-2 major histocompatibility complex, *Ann. Rev. Biochem.,* 50, 1025, 1981.
23. **Ploegh, H. L., Orr, H. T., and Strominger, J. L.,** Major histocompatability antigens: the human (HLA-A, -B, -C) and murine (H-2K, H-2D) class 1 molecules, *Cell,* 24, 287, 1981.
24. **Siegelman, M., Bond, M. W., Gallatin, W. M., St. John, T., Smith, H. T., Fried, V. A., and Weissman, I. L.,** Cell-surface molecule associated with lymphocyte homing is a ubiquitinated branched-chain glycoprotein, *Science,* 231, 823, 1986.
25. **Barondes, S. H.,** Lectins: their multiple endogenous cellular functions, *Ann. Rev. Biochem.,* 50, 207, 1981.
26. **Nicholson, G. L.,** The interactions of lectins with animal cell surfaces, *Int. Rev. Cytol.,* 39, 89, 1974.
27. **Roth, S.,** A molecular model for cell interactions, *Q. Rev. Biol.,* 48, 541, 1973.
28. **Weigel, P. H., Schnaar, R. L., Kuhlenschmidt, M. S., Schmell, E., Lee, R. T., Lee, Y. C., and Roseman, S.,** Adhesion of hepatocytes to immobilized sugars; a threshold phenomenon, *J. Biol. Chem.,* 254, 10830, 1979.
29. **Edelman, G. M.,** Cell adhesion molecules, *Science,* 219, 450, 1984.

30. **Rutishauser, U. and Gosidis, C.**, NCAM: the molecule and its genetics, *TIG*, 72, March 1986.
31. **Pytela, R., Pierschbacher, M. D., and Rouslahti, E.**, Identification and isolation of a 140 kD cell surface glycoprotein with properties expected of a fibronection receptor, *Cell*, 40, 191, 1985.
32. **Rouslahti, E. and Pierschbacher, M. D.**, Arg-gly-asp: a versatile cell recognition sequence, *Cell*, 44, 517, 1986.
33. **Horwitz, A., Duggan, K., Greggs, R., Decker, D., and Buch, C.**, The cell substrate attachment (CSAT) antigen has properties of a receptor for laminin and fibronectin, *J. Cell Biol.*, 101, 2134, 1985.
34. **Lesot, H., Kühl, U., and von der Mark, K.**, Isolation of a laminin-binding protein from muscle cell membranes, *EMBO J.*, 2, 861, 1983.
35. **Tervanova, V. P., Rao, C. N., Kalebic, T., Margulies, I. M., and Liotta, L.**, Laminin receptor on human breast cell carcinoma cells, *Proc. Natl. Acad. Sci. U.S.A.*, 80, 444, 1983.
36. **Liotta, L. A., Rao, C. N., and Wewer, U. M.**, Biochemical interactions of tumor cells with the basement membrane, *Ann. Rev. Biochem.*, in press.
37. **Lopes, J. D., dos Reis, M., and Brentani, R. R.**, Presence of laminin receptors in *Staphylococcus aureus*, *Science*, 229, 275, 1985.
38. **Gabius, H.-J., Springer, W. R., and Barondes, S. H.**, Receptor for cell binding site of Discoidin 1, *Cell*, 42, 449, 1985.
39. **Burger, M. M., Burkart, W., Weinbaum, G., and Jumblatt, J.**, Cell-cell recognition: molecular aspects. Recognition and its relation to morphogenetic processes in general, in *Cell-Cell Recognition*, Curtis, A. S. G., Ed., Cambridge University Press, London, 1978, 1.
40. **Bell, G. I.**, Models for the specific adhesion of cells to cells, *Science*, 200, 618, 1978.
41. **Dembo, M., Goldstein, B., Sobotka, A. K., and Lichtenstein, L. M.**, Histamine release due to bivalent penicilloyl haptens: the relation of activation and desensitization of basophils to dynamic aspects of ligand binding to cell surface antibody, *J. Immunol.*, 122, 518, 1979.
42. **Liebman, P. A. and Sitaramayya, A.**, Role of G-protein-receptor interaction in amplified phosphodiesterase activation of retinal rods, *Adv. Cyclic Nucleotide Protein Phosphate Res.*, 17, 215, 1984.
43. **Zhurkov, S. N.**, Kinetic concept of the strength of solids, *Int. J. Fract. Mech.*, 1, 311, 1965.
44. **McConnell, H.**, private communication.
45. **Jähnig, F.**, Thermodynamics and kinetics of protein incorporation into membranes, *Proc. Natl. Acad. Sci. U.S.A.*, 80, 3691, 1983.
46. **Hammer, D. and Lauffenberger, D. A.**, unpublished data.
47. **Evans, E. A.**, Detailed mechanics of membrane-membrane adhesion and separation. I. Continuum of molecular cross-bridges, *Biophys. J.*, 48, 175, 1985.
48. **Evans, E. A.**, Detailed mechanics of membrane-membrane adhesion and separation. II. Discrete kinetically trapped molecular cross-bridges, *Biophys. J.*, 48, 185, 1985.
49. **Mege, J. L., Capo, C., Benoliel, A. M., and Bongrand, P.**, unpublished data.
50. **Evans, E. A. and Kukan, B.**, Passive material behavior of granulocytes based on large deformation and recovery after deformation tests, *Blood*, 64, 1028, 1984.
51. **Singer, S. J.**, in *Surface Membrane Receptors*, Bradshaw, R. A., Frazier, W. A., Merrell, R. C., Gottlieb, D. I., and Hogue-Angletti, R. A., Eds., Plenum Press, New York, 1976, 1.
52. **Weis, R. M., Balakrishnan, K., Smith, B. A., and McConnell, H. M.**, Stimulation of fluorescence in a small contact region between rat basophilic leukemia cells and planar lipid membrane targets by coherent evanescent radiation, *J. Biol. Chem.*, 257, 6640, 1982.
53. **Capo, C., Garrouste, F., Benoliel, A. M., Bongrand, P., Ryter, A., and Bell, G. I.**, Concanavalin-A-mediated thymocyte agglutination: a model for a quantitative study of cell adhesion, *J. Cell Sci.*, 56, 21, 1982.
54. **McClosky, M. A. and Poo, M.**, Contact-induced redistribution of specific membrane components: local accumulation and development of adhesion, *Biophys. J.*, in press.
55. **Abney, J. R., Braun, J., and Owicki, J. C.**, Lateral interactions among membrane proteins. Implications for the organization of gap junctions, *Biophys. J.*, 52, 441, 1987.
56. **Bell, G. I., Dembo, M., and Bongrand, P.**, Cell adhesion: competition between nonspecific repulsion and specific bonding, *Biophys. J.*, 45, 1051, 1984.
57. **Atkins, P. W.**, *Physical Chemistry*, 2nd ed., W. H. Freeman, San Francisco, 1982.
58. **Bell, G. I.**, unpublished work.
59. **Bongrand, P. and Bell, G. I.**, Cell-cell adhesion: parameters and possible mechanisms, in *Cell Surface Dynamics, Concepts and Models*, Perelson, A. S., DeLisi, C., and Wiegel, F. W., Eds., Marcel Dekker, New York, 1984.
60. **Dembo, M. and Bell, G. I.**, The thermodynamics of cell adhesion, *Curr. Top. Membr. Transp.*, 29, 71, 1987.
61. **Dembo, M.**, unpublished data.
62. **Torney, D. C., Dembo, M., and Bell, G. I.**, Thermodynamics of cell adhesion. II. Freely mobile repellers, *Biophys. J.*, 49, 501, 1986.

63. **Bell, G. I.,** A theoretical model for adhesion between cells mediated by multivalent ligands, *Cell Biophys.*, 1, 133, 1979.
64. **Griffin, F. M., Jr., Griffin, J. A., and Silverstein, S. C.,** Studies on the mechanism of phagocytosis. II. The interaction of macrophages with anti-immunoglobulin IgG-coated bone marrow-derived lymphocytes, *J. Exp. Med.*, 144, 788, 1976.

Index

INDEX

9 780367 657390